Prairie Plants
OF THE
University of Wisconsin–Madison
Arboretum

over 360 *selected*
·Horsetails·Ferns·Rushes·Sedges·Grasses·Shrubs·
·Vines·Weeds·*and*·Wildflowers·

Theodore S. Cochrane • Kandis Elliot • Claudia S. Lipke

Copyright © 2006, Theodore S. Cochrane, Kandis Elliot, and Claudia S. Lipke.

ISBN-13 978-0-9789590-0-5

ISBN-10 0-9789590-0-0

First printing 2006 by the University of Wisconsin Press, 1930 Monroe Street, 3rd Floor, Madison, WI 53711-2059. Published by the University of Wisconsin–Madison Arboretum, 1207 Seminole Highway, Madison, WI 53711-3726.

All rights reserved. No part of this work covered by the copyright hereon may be reproduced or used in any form or by any means without the prior written permission of the authors. Fair use as defined by copyright law is granted for such purposes as criticism, comment, news reporting, teaching (including multiple copies for classroom use), scholarship, and research.

Printed in the U.S.A.

Contents

A Look at Prairies 2
    The Northern Temperate Grassland 3
    The Prairie Communities of Wisconsin 6
    The Arboretum Prairies. .8
    Arboretum Map 12
    About this Book 14

Glossary 16
    Flower Anatomy and Terminology 21
    Specializations in Flowers. 22

Species Treatments
    Prairie Flora List • *Families, Species, and Common Names*. 26
    Descriptions 36

Selected Bibliography. 344
Acknowledgments. 348
Index. 349

A Look at Prairies

Prairies filled with colorful flowers amidst waving fields of grasses are a joy to behold—Persian carpets not only aesthetically pleasing and mentally stimulating but also for the environmentalist gratifying to know that they still exist. In admiring the beauty and delicate design of leaf, flower, and fruit, we gain appreciation for the unending complexity of grassland communities, and by learning about their geographical distributions in Wisconsin and the world, a feeling for the ecological adaptations of their component grasses, herbs, and other organisms.

This book illustrates and describes 360 native and introduced species that grow and bloom on the Arboretum prairies, as well as briefly discussing or casually mentioning many additional species, infraspecific taxa, and hybrids. Through photographs and text, it aims to appeal to the eye and mind and to teach you to identify our prairie flora by understanding the structure of the plants and their flowers and fruits. By providing an enticing sample of the flora of the Arboretum and of the southern Wisconsin landscape, we hope it will increase your awareness and respect for our last remaining prairie remnants, motivate you to work for their preservation, and encourage you to grow native Wisconsin plants in your own yards and gardens. If nothing else, this volume should arouse your curiosity, stimulate you to learn plant names, and induce you to explore "Mother Nature."

prairie portrait

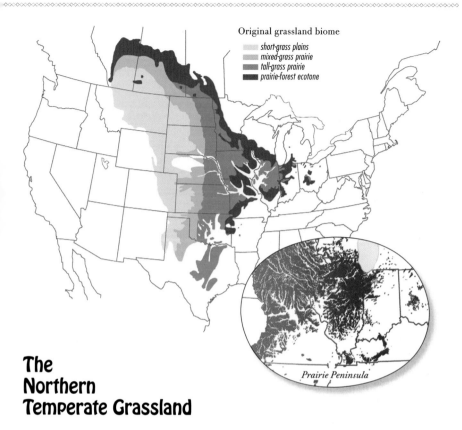

The Northern Temperate Grassland

The grasslands of North America once covered the vast plains from Texas through Iowa and Nebraska to Saskatchewan and Alberta and (apart from the Palouse Hills of Washington State and Great Valley of California) from the foothills of the Rocky Mountains to the oak-savanna margin of the eastern deciduous forest. From east to west, this region may be divided into three plant associations, recognizable by changes in climate and height of the vegetation: the **tall-grass prairies** (so-called because their most common grasses, big bluestem, Indian grass, and switch grass, grow in excess of 6 feet tall), the mixed-grass prairies, and the **short-grass plains**. In Wisconsin the tall-grass prairies predominate although on dry, southwest-facing hillsides, we find short-grass communities not unlike the mixed-grass or even short-grass prairies from out west.

At the time of European settlement prairies covered approximately 2.1 million acres of southern and western Wisconsin, but today they barely persist. Only a fraction of one percent of the original acreage survived untouched the advent of cow and plow.

A broad transition zone extends most of the length of the tall-grass prairie's eastern edge. Wisconsin's prairies lie along the northeastern boundary of this prairie-forest transition, gradually becoming broken into smaller and smaller patches as they extend eastward into the forest biome. Alternating with the prairies and oak forests are intermediate communities called oak woodlands and oak savannas (including oak barrens and oak openings). Ecologists and restorationists continue to debate the nature of savanna ecosystems, which are less well understood than are prairies and forests.

History A wedge-shaped extension of the tall-grass prairie, the "Prairie Peninsula," stretches from Arkansas and Minnesota into Ohio, encompassing southwestern Wisconsin and most of Illinois. The Prairie Peninsula developed during a warm dry period known as the Hypsithermal Interval, which followed the last glaciation and prevailed from about 8,000 to 3,500 years before present (yr B.P.). During that time species from all over, especially grassland species from farther west and southwest and even the Rocky Mountains, moved eastward into the deciduous forest of what is now the Upper Midwest, achieving their greatest extent around 5,500 yr B.P. These species tended to colonize upland areas and merged with a wet prairie flora that, together with forest species, had dispersed and migrated north from unglaciated refugia in the Southeast following the melting of the last, or Wisconsin, ice sheet. The intermingling of these with species of other floristic relationships created today's tall-grass prairie. By around 3,000 yr B.P. the climate of the Prairie Peninsula had become cooler and moister, and the reduced seasonal contrast in temperature allowed trees and shrubs to once again compete with and replace grassland vegetation from Minnesota to Illinois.

Climate The grassland formation is the product of climate. It is affected by limited precipitation, low humidity, extreme temperatures, and summer dry spells, as well as such weather phenomena as droughts, dust storms, hail, and blizzards. In the tall-grass region precipitation equals or exceeds evaporation, but in the mid- and short-grass regions rainfall is less. Evaporation and temperature vary in a similar manner. Cold-season precipitation, midsummer relative humidity, and frequency of drought are undoubtedly important in maintaining the prairies and savannas against advancing forest, which, however, seems to be extending into the tall-grass region at the present time. Prior to European settlement, fires, both natural and human caused, tilted the equation in favor of prairies, especially in the prairie-forest border region.

Arboretum prairie burn

Fire plays a major role in renewing prairies. It was a major component of the patterns and processes that existed on the grasslands for thousands of years, and prairie plants and animals ultimately became dependent on periodic fire. Plants do much better in a burned prairie than in an unburned one. Arboretum Ecologist M. K. Leach and University of Wisconsin–Madison Professor of Botany T. J. Givnish have documented long-term patterns of species loss in prairie remnants as the result of fire suppression. Successful maintenance of existing prairies and establishment of new restorations depends on the frequent use of fire, which prevents invasion by trees and shrubs, reduces competition with weeds by differentially damaging these competitors, removes accumulated mulch that smothers small plants and seedlings, and weakens the bluegrass sod so as to give native species a better chance to become established.

Biota The grassland biome is extremely complex, containing thousands of kinds of plants and animals, most of them too small to see with the naked eye. It is uniformly dominated by perennial grasses and originally by populations of grazing and burrowing animals. Five robust and adaptive native grasses in particular dominate the biome: big bluestem, little bluestem, Indian grass, switch grass, and prairie dropseed. Non-grass herbs, commonly called forbs, are important in all grasslands. Some shrubs, occasional tree seedlings, and under certain circumstances oak brush or "grubs" are present in prairie communities. There may be a layer of mosses, liverworts, ground lichens, and a component of saprophytes and parasites consisting mostly of fungi, including characteristic soil microfungi. The Arboretum prairies are wonderful places to experience this variety of life and witness the special adaptations that enable plants to survive wind, fire, drought, high light intensity, and early and late frosts.

The Prairie Communities of Wisconsin

Prairies are communities dominated by grasses on mineral soil. If trees are present, the average tree canopy cover is less than 10% of the area. Five broad prairie types were defined by J. T. Curtis and the Plant Ecology Laboratory, UW–Department of Botany, based on the behavior of prairie species with respect to soil moisture conditions: **wet, wet-medium, medium, dry-medium,** and **dry.** Percent organic matter and pH are also fundamentally important.

Lowland prairies Lowland prairies are divided into wet-medium prairies, which occur on level areas where surface water is present after heavy rains or flooding, and wet prairies, which have surface water present during winter and spring and the soil otherwise nearly always saturated. The intermediate or wet-medium prairies are dominated by big bluestem, blue-joint grass, slough grass, and Canada wild-rye. Among the typical forbs are New England aster, bottle gentian, eastern shooting-star, saw-tooth sunflower, thick-spike gay-feather, and mountain mint. The wet prairies are dominated by blue-joint grass, slough grass, big bluestem, and upland wild-timothy. They are much less diverse, and many of their prevalent forbs occur in medium prairies or other wet-soil communities as well.

Mesic or medium prairies This once very widespread community is found on flat or gently rolling landforms. The soils are deep, rich in nutrients, and generally well-drained, allowing for maximum species diversity and maximum plant height. The leading dominants are big bluestem, needle grass, little bluestem, prairie dropseed, and Leiberg's panic grass. Smooth aster, cream wild indigo, rattlesnake-master, yellow coneflower, prairie thistle, stiff sunflower, rough blazing-star, compass-plant, stiff goldenrod, and showy goldenrod are among the most characteristic forbs.

Short-grass or dry prairies Dry prairies occupy exposed sites that are well-drained, usually on steep, south- to west-facing, rocky or gravelly bluffs and slopes, but they also occur on flat uplands or plains, where dolomite, glacial drift, gravel, or sand lie close to the surface. Dry-medium prairies occur where the topographic position is lower and the soils are slightly deeper. The average height of the grasses is taller than in the dry prairies, and the floristic diversity approaches that of the medium prairies. Little bluestem, side-oats grama, big bluestem, prairie dropseed, and panic grasses dominate both the dry and dry-medium segments of the prairie continuum. Needle grass and June grass are prevalent in both types as are numerous other plants, including lead-plant, flowering spurge, purple prairie-clover, prairie tickseed, and prairie blue-eyed-grass. Other important plants of the dry-medium prairies like smooth sumac, common milkweed, black-eyed Susan, frost aster, and prairie violet scarcely occur in the driest prairies, on which such relatively well-known plants as pasqueflower, prairie-smoke, and silky aster are highly characteristic.

Related communities

include **fens,** in which over half of the dominance is contributed by grasses but which occur on peaty sites with a reliable internal flow of bicarbonate-rich water, and **sedge meadows,** which like wet prairies obtain their water supply from rain and surface drainage but have over half of the dominance contributed by sedges. Oak openings, oak barrens, and pine barrens are **savannas,** defined as having more than 10% but less than 50% canopy cover. The trees may be widely scattered or in groves. The oak openings occur on relatively heavy soils, whereas barrens occur on very sandy soils. These may be open and prairie-like (sand barrens), or they may contain black or Hill's oak (oak barrens) or jack pine (pine barrens). In Wisconsin the so-called **cedar glade** appears to be nothing more than dry prairie forested with red-cedar as a result of fire suppression.

PRINCIPAL PRAIRIE COMMUNITIES AND SOME OF THEIR MOST PREVALENT SPECIES

	Lowland Prairies Wet and Wet-medium poorly to very poorly drained silt loams and silty clay loams; muck from fibrous peat	**Medium Prairies** moderately well- to well-drained silt loams and silty clay loams	**Upland Prairies** Dry and Dry-medium well-drained, gravelly loam, silt loam, or fine sandy loam
Grasses	big bluestem blue-joint grass prairie cord grass Canada wild-rye prairie panic grass upland wild-timothy prairie dropseed Indian grass northern sweet grass fowl manna grass	big bluestem needle grass prairie dropseed prairie panic grass little bluestem Indian grass Canada wild-rye switch grass	little bluestem side-oats grama big bluestem prairie dropseed long-stalked panic grass needle grass few-flowered panic grass Indian grass Canada wild-rye June grass
Sedges	tussock sedge broad-leaved woolly sedge Bicknell's oval sedge Buxbaum's sedge common stiff sedge flat-stemmed spike-rush bald spike-rush tall nut-rush	Bicknell's oval sedge plains oval sedge long-awned bracted sedge field oval sedge broom sedge	early oak sedge Mead's sedge Richardson's sedge Bicknell's oval sedge plains oval sedge Pennsylvania sedge running savanna sedge
Forbs	tall meadow-rue mountain mint saw-tooth sunflower late goldenrod stiff cowbane New England aster field horsetail eastern shooting-star yellow star-grass thick-spike gay-feather	smooth aster rough blazing-star stiff sunflower yellow coneflower compass-plant common milkweed tick-trefoils field pussy-toes rattlesnake-master cream wild indigo	prairie-clovers pasqueflower flowering spurge old-field goldenrod whorled milkweed field sage-wort silky aster prairie tickseed heath aster sky-blue aster
Shrubs	prairie willow meadow willow beaked willow pussy willow smooth rose American hazelnut white meadowsweet elderberry	lead-plant New Jersey tea smooth sumac Carolina rose Arkansas rose smooth rose Allegheny blackberry	lead-plant smooth sumac Arkansas rose Carolina rose sand cherry
Weeds and Pioneers introduced (i) native (n)	reed canary grass (i) redtop (i) grass-leaved goldenrod (n) gray dogwood (n) honeysuckles (i) buckthorns (i) water-pepper (i) nut sedges (n) beggar-ticks (n) rough barnyard grass (n)	Kentucky bluegrass (i) smooth brome (i) sweet-clovers (i) leafy spurge (i) wild parsnip (i) bee balm (n) black-eyed Susan (n) gray dogwood (n) common evening-primrose (n) common milkweed (n)	quaking aspen (n) eastern red-cedar (n) black locust (i) honeysuckles (i) smooth sumac (n) common ragweed (n) Queen Anne's-lace (i) daisy fleabane (n) pepper-grasses (i,n) quackgrass (i)
Trees and Succession	bur oak quicky succeeds toward willows, aspens, ashes, elms, oaks	bur oak succeeds toward oaks, aspens, shagbark hickory, black cherry, elms, black walnut	black oak succeeds toward oak-hickory in the absence of management

The Arboretum Prairies

The six Arboretum prairies are located within the central portion south (**Curtis Prairie**) and east (**Juniper Knoll**) of the Visitor Center; along McCaffrey Drive opposite Longenecker Gardens (**Sinaiko Overlook Prairie**); south of the West Beltline Highway and east of Seminole Highway in the **Grady Tract** (East Knoll and West Knoll oak barrens, **Greene Prairie**); and along Monroe Street between its intersections with Glenway Street and Odana Road (**Marion Dunn Prairie**). The largest prairies, Curtis and Greene, range from dry to wet but differ in their soils, the former having loamy soils, the latter sandy soils, sedge peat, and clay. All except the Grady knolls were reestablished by transplanting plants and casting seeds collected from prairie remnants in southern Wisconsin. These restorations are works-in-progress as Arboretum staff and volunteers continue the process of converting them into reasonable facsimiles of the original community types.

Theodore S. Cochrane in Curtis Prairie

Curtis Prairie Curtis Prairie is evidently the world's first and oldest restored prairie; no one had ever attempted to re-create a large prairie before this project was undertaken. The land comprising the 73-acre restoration was variously plowed, mowed, or left undisturbed until 1927, after which all of it was grazed until 1932. Although the majority of the area was an abandoned horse pasture when it was acquired in 1933, apparently the eastern third, east of the north-south fire lane, was never plowed and only lightly grazed though the southern half was used as a mowing meadow. This irregularly shaped, partly wet, unplowed remnant is incorporated into the present-day prairie, which attempts to illustrate three of the five types of Wisconsin prairies.

The first plantings were experimental, carried out in 1935 by N. C. Fassett and two of his students with the help of the Civilian Conservation Corps. The principal planting effort began the following year under the direction of T. M. Sperry, and a second major planting program took place between 1950 and 1957 under D. Archbald. Many classic experiments on planting methods and management techniques were conducted during the 1930s, 1940s, and early 1950s. As a result of research on the effects of fire carried out by J. T. Curtis and his students, a schedule of prescribed burns was initiated in 1950. Further studies on prairie restoration and plant propagation were done in the 1960s and early 1970s. More recent studies have analyzed species richness in relation to such factors as edge effects, habitat fragmentation, storm water runoff, and invasive species. A research group led by J. B. Zedler is studying the interrelationships of hydrologic disturbance, wetland vegetation, and invasive species in Curtis and Greene prairies and Gardner Marsh.

Prominent in spring are white wild indigo, common spiderwort, and common blue violet. The most characteristic common species of summer—pale Indian-plantain, hemp-dogbane, rattlesnake-master, Canadian tick-trefoil, flowering spurge, wild bergamot, yellow and purple coneflowers—are succeeded by a wave of compositae: rosinweed, prairie dock, blazing-stars, sunflowers, and goldenrods. The dominant grasses, big bluestem and Indian grass, may reach a height of 4, 8, or even 10 feet in early fall. Another feature of the prairie is its high content of shrubs and the small groves of bur oak along the edges.

Grady Tract These restorations are all on old agricultural land. They include the Grady Dry Oak Woods, East Grady Woods, Greene Prairie, and the East and West Grady knolls, these a series of often-burned sand hills formed at the edge of a glacier and covered by a dry prairie groundlayer with scattered open-grown oaks and numerous oak grubs. Woody growth is denser on the East Knoll. In 1977 Arboretum Ecologist V. M. Kline initiated an experiment on the West Knoll to demonstrate the use of oak wilt fungus as a biological control on oak grubs. A convincing demonstration of the ruinous effects of uncontrolled invasive species can be seen on the West Knoll, where oriental bittersweet has made its way from the Southwest Grady Savanna onto the knoll and is spreading rapidly.

On the West Knoll dominant grasses are little bluestem and needle grass. Pussy-toes and rock cress, the earliest spring flowers, are soon followed by bird's-foot violet and puccoons, then spiderwort and lupine. The continuously changing flowering parade reveals a fairly rich flora as summer's legumes, milkweeds, penstemons, and evening-primroses yield to late summer and fall's sunflowers, goldenrods, and white- and blue-flowered asters. A few species, notably American pasqueflower and prairie larkspur, have been planted.

West Knoll

Greene Prairie This 50-acre restoration is on land that had been cultivated as recently as 1937, and in the case of the western addition, through 1946. The tract was first visited by botanist H. C. Greene in 1942, who the very next year introduced a few plants. Thus began his energetic, 15-year effort to re-establish a sand prairie. Working almost single-handedly until the early 1960s, and using a variety of techniques, he carefully replanted the 40-acre former corn field into a dry sand prairie in the north and a wet prairie in the south and southwest. His remarkable knowledge of the ecological requirements of individual species is confirmed by the dominance of native prairie grasses, by the diversity of native species including a number of rare and uncommon ones, by the relatively few weedy species, and until recently, by the limited invasion by woody species. Furthermore, the plants have hardly migrated within the prairie, staying put where he placed them.

Greene's artistry is still evident in the sprinkling of colors and balancing of textures. Showy species like Indian paintbrush, phloxes, prairie-clovers, gentians, and gay-feathers are effectively displayed against the backdrop of silphiums and shorter prairie grasses. Unfortunately, lower sections of this exquisite prairie are threatened by unmanaged storm water runoff from offsite, resulting in siltation and invasion by reed canary grass; while upper sections and the knolls are suffering from encroachment of trembling aspen and oaks.

hummingbird, bottle gentian

gay-feather, rattlesnake-master, compass-plant

Sinaiko (or Wingra) Overlook Prairie covers 5 acres that have not been subjected to as much management activity as the other Arboretum prairies. Landowners used the site as a pasture until the Arboretum acquired the property in 1935, after which it was maintained as an old field until 1969, when J. A. Schwarzmeier planted prairie-establishment research plots. Data were collected until 1971 by Schwarzmeier and subsequently by Arboretum Naturalist J. H. Zimmerman, who expanded the plantings during 1974–1976.

The prairie is fairly medium, given that it is developed on sandy loam with a low content of organic matter and half the topsoil has been lost. It is not known to what extent the prairie flora may have been reduced or removed during many years of grazing, how much it may have recovered on its own during the decades the site remained a field, or what species were planted by Zimmerman. In any case the area is reverting to prairie, the present species composition being a mixture of weeds, invasive woody species (which are abundant in the surrounding vegetation), and especially a conglomeration of typical dry, medium, and wet prairie species. If any one species is dominant today, it is Indian grass. The most conspicuous forbs include pale Indian-plantain, the two purple coneflowers, wild quinine, and along the road, hairy vetch.

Juniper Knoll This 8-acre, red-cedar-grassland demonstration area is developed on previously farmed, gently rolling terrain. The top and slopes of the hill were planted with red-cedar, common juniper, and horizontal juniper under Curtis's supervision, probably in the 1950s, in an effort to establish a "cedar glade" that grades into prairie. Management began in earnest in 1983, at which time an effort was initiated to remove not only honeysuckle, buckthorn, and black locust but also selected cedars. Prescribed burns have been carried out since the 1980s, and the cleared land has been planted with native prairie plants and seeds. Big and little bluestem and Indian grass now dominate the lower west and east slopes extending toward Curtis Prairie and the small wetland north of Teal Pond. These areas are slightly weedy and brushy, but groundlayer development is underway.

big bluestem

Marion Dunn Prairie This 4-acre, urban restoration is located in an area that was originally a sedge meadow, but that vegetation had long since disappeared by 1982, when part of the site was transformed into a storm water settling pond. Construction of the pond created several problems, not the least of which were loss of existing vegetation and drastic soil disruption. During the next two years, volunteers planted the large berm and graded areas with seeds of grasses and forbs collected mainly from other Arboretum prairies. An additional area north of the pond was planted in 1986-1987 and replanted in 1990. With the arrival of the construction crews again in 2004 to re-excavate and enlarge the pond, this area was changed again.

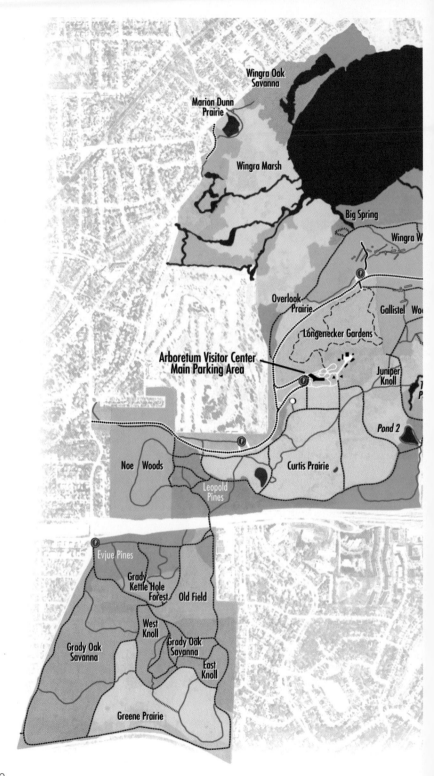

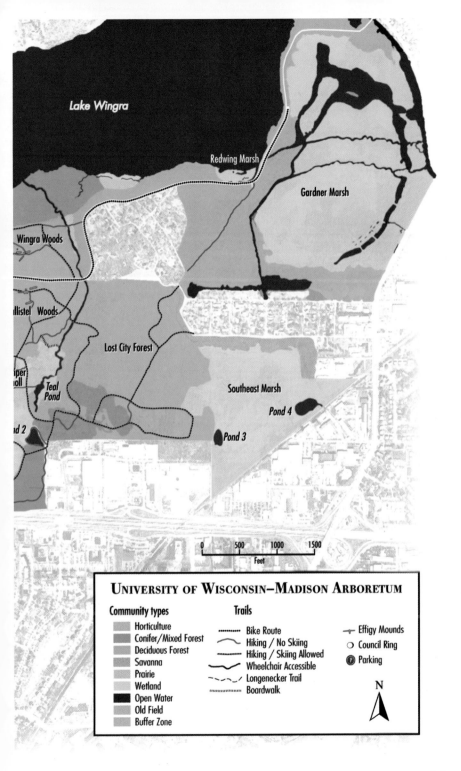

Sylvia Marek, Arboretum Naturalist, guiding a tour

About this Book

Plants are arranged according to the families to which they belong except for the ferns, which for convenience have been grouped together under "Ferns". However, you do not need to know plant families to use this book. Like ferns, plants in the same family will have similar structures and shapes, and the floral list thumbnails (pages 26-35) will help point you to the right section.

- **Use the photographs, text, and glossary together.** Identifying an unknown plant may be as simple as matching its flowers or other parts against the photographs. Confirm the identification by comparing the plant with the written description or the briefer description under "Similar species." The more characters you can observe, the easier it will be to correctly name each plant. Many species look superficially alike, and to distinguish them accurately, you must check their technical characters. A hand lens will be useful.

- **Inclusions and exclusions.** About 320 species of native vascular plants have been identified on Arboretum prairies, and these pages feature the majority of them—essentially all of the prevalent and common prairie species, as well as many non-native plants. This guidebook excludes many sedges, grasses, shrubs, and trees such as bur oak that were part of the original prairie landscape. Also omitted are several characteristic prairie wildflowers that have never become established on Arboretum prairies, such as grooved yellow flax (*Linum sulcatum*) and prairie buttercup (*Ranunculus rhomboideus*), and others that were once present in low numbers but have disappeared, including downy painted-cup (*Castilleja sessiliflora*), rough false pennyroyal (*Hedeoma hispida*), fringed puccoon (*Lithospermum incisum*), early buttercup (*R. fascicularis*), false-gromwell (*Onosmodium bejariense*), and upland white goldenrod (*Solidago ptarmicoides*). Finally, we did not include species that in the Arboretum are especially rare or far off-trail, such as short green milkweed (*Asclepias viridiflora*), kitten's-tails (*Besseya bullii*), and Canadian milk-vetch (*Astragalus canadensis*).

- **Nomenclature adopted.** This book generally follows traditional taxonomic delimitations. However, because modern research is showing familiar classification to be unsatisfactory in many respects, we have chosen to adopt some of the changes sanctioned in the Angiosperm Phylogeny Group (2003) system of classification. We do not follow the APG system throughout but at least mention places where their taxonomy differs from traditional systems. For example, we treat the Figwort Family in the customary sense instead of rearranging it into several families, and maintain the Milkweed Family separately from, rather than merging it with, the Dogbane Family. Likewise at the genus level, we accept *Packera* over *Senecio* but keep *Dodecatheon* in lieu of *Primula*. The result is a mixture of current APG and long-familiar names. We have also incorporated significant name changes if they have been accepted by the most important and influential national flora (i.e., the on-going Flora of North America project) and checklists (e.g., the USDA Plants Database, http://plants.usda.gov/). Examples include *Scirpus* rearranged into several genera, *Petalostemon* included in *Dalea*, and asters transferred into various other genera.

- **For additional information** or alternative treatments, ecological studies of prairies, and further background information on the Arboretum, you will want to consult references listed in the Selected Bibliography, especially *The Vegetation of Wisconsin*, the *Atlas of the Wisconsin Prairie and Savanna Flora*, and *Arboretum Prairies*, as well as the proceedings of prairie workshops and conferences and the more popular and technical articles published by the Wisconsin DNR and the Arboretum in its newsletter, *Arboretum News Leaf* (previously *Arboretum News*) and journal, *Ecological Restoration*.

Be respectful of our natural environment by staying on established trails and studying the flora non-destructively. Observe and enjoy the flowers, but please do not pick them. Wildflowers are sensitive to trampling, and besides causing damage that may diminish flowering or seed set, walking off trail even a few steps may leave a path that will last indefinitely. *Please observe Arboretum rules and trail etiquette.*

black and yellow Argiope

Glossary

Adventive • Spreading spontaneously from a native or introduced source but not well established.

Achene • Small dry, non-opening fruit, usually 1-seeded, technically derived from a single carpel but generally used for similar fruits (nutlets) derived from more than one carpel.

Alternate • Borne singly at the nodes, as leaves on a stem or branches in an inflorescence; also, situated between organs of another kind, as stamens alternate with the petals.

Anther • Terminal, pollen-producing part of a stamen, consisting of one or usually two pollen sacs and a connecting layer between them (see **Filament**).

Appressed • Lying flat or close against a stalk, surface, or margin.

Auricle • Lobe or appendage, often small and ear-shaped, at the base of an organ; in grasses, pointed appendages that occur laterally at the base of the leaf blade in some species and at the apex of the sheath in others.

Awn • Bristle-like appendage, usually the prolongation of a vein, extending from the apex or the outer surface of a structure, e.g., of a lemma or scale.

Axil • Angle between a main axis and a lateral organ, e.g., between a stem and a leaf stalk.

Bilaterally symmetrical • Divisible by only one plane into two similar sides (mirror-image halves). Most flowers in which the petals (or sepals) are dissimilar in form or orientation are bilaterally symmetrical.

Bract • Reduced or modified leaf associated with a flower or inflorescence, but not part of the flower; also applied to reduced, leaf- or scale-like structures on a stem or rhizome.

Bulb • Underground modified leaf bud bearing fleshy scaly leaves that function as food storage organs.

Chaffy • With thin dry scales.

Calcareous • Containing lime in a form available to plants.

Calyx (pl., **calyces**) • The sepals of a flower collectively, either fused or separate, comprising the outer floral envelope.

Capsule • Dry fruit that splits open at maturity, composed of more than one carpel.

Carpel • Fertile structure of a flowering plant, which encloses seeds and ripens to form a fruit. The carpel differentiates angiosperms from gymnosperms, which have "naked" seeds.

Clone • Colony of individuals (population of organisms) derived vegetatively from a single, sexually produced progenitor; all members of a clone are genetically identical.

Compound • Composed of multiple similar parts, e.g., a leaf made up of distinct leaflets, a pistil of more than 1 carpel; also, branched, e.g., a branched inflorescence (opposite of **Simple**).

Corolla • The petals of a flower collectively, either fused or separate, comprising the inner floral envelope.

Crown • Short tough persistent stem-base (botanical *caudex*), situated at or just beneath ground level, serving as the overwintering organ for an otherwise herbaceous perennial.

Dicot • Dicotyledon. A flowering plant having two cotyledons (twin leaves of germinating sprout) in the seed and net-veined leaves. Floral parts usually in 5s (compare **Monocot**).

Disk floret • Tiny tubular central flower of the composite head, the corolla with 5 lobes similar in size, shape, and orientation.

Glossary

Doctrine of signatures • The belief that a plant's utility to humans is recorded in the form of the plant. Thus, by the doctrine of signatures, the liver-shaped leaves of hepatica plants were taken to indicate that the plant is useful in healing liver ailments.

Drupe • Fleshy or pulpy fruit with the seed (or several seeds) permanently enclosed within a hard or stony inner layer, forming a pit or "stone," e.g., olive, peach, plum, cherry.

Elliptic • Structure with the shape of an ellipse, widest at the middle and rounded about equally to both ends (applied to flat structures; **Ellipsoid** is the equivalent term for solid structures).

Entire • With a continuous margin, not toothed, cut, or lobed.

Evapotranspiration • (Transpiration) Evaporation is passing of water into the air; transpiration is movement of water within a plant and subsequent loss of water as vapor from plant surfaces, mostly through stomata in the epidermis of leaves and stems.

Fen • Wetland community with impeded drainage dominated by smaller grasses and sedges; characterized by inflowing (calcareous-neutral to alkaline) groundwater.

Fertile • Bearing normal reproductive organs (opposite of **Sterile**).

Filament • Stalk of a stamen, usually thread-like but sometimes expanded.

Floral tube • Saucer-shaped, cup-like, or cylindrical structure formed from the union of sepals or petals (or tepals) and/or stamens. The floral tube may be free from the ovary, the ovary in this case superior, or fused to the ovary in flowers having inferior ovaries.

Floret • Reduced flower; in grasses, the lemma and palea with the enclosed flower. The floret may be bisexual, female, male, or neuter.

Follicle • Dry seed pod that opens along one side, derived from a single carpel.

Frond • Leaf of a fern, including both blade and stalk.

Genus (pl., **genera**) • Group of closely related species or sometimes a single species.

Glume • One of a pair of bracts or scales at the base of a grass spikelet (see **Lemma, Palea**).

Habitat • Kind of place or environment in which a particular species lives.

Head • A dense inflorescence of stalkless flowers clustered on a common receptacle.

Inflorescence • Flowering portion of a plant, including all associated stalks and bracts; the arrangement of the flowers on a plant.

Involucre • Bract or bracts (one or more series) surrounding the base of a flower or inflorescence.

Lemma • Lowermost of the two bracts or scales at the base of the grass floret (see **Glume, Palea**). The lemma typically has an odd number of veins.

Ligule • Appendage (membranous collar or fringe of hairs on the inner side) at the juncture of a leaf sheath and blade in many grasses and some sedges. In the Sunflower Family, the flattened part of the corolla of the ray floret.

Linear • Narrow and elongate with nearly parallel sides (applied to flat structures; **Cylindrical** is the equivalent term for solid structures).

Monocot • Monocotyledon. Any of various flowering plants, such as grasses, orchids, and lilies, having a single cotyledon (first leaf of germinating sprout) in the seed, and parallel-veined leaves. Floral parts in 3s (compare **Dicot**).

Glossary

Naturalized • Non-native species well established in the wild, whether locally at particular places or widespread.

Node • Point on a stem at which a leaf or branch arises. Nodes in many species are capable of sprouting roots and/or producing new shoots.

Oblong • Two or more times longer than broad with nearly parallel sides (applied to flat and solid structures).

Opposite • Borne in pairs directly across from one another at the same node, as in the branches, leaves, or leaflets of some plants; also, centered in front of another organ, as stamens opposite the petals.

Palea • Uppermost of the two bracts subtending the grass flower in the floret. The palea has its back to the axis and is usually 2-veined (see **Lemma**).

Palmate • Divided, hand-like, into a few diverging portions.

Panicle • Variously branched inflorescence, usually open, blooming from the base upward and theoretically capable of elongating indefinitely.

Pappus • Specialized calyx crowning the ovary (and achene) of the Sunflower Family, composed of scales, hairs, bristles, or plumes, or a mixture of these.

Perfoliate leaf • A leaf with tissue at the base of the blade fused around the stem, which appears to pass through the leaf.

Perigynium • Special bract (sac-like, flask-shaped, or flattened) that encloses the ovary (and achene) in *Carex*.

Petal • One of the innermost whorl of sterile floral leaves (when there are two series), often colored or white and modified to attract pollinators.

Pinnate • With two rows of parts (branches, leaflets, appendages) arising along each side of a common axis or midrib.

Pistil • Female organ of a flower, composed of one to many carpels and generally differentiated into an ovary, style, and stigma.

Raceme • Simple inflorescence consisting of a central axis bearing a number of stalked flowers (see **Spike, Panicle**).

Radially symmetrical • Wheel-shaped, with all structures symmetrical about the center (synonyms: actinomorphic, regular).

Ray • Flattened part of the bilaterally symmetrical, petal-like corolla in the Sunflower Family.

Ray floret • Petal-like flower at the outer rim of a head, the corolla strap-shaped, i.e., produced into a flattened expanded portion or "little tongue" (ligule) on one side.

Receptacle • Expanded end of a stalk to which the other flower parts are attached; in the Sunflower Family, the end of a stalk to which the flowers of the head are attached.

Rhizome • Underground stem, usually elongate and often branched, rooting at or producing new shoots from the nodes.

Rosette • Dense circular cluster of leaves or other organs from very short internodes, often basal.

Scale • Small bract, especially the one subtending an individual flower in a sedge.

Glossary

Schizocarp • A fruit that splits into separate carpels at maturity, e.g., fruits of the Carrot or Mallow families.

Sepal • One of the outermost whorl of sterile floral leaves, generally green and less conspicuous than the petals, but often petal-like (as in the genus *Anemone,* in which the petals are often lost, and the Lily Family, in which the petals are present but mirrored by the sepals).

Shapes

Egg-shaped • Shaped like a hen's egg in outline, one to two times as long as broad, widest toward the base and rounded at both ends. *Inverse-egg-shaped* • Egg-shaped but with the narrower end toward the attachment.

Elliptic • Structure with the shape of an ellipse, widest at the middle and rounded about equally to both ends.

Lance • Shaped like a lance-head, several times as long as broad, broadest toward the base and narrowed to the apex. *Inverse-lance-shaped* • Lance-shaped but broadest toward the apex and narrowed toward the attachment.

Linear • 10 or more times longer than broad with the sides nearly parallel.

Oblong • 2 or 3 times longer than broad with the sides nearly parallel.

Shape terminology may be applied to leaves, petals, and other plant anatomy; combinations of these terms are used to describe intermediate shapes.

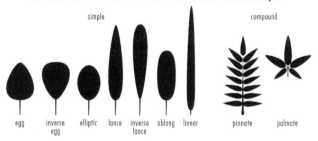

Sheath • Sheet of membranous tissue at the base of a leaf or bract encircling a stalk.

Simple • Not divided into distinct segments; stems not branched or leaves that are not divided into separate leaflets (opposite of **Compound**).

Sorus (pl., **sori**) • "Fruit dot," or cluster of spore-bearing structures (sporangia) on the underside of a fern frond.

Spathe • Modified leaf (bract) at the base of a monocot inflorescence, commonly with a reduced blade and expanded sheath.

Species • Discrete interbreeding population of similar plants, distinguishable by ordinary means from other such discrete interbreeding populations.

Spike • ± elongate inflorescence with flowers or spikelets attached directly (without a stalk) on an unbranched axis (see **Raceme**). Several to many spikes may be clustered into a panicle.

Glossary

Sporangium (pl., **sporangia**) • Spore case, or structure bearing spores; tiny, stalked, but usually occurring in visible clusters called sori.

Spore • Reproductive cell formed within a sporangium that germinates to give rise to the gametophyte generation.

Stalk • General term referring to the stem of any organ, such as a flower, fruit, or leaf.

Stamen • Male organ of a flower, consisting of a filament (usually) and an anther.

Sterile • Lacking functional reproductive organs, as in sterile florets; also used to describe lack of seed production (neuter) (opposite of **Fertile**).

Stigma • Portion of the female organ of the flower that receives pollen, usually located at the apex of the style.

Stipule • One of a pair of leaf-like appendages at the base of a leaf.

Stolon • Elongate stem ("runner") that creeps on or arches over the surface of the ground, rooting at the nodes and/or apex.

Style • Constricted (often slender and elongate) part of the female organ of the flower that connects the stigma(s) to the ovary.

Swale • Small, grassy or sedgy, open marshy spot that becomes seasonally dry.

Taxon (pl., **taxa**) • A taxonomic entity, irrespective of rank. For example, the genus *Scutellaria*, the species *Scutellaria parvula*, and the variety *Scutellaria parvula* variety *missouriensis* are taxa at three different ranks.

Tepal • Sepal or petal, used in describing flowers in which these organs are not differentiated in size, color, and texture (though usually distinguishable as a member of the outer or inner floral envelope).

Taxonomy • Science of classification of organisms.

Tuber • Thickened part of a rhizome or roots, storing food and often also bearing buds.

Umbel • Inflorescence in which the stalks of a cluster arise from a common point. Umbels may be simple (unbranched) or compound (umbels of umbellets), i.e., the primary stalks are again branched in umbellate fashion at the tip.

Whorl • Ring of three or more structures that emerge from a common node or point on the stem.

Winter annual • Annual plant whose seeds germinate in the fall, the first foliage leaves generally not appearing until the following spring.

Flower Anatomy and Terminology
Used in Identifying Species

Whorls of flower parts (all modified leaves)

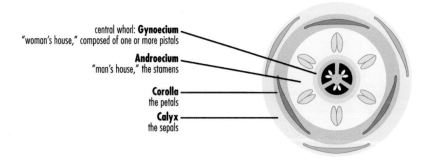

Whorls of flower parts in a generalized flower

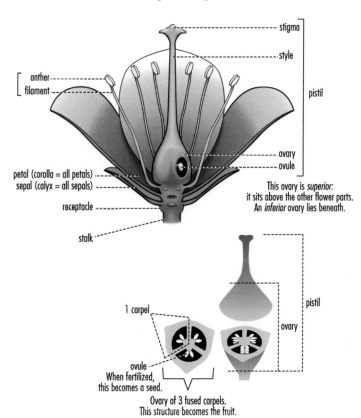

Specializations in Flowers

Flower parts are modified in many ways, from simple variations to highly complex forms in which the various parts may be difficult to identify.

Flower Heads

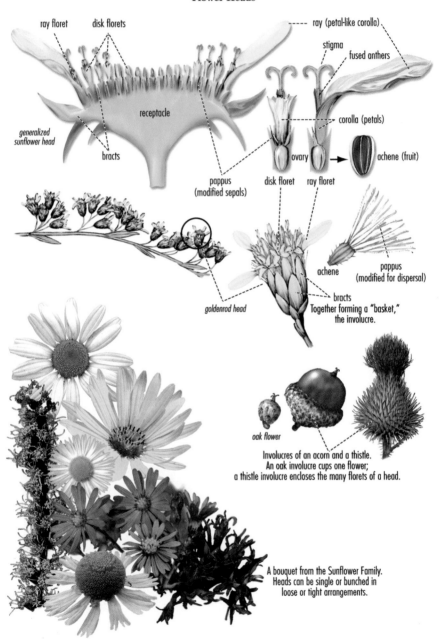

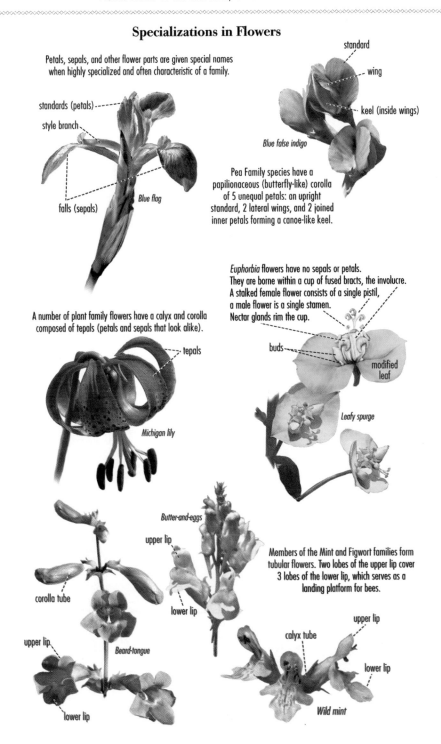

Specializations in Flowers

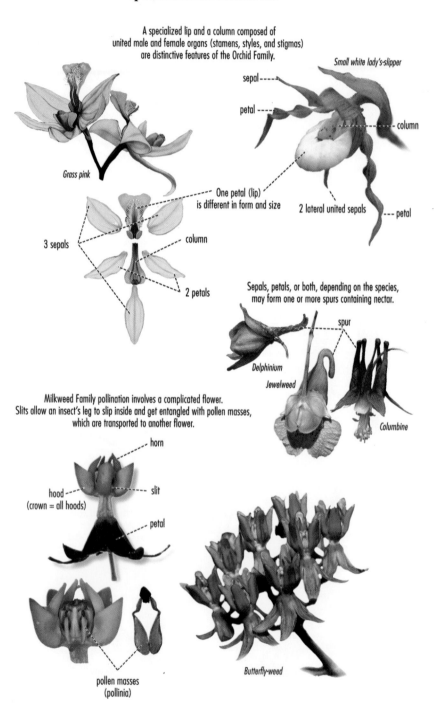

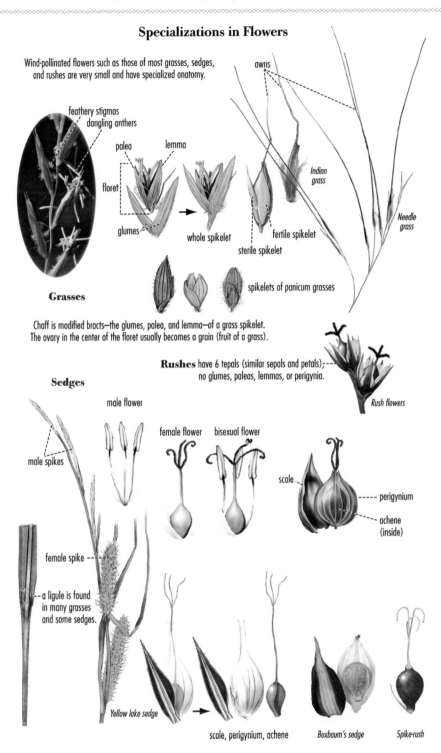

PRAIRIE PLANTS

Species in blue appear on an "Others" page after their respective families. These plants may be extremely rare, appear only intermittently, or are not visible from trails. Some may be primarily marsh, woodland, or woods-edge species, introduced or cultivated plants, or otherwise not part of the prairie community. Because visitors may encounter them, they warrant a brief note. All other species are fully described.

AMARANTHACEAE • AMARANTH FAMILY
Froelichia floridana • Cotton-weed.................................36

ANACARDIACEAE • CASHEW FAMILY
Rhus glabra • Smooth sumac 38
Rhus typhina • Staghorn sumac 39
Toxicodendron rydbergii • Poison-ivy 40

APIACEAE • PARSLEY OR CARROT FAMILY
Angelica atropurpurea • Angelica..................................42
Cicuta maculata • Water-hemlock..................................43
Daucus carota • Queen Anne's-lace................................44
Eryngium yuccifolium • Rattlesnake-master........................45
Oxypolis rigidior • Cowbane 46
Pastinaca sativa • Wild parsnip..................................47
Polytaenia nuttallii • Prairie-parsley...........................48
Taenidia integerrima • Yellow-pimpernel..........................49
Zizia aptera • Heart-leaved golden alexanders 50
Zizia aurea • Golden alexanders 51

APOCYNACEAE • DOGBANE FAMILY
Apocynum androsaemifolium • Spreading dogbane....................52
Apocynum sibiricum • Clasping dogbane 53

ASCLEPIADACEAE • MILKWEED FAMILY
Asclepias amplexicaulis • Clasping milkweed......................54
Asclepias incarnata • Swamp milkweed 55
Asclepias sullivantii • Prairie milkweed.........................56
Asclepias syriaca • Common milkweed..............................57
Asclepias tuberosa • Butterfly-weed..............................58
Asclepias verticillata • Whorled milkweed 59

ASTERACEAE • SUNFLOWER FAMILY
Achillea millefolium • Yarrow....................................60
Ageratina altissima • White snakeroot 61
Ambrosia artemisiifolia • Ragweed 62
Antennaria neglecta • Field pussy-toes 63
Antennaria plantaginifolia • Plantain pussy-toes.................64
Arnoglossum atriplicifolium • Pale Indian-plantain 65
Arnoglossum plantagineum • Prairie Indian-plantain...............66
Artemisia campestris • Field sage-wort 67

Artemisia ludoviciana • White sage 68
Bidens cernuus • Beggar-ticks 69
Cirsium discolor • Pasture thistle 70
Cirsium muticum • Swamp thistle 71
Conyza canadensis • Horseweed 72
Coreopsis palmata • Prairie tickseed 73
Echinacea pallida • Pale purple coneflower 74
Echinacea purpurea • Purple coneflower 75
Erigeron annuus • Annual fleabane 76
Erigeron pulchellus • Robin's-plantain 77
Erigeron strigosus • Daisy fleabane 78
Eupatorium perfoliatum • Boneset 79
Euthamia graminifolia • Grass-leaved goldenrod 80
Eutrochium maculatum • Spotted Joe-Pye-weed 81
Helenium autumnale • Sneezeweed 82
Helianthus grosseserratus • Saw-tooth sunflower 83
Helianthus hirsutus • Hairy sunflower 84
Helianthus occidentalis • Western sunflower 86
Helianthus pauciflorus • Stiff sunflower 87
Helianthus strumosus • Rough-leaved sunflower 85
Heliopsis helianthoides • Ox-eye 88
Hieracium longipilum • Long-haired hawkweed 89
Ionactis linariifolia • Flax-leaved aster 90
Krigia biflora • False-dandelion 91
Kuhnia eupatorioides • False boneset 92
Lactuca canadensis • Wild lettuce 93
Leucanthemum vulgare • Daisy 94
Liatris aspera • Rough blazing-star 95
Liatris cylindracea • Cylindrical blazing-star 96
Liatris ligulistylis • Northern Plains blazing-star 97
Liatris pycnostachya • Thick-spike gay-feather 98
Liatris spicata • Marsh gay-feather 99
Packera paupercula • Northern ragwort 100
Packera pseudaurea • Golden ragwort 101
Parthenium integrifolium • Wild quinine 102
Prenanthes racemosa • Glaucous white-lettuce 103
Pseudognaphalium obtusifolium • Sweet everlasting 104
Ratibida pinnata • Yellow coneflower 105
Rudbeckia hirta • Black-eyed Susan 106
Rudbeckia subtomentosa • Sweet coneflower 107
Rudbeckia triloba • Brown-eyed Susan 108
Silphium integrifolium • Rosinweed 109
Silphium laciniatum • Compass-plant 110
Silphium perfoliatum • Cup-plant 111
Silphium terebinthinaceum • Prairie dock 112
Solidago canadensis • Canadian goldenrod 113
Solidago gigantea • Late goldenrod 114
Solidago missouriensis • Missouri goldenrod 115
Solidago nemoralis • Old-field goldenrod 116
Solidago riddellii • Riddell's goldenrod 117
Solidago rigida • Stiff goldenrod 118
Solidago speciosa • Showy goldenrod 119

Symphyotrichum ericoides • Heath aster 120
Symphyotrichum firmum • Shining aster 121
Symphyotrichum laeve • Smooth aster 122
Symphyotrichum lanceolatum • Panicled aster 123
Symphyotrichum lateriflorum • Calico aster 124
Symphyotrichum novae-angliae • New England aster 125
Symphyotrichum oblongifolium • Aromatic aster 126
Symphyotrichum oolentangiense • Sky-blue aster 127
Symphyotrichum pilosum • Frost aster 128
Symphyotrichum sericeum • Silky aster 129
Vernonia fasciculata • Ironweed 130
Ambrosia trifida • Giant ragweed 131
Anthemis cotula • Stinking chamomile 131
Arctium minus • Burdock 131
Carduus acanthoides • Plumeless thistle 132
Centaurea stoebe • Spotted knapweed 132
Cichorium intybus • Chicory 132
Cirsium arvense • Canada thistle 132
Cirsium vulgare • Bull thistle 133
Doellingeria umbellata • Flat-top aster 133
Erechtites hieracifolius • Fireweed 133
Hieracium caespitosum • Field hawkweed 133
Lactuca serriola • Prickly lettuce 134
Solidago ulmifolia • Elm-leaved goldenrod 134
Sonchus asper • Sow-thistle 134
Symphyotrichum puniceum • Bristly aster 135
Taraxacum officinale • Dandelion 135
Tragopogon dubius • Goat's-beard 135

BALSAMINACEAE • TOUCH-ME-NOT FAMILY
Impatiens capensis • Orange jewelweed 136

BORAGINACEAE • BORAGE FAMILY
Lithospermum canescens • Hoary puccoon 137
Lithospermum caroliniense • Hairy puccoon 138
Hackelia virginiana • Stickseed 139

BRASSICACEAE • MUSTARD FAMILY
Arabis lyrata • Sand cress 140
Barbarea vulgaris • Yellow-rocket 141
Berteroa incana • Hoary-alyssum 142
Hesperis matronalis • Dame's rocket 142
Lepidium campestre • Fieldcress 142
Lepidium densiflorum • Small pepper-grass 142
Sisymbrium altissimum • Tall tumble mustard 142

CACTACEAE • CACTUS FAMILY
Opuntia humifusa • Eastern prickly-pear 143

CAESALPINIACEAE • SENNA FAMILY
Chamaecrista fasciculata • Partridge pea............................144

CAMPANULACEAE • BELLFLOWER FAMILY
Campanula rotundifolia • Harebell.................................145

CARYOPHYLLACEAE • PINK FAMILY
Dianthus armeria • Deptford pink 146
Silene latifolia • Bladder campion.................................147
Silene stellata • Starry campion...................................148
Cerastium fontanum • Mouse-ear chickweed..........................149
Saponaria officinalis • Bouncing-Bet 149

CELASTRACEAE • BITTERSWEET FAMILY
Celastrus orbiculata • Oriental bittersweet..........................150
Celastrus scandens • American bittersweet..........................151

CISTACEAE • ROCK-ROSE FAMILY
Helianthemum bicknellii • Bicknell's rock-rose......................152
Helianthemum canadense • Common rock-rose......................153
Lechea intermedia • Pinweed......................................154

COMMELINACEAE • SPIDERWORT FAMILY
Tradescantia ohiensis • Spiderwort 155

CONVOLVULACEAE • MORNING-GLORY FAMILY
Calystegia sepium • Hedge bindweed..............................156
Convolvulus arvensis • Field bindweed 156

CORNACEAE • DOGWOOD FAMILY
Cornus racemosa • Gray dogwood 157
Cornus amomum • Silky dogwood 158
Cornus stolonifera • Red osier dogwood 158

CYPERACEAE • SEDGE FAMILY
Carex bicknellii • Bicknell's sedge..................................160
Carex buxbaumii • Buxbaum's sedge 161
Carex haydenii • Hayden's sedge 162
Carex molesta • Field oval sedge..................................163
Carex pensylvanica • Pennsylvania sedge...........................164
Carex siccata • Dry-spiked sedge 165
Carex stricta • Common tussock sedge 166
Carex tenera • Marsh straw sedge.................................167
Carex tetanica • Stiff sedge 168
Carex vulpinoidea • Fox sedge....................................169
Eleocharis compressa • Flat-stemmed spike-rush 170
Scleria triglomerata • Nut-rush 171

Carex annectens • Yellow-headed fox sedge..........................172
Carex blanda • Wood sedge 172
Carex brevior • Fescue sedge......................................172
Carex conoidea • Prairie gray sedge...............................172
Carex interior • Inland sedge 173
Cyperus species • Umbrella sedge 173
Schoenoplectus tabernaemontani • Soft-stem bulrush................173

Equisetaceae • Horsetail Family
Equisetum arvense • Common horsetail 174
Equisetum laevigatum • Smooth scouring rush 175

Euphorbiaceae • Spurge Family
Euphorbia corollata • Flowering spurge............................176
Euphorbia esula • Leafy spurge....................................177

Fabaceae • Pea or Bean Family
Amorpha canescens • Lead-plant 178
Amphicarpaea bracteata • Hog-peanut 179
Baptisia alba • White wild indigo.................................180
Baptisia bracteata • Cream wild indigo............................181
Dalea candida • White prairie-clover 182
Dalea purpurea • Purple prairie-clover............................183
Desmodium canadense • Canadian tick-trefoil 184
Desmodium illinoense • Illinois tick-trefoil 185
Lathyrus palustris • Marsh pea 186
Lathyrus venosus • Veiny pea......................................187
Lespedeza capitata • Round-headed bush-clover 188
Lupinus perennis • Wild lupine....................................189
Melilotus alba • White sweet-clover 190
Melilotus officinalis • Yellow sweet-clover.......................190
Pediomelum esculentum • Prairie-turnip............................192
Tephrosia virginiana • Goat's-rue 193
Vicia americana • American vetch..................................194
Vicia villosa • Hairy vetch 195
Coronilla varia • Crown-vetch 196
Lotus corniculatus • Bird's-foot trefoil 196
Medicago lupulina • Black medick 196
Trifolium hybridum • Alsike clover................................197
Trifolium pratense • Red clover...................................197
Trifolium repens • White clover...................................197

Ferns
Onoclea sensibilis • Sensitive fern 198
Osmunda regalis • Royal fern 199
Pteridium aquilinum • Bracken fern 200
Thelypteris palustris • Marsh fern 201

GENTIANACEAE • GENTIAN FAMILY
Gentiana alba • Cream gentian 202
Gentiana andrewsii • Bottle gentian 203
Gentiana puberulenta • Downy gentian 204
Gentianella quinquefolia • Stiff gentian.........................205
Gentianopsis crinita • Fringed gentian 206

GERANIACEAE • GERANIUM FAMILY
Geranium maculatum • Wild geranium.........................208

HYPERICACEAE • ST. JOHN'S-WORT FAMILY
Hypericum kalmianum • Kalm's St. John's-wort 209
Hypericum perforatum • Common St. John's-wort 210
Hypericum pyramidatum • Great St. John's-wort 211

IRIDACEAE • IRIS FAMILY
Iris virginica • Blue flag..212
Sisyrinchium campestre • Prairie blue-eyed-grass 213

JUNCACEAE • RUSH FAMILY
Juncus tenuis • Path rush 214
Juncus dudleyi • Dudley's rush 215

LAMIACEAE • MINT FAMILY
Lycopus americanus • Water-horehound......................216
Mentha arvensis • Wild mint 217
Monarda fistulosa • Bergamot 218
Monarda punctata • Dotted horsemint........................219
Prunella vulgaris • Heal-all...220
Pycnanthemum virginianum • Mountain mint 221
Scutellaria leonardii • Leonard's skullcap 222
Stachys palustris • Hedge-nettle 223
Teucrium canadense • American germander.................224
Agastache scrophulariaefolia • Giant hyssop 225
Glechoma hederacea • Creeping-Charlie......................225
Leonurus cardiaca • Motherwort.................................225
Nepeta cataria • Catnip..225

LILIACEAE • LILY FAMILY
Allium canadense • Wild onion 226
Allium cernuum • Nodding wild onion........................227
Camassia scilloides • Wild-hyacinth.............................228
Hypoxis hirsuta • Yellow star-grass...............................229
Lilium michiganense • Michigan lily............................230
Lilium philadelphicum • Wood lily 231
Zigadenus elegans • Death camas 232
Asparagus officinalis • Asparagus.................................233
Maianthemum racemosum • Solomon's-plume.............233
Polygonatum biflorum • Solomon's-seal 233
Smilax herbacea • Carrion-flower 233

Lobeliaceae • Lobelia Family
Lobelia silphilitica • Great blue lobelia..............................234
Lobelia spicata • Pale-spike lobelia................................235
Lobelia cardinalis • Cardinal-flower...............................236

Lythraceae • Loosestrife Family
Lythrum alatum • Winged loosestrife.............................237

Malvaceae • Mallow Family
Napaea dioica • Glade mallow....................................238
Malva neglecta • Cheeses...239

Melastomataceae • Meadow-beauty Family
Rhexia virginica • Virginia meadow-beauty 240

Onagraceae • Evening-primrose Family
Gaura biennis • Biennial gaura...................................241
Oenothera biennis • Common evening-primrose 242
Oenothera clelandii • Sand evening-primrose 243
Oenothera perennis • Small evening-primrose 244
Epilobium coloratum • Eastern willow-herb........................245

Orchidaceae • Orchid Family
Calopogon oklahomensis • Oklahoma grass pink 246
Cypripedium candidum • Small white lady's-slipper 247
Cypripedium parviflorum var. *makasin* • Small yellow lady's-slipper......248
Platanthera flava • Pale green orchid..............................249
Platanthera leucophaea • Prairie white fringed orchid..................250
Spiranthes magnicamporum • Great Plains lady's-tresses 251

Oxalidaceae • Wood-sorrel Family
Oxalis violacea • Violet wood-sorrel...............................252
Oxalis dillenii • Dillenius's oxalis 253
Oxalis stricta • Common yellow oxalis............................253

Plantaginaceae • Plantain Family
Plantago patagonica • Woolly plantain 254
Plantago rugelii • Rugel's plantain 255
Plantago lanceolata • English plantain 256
Plantago major • Common plantain 256

Poaceae • Grass Family
Andropogon gerardii • Big bluestem..............................258
Bouteloua curtipendula • Side-oats grama.........................259
Bouteloua hirsuta • Hairy grama 260
Calamagrostis canadensis • Blue-joint grass 261
Dichanthelium leibergii • Leiberg's panic grass 262

Dichanthelium oligosanthes • Scribner's panic grass 263
Dichanthelium ovale ssp. *praecocius* • Stiff-leaved panic grass 263
Elymus canadensis • Canada wild-rye.................264
Hesperostipa spartea • Needle grass.................265
Koeleria macrantha • June grass.................266
Panicum virgatum • Switch grass.................267
Phalaris arundinacea • Reed canary grass.................268
Schizachyrium scoparium • Little bluestem.................269
Sorghastrum nutans • Indian grass.................270
Spartina pectinata • Prairie cord grass.................271
Sporobolus heterolepis • Prairie dropseed.................272
Agrostis gigantea • Redtop.................273
Agrostis hyemalis • Bent grass.................273
Elymus hystrix • Bottlebrush grass.................273
Elytrigia repens • Quackgrass.................273
Eragrostis spectabilis • Purple love grass.................274
Phleum pratense • Timothy.................274
Poa compressa • Canada bluegrass.................274
Poa pratensis • Kentucky bluegrass.................274
Bromus ciliatus • Fringed brome.................275
Bromus inermis • Smooth brome.................275
Bromus kalmii • Kalm's brome.................275

POLEMONIACEAE • PHLOX FAMILY
Phlox glaberrima • Smooth phlox.................276
Phlox pilosa • Prairie phlox.................277
Polemonium reptans • Jacob's-ladder.................278

POLYGALACEAE • MILKWORT FAMILY
Polygala polygama • Purple milkwort.................279
Polygala sanguinea • Field milkwort.................280
Polygala senega • Seneca snakeroot.................281
Polygala verticillata • Whorled milkwort.................282

POLYGONACEAE • SMARTWEED FAMILY
Persicaria amphibia • Water smartweed.................283
Persicaria hydropiper • Water-pepper.................284
Persicaria sagittata • Arrow-leaved tear-thumb.................285
Rumex acetosella • Sheep sorrel.................286
Persicaria maculosa • Heart's-ease.................287
Rumex crispus • Curly dock.................287

PRIMULACEAE • PRIMROSE FAMILY
Dodecatheon meadia • Shooting-star.................288
Lysimachia quadriflora • Narrow-leaved loosestrife.................289
Lysimachia ciliata • Fringed loosestrife.................290

RANUNCULACEAE • BUTTERCUP FAMILY

Anemone canadensis • Canada anemone 291
Anemone cylindrica • Candle anemone 292
Anemone patens • Pasqueflower 293
Anemone virginiana • Thimbleweed 294
Aquilegia canadensis • Wild columbine 295
Clematis virginiana • Virgin's-bower 296
Delphinium carolinianum • Carolina larkspur 297
Thalictrum dasycarpum • Tall meadow-rue 298
Caltha palustris • Marsh-marigold 299
Ranunculus pensylvanicus • Bristly buttercup 299

RHAMNACEAE • BUCKTHORN FAMILY

Ceanothus americanus • New Jersey tea 300

ROSACEAE • ROSE FAMILY

Fragaria virginiana • Wild strawberry 301
Geum triflorum • Prairie-smoke 302
Pentaphylloides floribunda • Shrubby cinquefoil 303
Potentilla argentea • Silvery cinquefoil 304
Potentilla arguta • Tall cinquefoil 305
Potentilla norvegica • Rough cinquefoil 306
Potentilla recta • Sulphur cinquefoil 307
Potentilla simplex • Old-field cinquefoil 308
Rosa arkansana • Prairie rose 310
Rosa blanda • Smooth rose 311
Rosa carolina • Carolina rose 312
Rosa setigera • Climbing prairie rose 313
Spiraea alba • White meadowsweet 314
Agrimonia gryposepala • Common agrimony 315
Geum aleppicum • Yellow avens 315
Geum canadense • White avens 315
Rubus allegheniensis • Common blackberry 316
Rubus occidentalis • Black raspberry 316

RUBIACEAE • BEDSTRAW OR MADDER FAMILY

Galium boreale • Northern bedstraw 317
Galium obtusum • Wild madder 318
Houstonia caerulea • Bluets 319

SANTALACEAE • SANDALWOOD FAMILY

Comandra umbellata • Bastard-toadflax 320

SAXIFRAGACEAE • SAXIFRAGE FAMILY

Heuchera richardsonii • Prairie alumroot 321
Saxifraga pensylvanica • Swamp saxifrage 322

Scrophulariaceae • Figwort Family
Aureolaria grandiflora • Yellow false foxglove 323
Castilleja coccinea • Indian paintbrush 324
Chelone glabra • Turtlehead.......................................325
Pedicularis canadensis • Wood-betony 326
Penstemon digitalis • Tall beard-tongue 327
Penstemon gracilis • Slender beard-tongue 328
Penstemon grandiflorus • Large-flowered beard-tongue...............329
Veronicastrum virginicum • Culver's-root..........................330
Linaria vulgaris • Butter-and-eggs.................................331
Mimulus ringens • Monkey-flower 331
Verbascum thapsus • Mullein 331

Solanaceae • Nightshade Family
Physalis heterophylla • Clammy ground-cherry......................332
Physalis virginiana • Virginia ground-cherry 333
Solanum dulcamara • Deadly nightshade..........................334
Solanum ptychanthum • Nightshade...............................334

Typhaceae • Cat-tail Family
Typha latifolia • Broad-leaved cat-tail 335

Valerianaceae • Valerian Family
Valeriana edulis • Edible valerian..................................336

Verbenaceae • Vervain Family
Verbena hastata • Blue vervain 337
Verbena stricta • Hoary vervain...................................338
Verbena urticifolia • White vervain 339

Violaceae • Violet Family
Viola pedata • Bird's-foot violet 340
Viola pedatifida • Prairie violet 341
Viola sagittata • Arrow-leaved violet 342
Viola sororia • Common blue violet...............................343

35

Cotton-weed
Florida snake-cotton, plains snake-cotton, large cotton-weed
Froelichia floridana

This casual pioneer of disturbed sandy soil makes little or no claim to beauty, having small inconspicuous flowers. What color it has resides in the covering of white or brownish, silky-woolly hairs (stems, leaves, and calyx) and straw-colored or blackish bracts of the cottony, raceme-like spike. **Erect annual, 6–39 inches tall, from a taproot.**

Leaves opposite, the main ones inverted-lance- to somewhat spoon-shaped, 1–3¾ inches long, ¼–1½ (averaging more than ⅜) inches wide, entire. **Stem** solitary, simple or with a few erect branches.

Flowers bisexual, stalkless, each subtended by a membranous bract and 2 concave overlapping bractlets, in mostly terminal, interrupted spikes ⅜–3 or more inches long and about ½ inch thick; sepals 5, united most of their length into a flask-shaped tube ¼ inch long, bearing 2 lateral, nearly entire to deeply toothed crests or wings, these largely hidden by copious wool; petals none; staminal filaments united into a membranous tube as long as the calyx tube, bearing 5 anthers alternating with 5 prolonged points between the anthers. Bractlets somewhat circular, thin, dry, and chaffy, shorter than the calyx, enclosing and falling with the flower. *Mid July–beginning of September*

Fruits bladder-like, non-opening, small, membranous, included within the filament tube and the whole surrounded by the hardened calyx; seed 1, dark reddish-brown, lens-shaped. *Mid August–November*

WI range and habitats Local southwest, mostly on sand terraces along major rivers and sand plains in central Wisconsin, in disturbed sandy fields, sandy prairies, and sand blows, less frequently on edges of sandy or rocky woodlands. *West Knoll*

Similar species Slender cotton-weed (*Froelichia gracilis*) is slenderer and narrower-leaved (leaves averaging less than ⅜ inch wide) and has stalkless lateral spikes and smaller flowers. Moreover, the crests of the mature calyx consist of 2 lateral rows of distinct spines. One reference states that slender cotton-weed and large cotton-weed are distinct with no known evidence of hybridization, whereas another says that on the Great Plains, hybrids between them are "all too frequent." Adventive from west of the Mississippi River, the former occurs in our range but is not known from the Arboretum.

Greene Prairie in August

Anacardiaceae • Cashew Family

SUMACS are famous for brightly colored autumn foliage, and staghorn sumac is sometimes cultivated for this reason. Fruits persist over winter and through the following spring, providing excellent winter food for birds. *Erect shrubs or rarely small trees with a few thick branches, forming dense colonies.*

Leaves alternate, pinnately compound, with 11–23 leaflets, occasionally more. Sharp-toothed leaflets taper to a long point. Branchlets, leaf stalks, and inflorescence axes of smooth sumac are hairless and covered by a whitish waxy bloom; those of staghorn sumac are velvety-hairy, covered with long, straight soft hairs.

Flowers numerous, cream-colored to greenish-yellow, in a large dense terminal panicle. Individual plants bear both bisexual and unisexual flowers (polygamous), but whole panicles are often unisexual. *Early June–mid July*

SMOOTH SUMAC
Rhus glabra

Abnormal growth due to disease

Fruits in both species are small, berry-like drupes covered with a persistent red, sticky-hairy outer layer, resinous middle layer, and stony inner layer. The fruit hairs of smooth sumac are very short and fleshy; those of staghorn sumac are long, shaggy, and needle-like.
Late July (smooth sumac) or early August (staghorn sumac)–November

WI range and habitats

Smooth sumac: Southwest, abundant in dry open ground: prairies, sand barrens, abandoned fields, borders of woods and clearings, roadsides, railroads, and fencerows.

Staghorn sumac: Common throughout, in oak barrens, thin woods, thickets, borders, clearings, and wooded or brushy slopes, bluffs, banks, fields, roadsides, and excavations, also on moist shores, rarely in prairies. ***Both throughout the Arboretum***

These two species frequently hybridize, producing a polymorphic series of hybrids and hybrid segregates lumped under the collective name hybrid sumac (*Rhus ×pulvinata;* synonym: *R. ×borealis*). Often, specimens have sparsely hairy herbage and fruits with the two hair types intermixed; others have hairless or sparsely hairy herbage and drupes with long needle-like hairs or densely hairy herbage with extremely minute, nipple-shaped fruit hairs. Hybrids appear to be about half as frequent as smooth sumac, but they are difficult to identify without fruit and are undoubtedly more common than is generally realized.

STAGHORN SUMAC
Rhus typhina
Synonym *Rhus hirta*

Poison-Ivy

Rydberg's poison-ivy, western poison-ivy
Toxicodendron rydbergii
Synonym *Rhus radicans*
Poisonous

Common northern, non-climbing poison-ivy is often lumped into the collective species *Rhus radicans*. The old saying "Leaflets three, let them be" is as applicable as ever. **Small semishrub, 4–36 inches tall, often forming patches from branched subterranean rhizomes.**

Leaves near summit of stem, compound, the 3 leaflets small to large, variable in shape (egg-shaped, elliptic, diamond-shaped, or inversely egg-shaped), $1\frac{1}{4}$ –$7\frac{3}{4}$ inches long, the margins entire, unevenly wavy, coarsely toothed, or bluntly lobed, green and usually shiny in spring and red or russet in fall. Lateral leaflets are somewhat smaller than the terminal one. Lower stem is brown, woody, the few branchlets (if present) upright and hairless.

Flowers in slender axillary, raceme-like panicles that are often hidden by the leaves. *Early June–August*

Fruit a globose drupe, $\frac{1}{4}$ inch or a little less in diameter, smooth and hard. The outer layer of the fruit wall is whitish or cream-colored (light yellowish-gray late in the season) and papery, the inner layer stony. *Early July–winter*, drupes persisting until the following June

WI range and habitats Common to abundant throughout, in almost all habitats: woods, dry prairies, old fields, fencerows, and shores, invading cut-over woodlands, roadsides, railroads, and waste ground. *Throughout the Arboretum*

The entire plant, at all times of the year, is **poisonous to the touch,** causing severe dermatitis in people who are sensitive to its oily sap. Besides direct contact, poisoning cases are caused by handling clothing, garden implements, and pets that have been contaminated, and by smoke from burning vines. Any time you have handled poison-ivy or brushed against it, wash thoroughly with strong soap as soon as possible. Severe cases may require medical help.

Similar species Climbing poison-ivy *(Toxicodendron radicans* subspecies *negundo)*, a southern, high-climbing, woody vine that extends along river valleys in southern Wisconsin, forms clinging aerial roots, is much branched, and has leaves scattered along the branches. It prefers damp to wet forests and bottomlands but can also grow in upland woods. Both types of poison-ivy occur in the Arboretum.

Anacardiaceae • Cashew Family

Look-alikes

"Leaflets three" is a common leaf pattern in the plant world. Shown here are a few familiar species that are often mistaken for poison-ivy.

Virgin's-bower, *Clematis virginiana*

Raspberries, blackberries, *Rubus* species

This is the real thing!

Hog-peanut, *Amphicarpaea bracteata*
This vine is a remarkable look-alike! However, the leaves are not glossy, and the flowers are pea-like.

Box elder, *Acer negundo*
Seedlings and young shoots have 3 or 5 leaflets very similar to poison-ivy in shape and texture.

Virginia creeper and woodbines, *Parthenocissus* species
Five leaflets are obvious; nonetheless these vines are regularly confused with poison-ivy.

Wild strawberry, *Fragaria virginiana*

Angelica
Common great angelica
Angelica atropurpurea

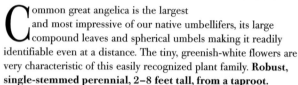

Common great angelica is the largest and most impressive of our native umbellifers, its large compound leaves and spherical umbels making it readily identifiable even at a distance. The tiny, greenish-white flowers are very characteristic of this easily recognized plant family. **Robust, single-stemmed perennial, 2–8 feet tall, from a taproot.**

Leaves compound, the lower ones 2 or 3 times divided into 3s, the uppermost ones bladeless or nearly so. Leaflets coarse, oblong-lance- to egg-shaped, 1½–4 inches long, cleft or sharply toothed. The stalks have inflated, coarsely veiny sheaths. Stems very stout, often purplish.

Flowers in compound umbels, the principal branches radiating in all directions to form a spherical inflorescence. *Late May–July*

Fruits dry schizocarps separating at maturity into two 1-seeded portions (called mericarps, meaning "split fruits"), each strongly flattened, with thin marginal wings and prominent lengthwise ridges on the back. *Late June–early October*

WI range and habitats
Throughout except northeast, fairly local in marshes, shores, sedge meadows, fens, swales, wet thickets, and swamps, particularly characteristic of springy habitats like shaded seepage slopes; full sun to partial shade. *Curtis Prairie, Marion Dunn Prairie, Wingra Oak Savanna*

Common great angelica used to be grown in kitchen and herb gardens, but it has very little medicinal value compared to the European *Angelica archangelica*.

Water-hemlock

**Common water-hemlock,
spotted water-hemlock**
Cicuta maculata
Poisonous

Compound leaves, flat-topped compound umbels, and tiny white to greenish flowers help identify common water-hemlock. **Relatively stout perennial, 2–7 feet tall, with finger-like roots the size and shape of small sweet potatoes.**

Leaves 2 or 3 times compound, the stalks sheathing, the leaflets pinnately arranged on a common axis, lance- to oblong-lance-shaped, $3/4$–$4\ 3/4$ inches long, sharply and coarsely toothed. Unlike most plants, the *veins end at the sinuses between the teeth* instead of running to the tip of the teeth.

Flowers in compound umbels, the primary stalk of which is longer than the leaves and without bracts at the base, whereas the numerous secondary stalks are ± of even lengths and are subtended by linear to lance-shaped bractlets. Individual flowers are on stalks and have 5 white petals that soon fall. *Late June–early September*

Fruits of 2 oval to almost-circular halves (mericarps), distinctly broader than thick, with prominent corky low ribs alternating with reddish-brown grooves. *Late July–mid October*

WI range and habitats Common throughout, in marshes, sedge meadows, wet-mesic prairies, moist to wet woods and thickets, shores, swales, sloughs, and ditches. *Curtis Prairie, Greene Prairie*

Water-hemlocks, like cowbanes, may be harmless-looking, but they are **deadly poisonous.** *Cicuta* is the ancient Latin name of a European relative, poison-hemlock, the herb that was given to the great Athenian philosopher Socrates.

Similar species include bulblet water-hemlock *(Cicuta bulbifera)*, hemlock-parsley *(Conioselinum chinense)*, and poison-hemlock *(Conium maculatum)*, but their leaves are finely divided into linear segments. The first of these has small bulblets in its leaf axils. The other two are rare in Wisconsin and not known from the Arboretum. Water-parsnip *(Sium suave)* has corrugated stems, very thin, finely divided, submersed leaves, and once-compound, copiously toothed, above-water leaves.

Queen Anne's-Lace
Wild carrot, devil's plague
Daucus carota
Invasive weed

Finely divided leaves and delicate inflorescences readily identify this regal Cinderella. Those who appreciate its loveliness use its royal common name, but the gardener or farmer might prefer devil's plague. The tiny flowers are very characteristic of this large and important plant family. **Biennial, 1–5 feet tall, from a fleshy, cone-shaped taproot.**

Leaves more than once pinnately compound, to 6 inches long, the ultimate divisions small and narrow, the stalk with a sheathing base.
Stems solitary, bristly-hairy, the sparse to dense hairs divergent or bent downward.
Flowers in a flat lacy mass to 4 inches in diameter, white except for a central purple to black flower in each umbellet. The primary flower cluster is subtended by feathery bracts, easily seen when the aging inflorescence curls into a basket-like "bird's nest" over the ripening fruits. *Late June–early September*

Fruits dry schizocarps splitting when ripe into halves bearing 4 rows of stiff barbed prickles. These catch on a multitude of dispersing agents, from animals to highway mowers, spreading Queen Anne's-lace seed hither and yon. *Late July–mid October*

WI range and habitats Spreading southeast to northwest, abundant on roadsides and in fields, pastures, and waste places, invading prairies, woods, and other disturbed habitats. *Throughout the Arboretum*

This pernicious, albeit lovely weed is the Eurasian wild ancestor of the common garden carrot and a relative of parsnips, celery, parsley, coriander, caraway, and dill. It was brought to America by the first settlers because of its medicinal properties. The long tap roots are soft enough to be edible their first year, and they can be roasted and ground for a coffee substitute. The leaves were used to flavor soups and stews. The black swallowtail butterfly makes Queen Anne's-lace its food plant.

Of several stories about the best-known common name for this plant, one commemorates an incident in the life of Anne of Denmark, Queen of James I of England, who so admired the wild carrot of herb gardens that she challenged the ladies of the court to produce a lace with a pattern as lovely as its inflorescence. (Fancy adds that the queen, in her haste, pricked her finger and bled a single drop into the center of her work. Hence the one dark flower in the center of the inflorescence.)

Similar species Caraway, *Carum carvi*, an occasional weed of disturbed ground, has less-divided leaves. Inflorescences of Japanese hedge-parsley *(Torilis japonica)*, an ecologically invasive, new exotic to Wisconsin, resemble those of Queen Anne's-lace but are smaller, somewhat less dense, and are subtended by 2–several bracts.

Apiaceae • Parsley or Carrot Family

RATTLESNAKE-MASTER
Eryngium yuccifolium

This remarkable species is unlike any other native plant in our area, having yucca-like leaves and tiny white flowers packed into dense prickly, ball-like heads. **Stiffly erect perennial, 1–3 feet tall, from a bundle of tuberous woody roots.**

Leaves strap-shaped, leathery, with parallel venation, remotely spaced, marginal bristles, and sheathing stalks. Basal leaves rigid; stem leaves reduced above.

Flowers in dense heads (really contracted umbels), the heads several in an irregularly branched open cluster, the central head blooming earliest. *Early July–late August*

Fruits stalkless schizocarps separating at maturity into 2 seed-like mericarps covered with little flattened scales. *Mid October–mid November*

WI range and habitats Locally common only in the southern two tiers of counties, in dry-mesic to wet-mesic prairie remnants. *Curtis, Greene, and Sinaiko Overlook prairies, Juniper Knoll, Visitor Center parking lot*

Eryngium comes from a Greek name for sea holly *(Eryngium vulgare)*. The common name refers to an old belief that the roots could be used to heal rattlesnake bites.

Stiff Cowbane
Water drop-wort
Oxypolis rigidior

Slender and scarcely branched, stiff cowbane is one of the less conspicuous umbellifers in our area. The few, firm-textured leaflets of the once-compound leaves suffice to identify this species even in the absence of flowers or fruits. **Slender perennial, 2–4 feet tall, from a bundle of spindle-shaped tubers.**

Leaves few, once-pinnately compound, to 12 inches long and 10 inches wide, the stalks sheathing, the usually 5–9 stalkless leaflets oblong to lance-shaped, somewhat leathery, entire or less often with a few coarse, irregularly spaced teeth. Plants entirely hairless.

Flowers white, tiny, each on a slender spreading stalk, in compound umbels with as many as 45 spreading branches of almost equal length, the leading umbel with 0–2 primary bracts but the umbellets with a few linear bractlets. *Late July–September*

Fruits oval to egg-shaped, $1/8$–$1/4$ inch long, flattened, appearing 5-ribbed. *Mid August–mid October*

WI range and habitats Infrequent south, in marshes, mesic to wet prairies, fens, borders of swampy woods and thickets, swales, and ditches, often near springs. *Curtis and Greene prairies*

The genus name is from the Greek *oxys*, sharp, and *polios*, white, probably alluding to the awl-shaped bracts and white petals. Cowbane superficially resembles water-parsnip *(Sium suave)*, which differs in its finely and uniformly toothed leaf margins, numerous bracts and bractlets, and larger fruits.

Wild Parsnip
Pastinaca sativa
Poisonous invasive weed

Large, coarsely divided leaves and large, flat-topped, yellow-flowered umbels make wild parsnip readily identifiable even at a distance. It can form extensive populations that turn roadsides and fields into a carpet of yellow. **Tall branching biennial, 1–5 feet tall, from a long taproot.**

Leaves once pinnate-compound, the lower long-stalked, the upper ones with conspicuously wide sheaths, 3–10 inches long, made up of 5–15 oblong leaflets that are usually more than $3/8$ inch wide. The stout stem is hollow and angular or grooved, the herbage hairless.

Flowers in large compound umbels of 15–25 primary branches, each supporting 7–10 umbellets consisting of many tiny yellow flowers. ***Late June–early September***

Fruits dry schizocarps separating at maturity into 2 seed-like oval mericarps, each $1/4$ inch long, very much flattened, the lateral ribs extended into thin (not corky) wings. ***Late July–mid October***

WI range and habitats Abundant south and central, scattered northward, on roadsides, abandoned fields, pastures, and waste places, invading prairies, woods, and other disturbed habitats. *Throughout the Arboretum*

Like many Carrot Family species, parsnip is a host plant for the black swallowtail butterfly caterpillar.

This pernicious Eurasian weed, originally introduced as a garden food crop or possibly accidentally, is widely naturalized in North America and ubiquitous in disturbed ground in southern Wisconsin. Chemical compounds called furocoumarins in the **stem and leaf sap can cause phytophotodermatitis**: red blotches, rashes, and blisters will appear on skin within hours and usually worsen during following days. This is a *phototoxic* reaction, not an allergic dermatitis. Tissue injury from plant sap ("phyto") develops only after light exposure ("photo"). Wear gloves, long sleeves, and long pants when working among wild parsnip.

Prairie-Parsley
Polytaenia nuttallii
Wisconsin threatened species

Prairie-parsley is a distinct native plant in our area and, though possessing the characteristics of its easily recognized plant family, is nonetheless unmistakable in all stages of life. **Stout, single-stemmed perennial, 2–3 feet tall, from a taproot.**

Leaves mostly alternate and pinnate-dissected, the leaflets stalkless or confluent, the ultimate segments narrowly egg-shaped, coarsely few-toothed. Upper leaves are opposite and 3-cleft and have conspicuously dilated sheaths.

Flowers in compound umbels, the 5 yellow petals with abruptly in-turned tips. Flower stalks and bractlets hairy. *Late May–June*

Fruits dry corky schizocarps that split lengthwise from the base into stalked pairs of seed-like mericarps, these broadly oval, strongly flattened. *Mid July–early November*

WI range and habitats Rare and local southeast, reappearing in Monroe County, in dry to wet-mesic prairie remnants, oak barrens, and cut-over oak forests, and on outcrops and cliffs in the Baraboo Hills.

Greene Prairie

The species name honors its discoverer, Thomas Nuttall (1786–1859), who described for the first time many of our prairie species while accompanying the Astoria Party in its 1811 exploration of the upper Missouri River.

Yellow-Pimpernel
Taenidia integerrima

An easily recognized member of the Carrot Family, this elegant species is our only umbellifer with both yellow flowers and strictly entire leaflets. The few, well-spaced leaves and tiny yellow flowers in a loose open circular umbel give it an airy appearance. **Slender branching perennial, 12–44 inches tall, from a deep taproot.**

Leaves divided into 3 divisions, these in turn once or twice divided into 3s, the basal on long sheathing stalks, the upper ones stalkless; leaflets stalkless, egg-, elliptic-egg-, or inverted-egg-shaped, $3/8$–$1 1/2$ inches long, with a tiny short abrupt point and translucent margins. Entire plant hairless, often whitened with a waxy bloom, with a light, celery-like odor.

Flowers in compound umbels 2–5 inches across (to 7 inches in fruit) on stalks 2–8 inches long, the 9–20 primary branches spreading or ascending, of unequal lengths, lacking bracts and bractlets; calyx wholly fused to the ovary, the teeth (sepals) obsolete; style bases not expanded into a disk. Central flowers of umbellets male (with reduced pistil); marginal flowers female and fertile (with reduced or no stamens). *Early May–early July*

Fruits separating at maturity into 2 seed-like mericarps, these broadly oblong to oval, $1/8$ inch long, slightly compressed, pentagonal in cross-section, the 5 longitudinal ribs brown, inconspicuous, the intervals between them nearly black. *Late June–August*

WI range and habitats Infrequent to locally abundant, mostly southeast, in deciduous woodlands (especially under oaks and aspen), oak savannas, and thickets (often along streamsides), less often on dry to medium prairies, where thriving in transition zones of mixed prairie and woody plants, sporadic northward on jack pine plains and in medium forests; generally on wooded slopes, banks, and outcrops, invading fencerows, railroad embankments, and roadsides; basic soils. *West Curtis Prairie, East Greene Prairie*

Flower-visiting insects include bees (both short- and long-tongued), wasps, flies, and beetles, all of which suck nectar. Yellow-pimpernel is a larval host plant for the black swallowtail butterfly.

Like starry campion (*Silene stellata*), yellow-pimpernel is shade-tolerant, and if present in the soil seed bank, it will reappear immediately in oak savanna remnants in which fire is reintroduced as a management technique.

Apiaceae • Parsley or Carrot Family

Golden Alexanders

The attractive compound foliage of these plants comes up in spring, looking less divided and firmer-textured than other Carrot Family members, soon to be accompanied by tiny yellow flowers in dense, flat-topped umbels. **Perennials, 1–3 feet tall, from bunched roots.**

Basal leaves long-stalked, $3/4$–4 inches long, toothed. **Stem leaves** few, compound, irregularly toothed, pointed. Most or all basal leaves of heart-leaved golden alexanders are undivided and heart-shaped, whereas those of golden alexanders are once or twice divided into 3s. **Stems** and leaf sheaths relatively slender.

Flowers in dense, flat-topped, compound umbels 5–6 inches across with 6–18 spreading-ascending, mostly somewhat equal branches. The central flower and fruit of each umbellet is fertile and stalkless. Bracts lacking, but umbellets are subtended by a few inconspicuous linear bractlets shorter than or equaling the flower stalks. *Early May through July*

Fruits dry schizocarps separating at maturity into 2 oblong-egg-shaped mericarps, these $1/8$ inch long, compressed laterally, the low ribs threadlike (not winged). *Late June–early September*

Heart-leaved Golden Alexanders
Zizia aptera

Apiaceae • Parsley or Carrot Family

Wisconsin ranges and habitats

Heart-leaved golden alexanders: South, locally common in moist to mesic prairies, drained or burned marshland, bur oak groves, and (rarely) dry lime prairies, also rare in northwestern sandy pine barrens. *Curtis Prairie, Greene Prairie*

Golden alexanders: Common south, less frequent north, mesic to wet prairies, fens, old fields, roadsides, openings and thickets along streams, rights-of-way, and fencerows, less often in dry prairies and open spots in deciduous woods. *Curtis Prairie, East Knoll, Greene Prairie, Sinaiko Overlook Prairie*

These two species strongly resemble not only one another but also meadow-parsnip (*Thaspium trifoliatum*), which has almost stalkless leaves that are finely toothed. In meadow-parsnip, all flowers are stalked, and the fruits have thin wings instead of low ribs. The three species differ in their basal leaves, which in heart-leaved golden alexanders are simple (not divided), egg-shaped or almost circular, the leathery blades with rounded teeth. In golden alexanders both basal and stem leaves are alike, all divided, stalked, and sharply toothed, the leaflets egg- to lance-shaped, membranous in texture. Golden alexanders occurs in large patches in late spring along roads in wet ditches and is often confused with wild parsnip (*Pastinaca sativa*). All species are common throughout the tall-grass prairie region, but meadow parsnip does not occur in the Arboretum. Golden alexanders was once reputedly used to heal wounds and relieve headaches, fevers, and syphilis.

GOLDEN ALEXANDERS
Zizia aurea

SPREADING DOGBANE
Apocynum androsaemifolium

Low, reddish stems with spreading forked branches, drooping leaves, and pink-striped flowers distinguish this shrub-like herb. **Perennial, 8–24 inches tall, with milky sap, spreading by rhizomes.**

Leaves opposite, short-stalked, spreading or drooping, dark green above, egg- to oblong-lance-shaped, with a short abrupt projection at the tip.

Flowers in showy clusters (cymes), slightly nodding and fragrant, nearly white or pink with pink-veins, the corolla bell shaped, $3/16$ inch long and broad, tubular, the lobes flaring or recurved. The inflorescences flower simultaneously, the uppermost one usually overtopping the foliage. *Mid June–August*

Fruits paired slender pods (follicles) characteristic of the family, $2 1/4$ – 6 inches long, splitting open down one side; seeds numerous, spindle-shaped, $1/8$ inch long, tipped with a parachute of long silky hairs. *Late August–early October*

WI range and habitats Common throughout on edges and openings of upland forests and pine and oak savannas, sometimes in prairies, but more often along roadsides, railroads, and old fields, occasionally in damp sites; in diverse sunny and grassy or partly shady and brushy habitats. *Curtis Prairie, East and West knolls*

The genus name comes from the Greek words *apo,* far from, and *cyon,* a dog, and reflects the allegation that dogbanes are especially toxic to dogs.

Close relatives of the milkweeds, dogbanes also produce an acrid milky sap (latex) throughout the plant. This juice contains several glycosides, which have medicinal effects similar to those of digitoxin, one of the drugs most commonly used as a heart stimulant. It causes severe skin blisters in some people, whereas others can handle the plant with impunity. The roots were used as an emetic. When dried and peeled, the bark provides a strong fiber that was once used by many Native American tribes for thread for sewing and cord for fishing and trapping. The Menominee used these fibers to make their bow strings, and orioles often use them to build nests.

Clasping Dogbane
Indian hemp
Apocynum sibiricum

Flowers of this relatively large plant are delicate and fragrant, like those of spreading dogbane. The leaves are reminiscent of those of milkweeds and flowering spurge. **Perennial, 1–5 feet tall, with milky juice, reproducing freely from bud-bearing roots to form patches.**

Leaves simple, opposite, ascending or slightly spreading, from stalkless or nearly so (those of the main stem) to short-stalked (on upper branches), oval to oblong-lance-shaped, entire, always hairless. The uppermost inflorescence is usually overtopped by foliage.

Flowers small, in terminal and axillary round clusters (cymes) of 2–10, milky white to yellowish or greenish, the corolla urn shaped. *June–mid August*

Fruits usually paired slender elongate pods, 3–8 inches long, hanging at maturity, opening along a suture to reveal many seeds furnished with long tufts of white silky hairs to carry them on the wind. *September*

WI range and habitats Frequent in prairies, fens, and similar grassy areas, also abandoned fields, marsh edges, and fencerows, readily colonizing railroads, roadsides, and waste places; in dry to moist, usually open habitats. *Curtis and Greene prairies*

Like spreading dogbane, clasping dogbane produces toxic, glycoside-containing latex and consequently was much used in local medicine. The tough fibrous bark of its stems yields a strong thread that was braided into cord by Native Americans to make nets, string bags, and coarse fabric, hence the common name.

Similar species Indian hemp and hemp dogbane *(Apocynum cannabinum)* are readily confused with each other. Hemp dogbane is less common and not as weedy, with leaves noticeably stalked, narrowed to broadly rounded at the base, and hairy to hairless. The species are variable, and some plants are intermediate, making it doubtful whether they should be maintained as separate species. All three dogbanes apparently hybridize with each other.

Clasping Milkweed
Sand milkweed, blunt-leaf milkweed
Asclepias amplexicaulis

Milkweeds constitute a large and interesting genus characterized by showy umbels of distinctive, extremely complex flowers on usually simple, strong straight stems. Blunt-leaf milkweed is so-called from the broad leaves that are blunt to rounded at the apex. **Perennial, 12–30 inches tall, from a crown at the surface of the ground and deep rhizomes.**

Leaves stalkless, broadly oval to broadly oblong, to 5 inches long, with broadly rounded to heart-shaped bases and wavy margins, firm, hairless (except margins), with a waxy cast.

Flowers in a solitary terminal spherical cluster (umbel) on an erect elongate stalk; flower stalks mostly 1–2 inches long, minutely hairy; flowers greenish-purple. In this genus the 5 corolla segments are bent downward when in full bloom and conceal the calyx. There is also a prominent, 5-parted crown borne on a stout column, made up of cup- or hood-like structures called hoods, each often bearing a slender incurved horn (the function of the horn is unknown). *Mid June–mid July*

Fruits narrowly spindle-shaped, erect on curved or contorted stalks, 3–6 inches long, minutely hairy to hairless, with a waxy whitish cast; seeds broadly egg-shaped, $5/8$ inch long, bearing a tuft of long silky hairs.

WI range and habitats Rather uncommon south and central, also Oconto County, in dry prairies, recovering old fields, savannas, clearings in oak woods, roadsides, and railroads; dry open sandy ground. *West Knoll*

See notes under swamp milkweed.

Similar species Of the 14 milkweed species in our area, clasping milkweed is most similar to prairie milkweed (*Asclepias sullivantii*), a moist-prairie species often having 2 or 3 umbels, hairless flower stalks, and the horns shorter than the hoods (and largely concealed).

Swamp Milkweed
Asclepias incarnata

This is our tallest and showiest milkweed. As its common name indicates, it is a denizen of marshes and wet meadows, its beautiful rose-pink, relatively small flowers initiating a succession of late summer bloom. **Fairly stout perennial, 1–4 feet tall.**

Leaves opposite, rather numerous, short-stalked, elliptic-egg- to lance-shaped, to 4 inches long, entire. Sap less milky than in other milkweeds.

Flowers in several umbrella-shaped umbels, $1/4$ inch wide, with 5 calyx lobes and 5 petals, all downward curving, and just above the petals and colored like them, a crown of 5 scoop-shaped hoods, each surrounding a little incurved horn. Flowers delicately and sweetly fragrant. *Mid June–mid August*

Fruits erect, lance-spindle-shaped pods 2–3$1/2$ inches long, opening along one side to reveal 30–60 flat brown seeds, each bearing a tuft of white silky hairs. Seeds downwardly overlapping, filling the interior of the ovary.

WI range and habitats Ubiquitous in wet open habitats, such as edges of swamps, bogs, woods, and thickets, shores, fens, prairies, swales, and ditches. ***Curtis Prairie, Greene Prairie, Teal Pond***

Swamp milkweed is a host and nectar plant for monarch butterflies and some of the hairstreaks and blues.

Each milkweed flower has slits at the base of the crown that allow an insect's leg to slip inside, where it tends to catch on clips that connect bands to which the pollen masses are attached. Extricated by the legs of struggling insects, pairs of pollen masses are unwittingly carried away, to be deposited near the stigmatic chamber of the next flower visited. This complicated mechanism means that only a few flowers are pollinated. The flowers become fatal traps to small insects whose legs become so firmly lodged that they can not escape.

Following the Angiosperm Phylogeny Group (APG) system, most current taxonomists now include the Milkweed Family in the Dogbane Family (Apocynaceae).

Prairie Milkweed
Smooth milkweed, Sullivant's milkweed
Asclepias sullivantii
Wisconsin threatened species

This species is quite similar to other milkweeds having broad thick leaves and relatively large, purplish to pinkish-rose flowers. These are not easily distinguished without resorting to technical characters of the extraordinary flowers, which in prairie milkweed are faintly scented and have tall hoods largely concealing the horns. **Coarse perennial, 2–5 feet tall, from a deep rhizome.**

Leaves opposite, oblong to oblong-egg-shaped, $2\frac{1}{2}$–7 inches long, on a stalk $\frac{1}{8}$ or less inch long, with a rounded to heart-shaped, non-clasping base. Plants hairless, oozing thick white sap from any cut or broken surface.

Flowers $\frac{1}{2}$–$\frac{3}{4}$ inch long, on stalks about 1 inch long, in a solitary umbel at the end of the stem or a few, globe-shaped umbels in the axils of upper leaves, each umbel to 2 inches in diameter; calyx lobes 5; corolla lobes 5, downward-curving; crown of 5 upward-directed, somewhat fleshy, pouch-like hoods attached to the column and surpassing it in height, each hood with a sickle-shaped horn that is sharply incurved over the anther head. *Late June–mid August*

Fruits lance-egg-shaped pods 3–4 inches long, 1 inch thick, minutely hairy to hairless, obscurely warty toward the tip; seeds numerous, broadly oval, $\frac{1}{4}$ inch long, flat, each with a white tuft of silken hairs $1\frac{3}{4}$ inches long. As they escape from the pods, the seeds are lifted into the air and carried away on the wind.

WI range and habitats Once more common, now local in a few low to medium prairie relicts and fens in 8 southeastern counties, possibly still surviving as isolated plants in prairie habitats along railroads and roadsides even though the native vegetation may have largely disappeared; avoids permanently saturated soil, intolerant of disturbance. *Greene Prairie*

Similar species Among Wisconsin species, prairie milkweed is most closely related to purple milkweed (*Asclepias purpurascens*), a rare species of dry prairies and savannas that does not grow in the Arboretum. Purple milkweed and common milkweed differ from prairie milkweed in having leaves and pods that are coated with grayish down. Common milkweed pods are usually abundantly covered with pointed, wart-like projections.

COMMON MILKWEED
Asclepias syriaca

This pleasant plant is so common that its main characteristics are well known: milky juice; heavily scented, 2-inch clusters of cream-, green-, or purplish-tinged pink flowers; large pods; and seeds tipped with silky parachutes. **Coarse perennial, 3–6 feet tall, spreading by horizontal rhizomes.**

Leaves opposite, broadly oval to lance-oblong, large, often 4–6 inches long, finely and softly grayish-downy beneath.

Flowers in few to several umbels at upper nodes, relatively large and showy, with 5 downward-curving petal lobes and a 5-part crown of upward-directed, hood-like nectar cups, each with a claw-like horn arising from the cavity. *Early June–early August*

Fruits warty green pods, single or paired, erect on downward-bent stalks, slenderly egg-shaped, 3–5 inches long, covered with whitish woolly hairs, eventually turning brown and splitting open to expose numerous flat brown seeds, overlapped like the scales of a fish, each with a parachute-like tuft of silken hairs.

WI range and habitats
Extremely common throughout, in prairies, fields, pastures, roadsides, and railroads, also dunes near Lake Michigan; weedy in disturbed, sunny habitats, including city lots. *Curtis Prairie, Greene Prairie, grassy areas throughout the Arboretum*

Milkweeds were considered outstanding medicinal plants by many Native American tribes. The sticky milky latex contains potent compounds, including cardiac glycosides, and protects the milkweeds from the ravages of many leaf-eating insects. One insect, however, the monarch butterfly, has co-evolved with the milkweeds, which are the exclusive food of the caterpillars and a good nectar source for the adults. The caterpillars ingest the toxic sap with impunity, retaining the poison as adults and making them unpalatable to birds, which soon learn to avoid monarch butterflies and their look-alikes, the viceroys.

Similar species Purple milkweed (*Asclepias purpurascens*) holds its umbels above the leaves and has pinkish-red to purple corolla lobes and hoods that average somewhat longer than in common milkweed. It favors savanna borders and prairie thickets but is not found in the Arboretum.

BUTTERFLY-WEED
Orange milkweed
Asclepias tuberosa

Clusters of vivid yellow-orange to orange-red flowers make this is a spectacular plant in the wildflower garden as well as prairies. The erect to sprawling stems arise in clusters and bear many narrow leaves and broad umbels at the branch tips. **Rather stout-stemmed perennial, 15–30 inches tall, from a deep tuberous root.**

Leaves mostly alternate (only in small part opposite), unstalked or slightly stalked, linear to oblong-lance-shaped, mostly 2–4 inches long, covered with short coarse hairs, particularly beneath. **Stems** conspicuously rough-hairy.

Flowers moderately large for a milkweed, in 2 to several umbels at the top of the stem, with greenish-orange to red or yellow corolla divisions and orange to red or (in a color form) pale yellow hoods. *Mid June–August*

Fruits narrow spindle-shaped pods, erect but on deflexed stalks, about 3–5 inches long, with broadly oval seeds about $1/4$ inch long, their tufts of hairs $1-1\frac{1}{2}$ inches long.

WI range and habitats Frequent southwest, on dry-medium prairies, steep limy prairies, fields, and roadsides; particularly common in sandy areas of central Wisconsin; requires excellent drainage and full sun. *East Knoll, West Knoll, Greene Prairie*

This milkweed is unique in our flora in having leaves that are not opposite, juice that is not obviously milky, and orange flowers. Butterfly-weed is so-called from the number of butterflies (especially the monarch) that are almost always to be seen on its flowers. It should not be picked or dug, as it has become increasingly rare in many areas (this species grows readily from seed).

Butterfly-weed was one of the most important medicinal herbs for Native Americans. The tubers were grated and applied to cuts and bruises, chewed to treat pleurisy, and brewed into a tea to treat heart trouble, diarrhea, inflammation of the lungs, and other diseases.

Asclepiadaceae • Milkweed Family

WHORLED MILKWEED
Asclepias verticillata

This is a distinctive slender plant with numerous, whorled or opposite, almost needle-like leaves and tiny white or greenish flowers in several small umbels from the upper nodes. **Slender perennial, 8–20 inches tall, from crowns with shallow roots and deep rhizomes, forming patches.**

Leaves opposite or whorled, 3–6 per whorl, stalkless, narrowly linear, $^3/_4$–2 inches long, $^1/_{16}$ inch or less wide, the margins rolled toward the underside.

Flowers in several small round umbels from the upper nodes, $^1/_8$ inch wide, with 5 narrow angled calyx lobes, 5 downward-curving corolla divisions, and long, claw-shaped horns arching gradually over the anther head. *July–early September*

Fruits erect pods on erect stalks, slenderly spindle-shaped, about 3–4 inches long; seeds oval, about $^3/_{16}$ inch long, the white hair tuft about 1 inch long.

WI range and habitats Very common south, in medium to dry prairies and open woodlands with prairie flora, weedy in abandoned fields, pastures, and roadsides; nutrient-poor, often sandy soils. *Curtis Prairie, West Knoll*

The name of the genus honors Asklepios, the Greek god of medicine, in acknowledgment of the use of many milkweed species in folk medicine. Several species have been reported to be poisonous to grazing animals, including whorled milkweed, common milkweed (*Asclepias syriaca*), and butterfly-weed (*A. tuberosa*).

One of the other rarer prairie species you might spot on the Arboretum's West Knoll is short green milkweed (*Asclepias viridiflora*), which has few (6–12 pairs) oval to oblong, thick-textured leaves; densely flowered, globose umbels that are essentially unstalked; and moderately large, pale green flowers.

Common Yarrow
Milfoil
Achillea millefolium

A familiar roadside companion, yarrow is distinctive for leaves even lacier than those of Queen Anne's-lace and dull white heads in flat-topped flower clusters. The tiny flowers are closely grouped into small heads often mistaken for single flowers. **Rhizomatous perennial, 8–30 inches tall, often growing in patches.**

Leaves dull green, fernlike, 1, 2, or 3 times pinnate-parted and -dissected into fine segments, to 6 inches or more long, with a pungent scent reminiscent of the mouthwash Listerine. **Stems** unevenly cobwebby with long soft hairs.

Flower heads numerous, in rather dense, flat-topped clusters at the top of the plant, each $1/4$ inch or less across, with a ring of 3–5 white (occasionally pink to rose purple), petal-like, female ray florets surrounding a central group of 10–30 whitish bisexual disk florets. Each floret has an inferior ovary below the tubular corolla formed by 5 joined petals, which in ray florets becomes one large ray. ***June–mid October***

Achenes oblong to inverted-egg-shaped, $1/16$ inch long, obliquely flattened and 3-edged, hairless. **July–mid October**

WI range and habitats Almost ubiquitous wherever it is not too wet or shady: prairies, abandoned fields, open woods, pastures, roadsides, railroads, shores, waste places, and other disturbed habitats. ***Throughout the Arboretum***

The genus is named after Achilles, who is said to have carried the plant into battle to treat wounds. The species epithet means "thousand-leaved." Such common names as *soldier's woundwort, sanguinary, nosebleed,* and *staunchweed* refer to its pharmaceutical properties, whereas *milfoil* and *thousand-leaf* allude to the many divisions of the leaves. Many cultures have used yarrow not only in folk medicine but also in magic and divination. It enhances many a home garden and dried flower arrangement.

Yarrow is a polyploid complex throughout far northern regions. Both native and Eurasian taxa occur in North America, and it is often uncertain whether a particular specimen is native or introduced.

Asteraceae • Sunflower Family

WHITE SNAKEROOT
Ageratina altissima
Synonym *Eupatorium rugosum*

Although considerably smaller than our other species of Joe-Pye-weed and weedy besides, white snakeroot is nonetheless a handsome plant. It produces emerald green leaves and velvety-looking, white heads in small dense clusters that brighten up shady spots late in the season. Like virtually all members of the Sunflower Family, the fruits are dispersed by wind. **Perennial, 1–4 feet tall, from a small crown.**

Leaves opposite, simple, triangular-egg-shaped, $2\frac{1}{2}$–6 inches long, on distinct stalks $\frac{3}{8}$–$1\frac{1}{2}$ inches long, rather coarsely and evenly cut-toothed, 3-veined basally. **Stems** solid, green, without a pale waxy bloom.

Flower heads in flattish to slightly dome-shaped clusters, each consisting of 10–24 (mostly 12–20) bisexual disk florets only; involucre $\frac{1}{8}$–$\frac{3}{16}$ inch high, its bracts in 2 overlapping ranks of nearly the same length, mostly green with narrow pale margins; corollas white, $\frac{1}{8}$ inch long, with long style branches protruding far out of the corolla tube. **Late July–September**

Achenes blackish, $\frac{3}{16}$–$\frac{1}{4}$ inch long, 5-angled and -veined, glandless, with a well-developed pappus of hair-like bristles in a single whorl. *(Late July) August–early October*

WI range and habitats Abundant throughout except the extreme northwest, in dry to moist forests, woodlands, savannas, thickets, shady ravines, riverbanks, and floodplains, especially in disturbed areas, openings, and trails; prefers deep soil and light shade; weedy, including in urban areas. *Throughout the Arboretum*

This is a widespread, abundant, and variable species. Its stems and leaves contain glycosides and the complex alcohol tremetol, which causes "trembles," a fatal disease of livestock. Tremetol is soluble in milk fat and is transmitted to other animals or humans when they drink poisonous milk from cows that have grazed on white snakeroot tops. "Milk sickness" became a common condition and one of the leading causes of human death in parts of the East and Midwest during early colonial times. It is said to have killed Nancy Hanks Lincoln, mother of Abe, when the future president was nine years old.

Common Ragweed
Annual bur-sage, short ragweed
Ambrosia artemisiifolia

This homely herb is unmistakable owing to the deeply cut leaves, from which it gets both its common name and species epithet, and inconspicuous greenish heads with reduced florets, which are specialized for wind pollination. **Low slender annual, 3–28 (40) inches tall, from a short taproot.**

Leaves opposite below, alternate above, once to thrice pinnate-divided into toothed segments that may themselves be lobed (uppermost leaves occasionally unlobed), 2–4 inches long, ± smooth above. Plants coarsely hairy, resinous, and aromatic.

Flower heads unisexual, $1/8$ inch across, both male and female borne on the same plant, the former in 1–several, elongate, spike-like racemes, the female clustered in axils of leaves and bracts below. Male heads with only a few involucral bracts in one saucer-shaped series. Female heads 1- or 2-flowered. *Late July–late September*

Achenes enclosed within the involucres, the whole forming inversely egg-shaped burs commonly with reddish-brown to black mottling, $1/8$ inch long, each with 4–7 or more thorn-like protuberances below the base of the awl-shaped beak. *Early September–late October*

WI range and habitats Abundant throughout except a few north-central counties, in disturbed spots in dry prairies and woodlands and open waste ground generally: fields, pastures, roadsides, railroads, city lots and streets; a serious weed, aggressive in any disturbed, over-grazed, or sterile soils. *Throughout the Arboretum*

Ecologically speaking, all the ragweeds are "weeds," being adapted to open or disturbed soil. Fields, roadsides, and gardens are weedy precisely because we are constantly cultivating or otherwise disturbing the ground, creating a bed for ever-present, rapidly growing seeds.

Probably few hay fever victims recognize common ragweed, even though it is the primary culprit producing the irritating, air-borne pollen that torments them.

Similar species Western ragweed (*Ambrosia psilostachya*) is perennial, forming patches from runner-like roots. Its leaves are less divided and thicker in texture than those of common ragweed. Giant ragweed (*A. trifida*) is usually taller (to 9 feet) and has larger, palmately 3–5-lobed or sometimes unlobed leaves (never compound or pinnate-cleft).

Asteraceae • Sunflower Family

Field Pussy-toes
Cat's-foot
Antennaria neglecta complex

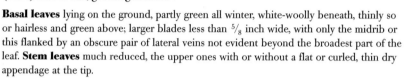
female flower heads

Pussy-toes bear puffs of small whitish heads, looking like furry cats' paws, at the summit of semi-naked stems that shoot up from basal rosettes interconnected by leafy creeping runners. Field pussy-toes is distinguished from the other main group of pussy-toes by its small, 1- or obscurely 3-veined leaves. **Perennial, $1\frac{1}{2}$–16 (female) or $\frac{3}{4}$–5 (male) inches tall, growing in mats.**

Basal leaves lying on the ground, partly green all winter, white-woolly beneath, thinly so or hairless and green above; larger blades less than $\frac{5}{8}$ inch wide, with only the midrib or this flanked by an obscure pair of lateral veins not evident beyond the broadest part of the leaf. **Stem leaves** much reduced, the upper ones with or without a flat or curled, thin dry appendage at the tip.

Flower heads 1–15, clustered, of disk florets only; involucre of female heads $\frac{3}{16}$–$\frac{3}{8}$ inch high; involucral bracts white (often pinkish toward the base), dry, thin, and papery; male heads smaller with thicker tubular corollas, undivided styles, and meager pappus; female heads with threadlike tubular corollas, deeply 2-cleft styles, and copious pappus of fine white hairs. Individual clones bear either male or female heads but not both. *Mid April–late June*

Achenes minute, veinless. *Late May–July*

WI range and habitats Common south and central, sporadic north, in pastures, oak openings, oak and pine barrens, dry oak or pine woods, cedar glades, and dry to wet-medium prairies. *Curtis Prairie, Upper Greene Prairie, Juniper Knoll, West Knoll, Sinaiko Overlook Prairie*

Classification in this taxonomically difficult genus is complicated by polyploidy (multiple chromosome complements in each cell) and apomixis (setting of seed without fertilization), resulting in semi-distinct races. The conservative treatment followed here recognizes two highly polymorphic species, field pussy-toes and plantain pussy-toes (*Antennaria plantaginifolia*). Field pussy-toes in the narrowest sense is the smallest of several Wisconsin taxa having short stems and small leaves. One of our two sexual diploid species, it intergrades with three mainly polyploid and apomictic taxa that have been treated as separate species, as varieties, as subspecies of *A. neglecta*, or as subspecies of Howell's pussy-toes (*A. howellii* complex).

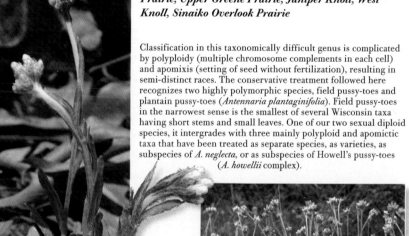

Plantain Pussy-toes
Plantain-leaved pussy-toes, woman's-tobacco
Antennaria plantaginifolia complex

Plantain pussy-toes is characterized by relatively large leaves that are covered with white woolly hairs and have 3–7 prominent long ribs beneath. Low perennial with leafy runners, 4–36 (female stems) or 1–12 (male) inches tall, mat-forming or colonial.

Basal leaves lying on the ground, partly evergreen, very broadly oval, 1–3 inches long, $^5/_8$ or more inch wide, the midrib flanked by 1 or 2 pairs of well–developed lateral veins. Runners ascending and short, or prostrate and elongated (depending on subspecies). **Stem leaves** to 2 inches long, sharp-pointed to bristle-tipped.

Flower heads 4–22, the male with disk florets only, the female heads with central disk florets and marginal florets with reduced corollas; female involucre $^3/_{16}$–$^3/_8$ inch high; male heads smaller, with undivided styles and meager pappus; female heads with threadlike tubular corollas, deeply 2-cleft styles, and copious pappus bristles. Individual clones male or female, never both. *Early May–mid July*

Achenes minute, veinless. *Early June–mid July*

WI range and habitats More common south than north, mostly in dry to medium prairies, oak openings, sand barrens, oak woods (grazed and ungrazed), and pastures, infrequent in bedrock glades and pine barrens. *Curtis Prairie, Juniper Knoll, Grady knolls, Greene Prairie, Sinaiko Overlook Prairie*

In some areas male plants may be rare or absent.

Several factors allow pussy-toes to compete successfully in many plant communities. Among these are drought resistance; evergreen basal leaves, which can photosynthesize while other plants are dormant; and the secretion of toxic chemicals that inhibit the growth of other species in the vicinity.

Plantain pussy-toes in the narrow sense is our other sexual diploid species, that is, it requires cross-pollination to produce viable seed, and its cells have the basic two sets of chromosomes. Its distribution may be limited to the Driftless Area. Many specimens, however, are borderline. These have been variously treated as separate species, as varieties of plantain pussy-toes, or as subspecies of Parlin's pussy-toes (*Antennaria parlinii*).

Pale Indian-plantain
Arnoglossum atriplicifolium
Synonym *Cacalia atriplicifolia*

Tall stems and characteristic leaves—triangular-egg-shaped, palmately veined, and whitened beneath—distinguish this statuesque plant whether or not it is blooming. In flower, the compound, flat-topped inflorescences of whitish, few-flowered heads are unmistakable. **Stout perennial, 3–6 (occasionally to 8) feet tall, from a short thick crown.**

Leaves alternate, the lower ones long-stalked, the middle and upper ones few, progressively reduced; blades mostly triangular-egg-shaped, about as wide as long, the margin shallowly lobed or coarsely toothed, the underside strongly whitened by a waxy bloom, with 3–5 (7) main veins diverging from a common point. **Stem** simple (except at the extreme top), round in cross section or grooved, with a whitish waxy cast. Plants hairless.

Flower heads in a large loose, flat-topped cluster at the top of the plant; involucre (whorl of bracts) yellow-green, urn-shaped, to $3/8$ inch long; ray florets absent; disk florets 5, bisexual, pale yellowish or nearly white, the expanded tip of the corolla deeply and equally 5-toothed. *Mid June–September*

Achenes dull blackish-brown, narrowly ellipsoid, barely over $1/8$ inch long, 10-ribbed, crowned by numerous fine white bristles that fall off early. *Mid August–mid October*

WI range and habitats Conspicuous but rare southeast, in dry to medium or even swampy open woods, oak openings and borders, thickets, dry to wet prairies, and railroad embankments, rarely weedy in fields and roadsides. *Curtis Prairie, Greene Prairie, Pond 2, Sinaiko Overlook Prairie, Visitor Center parking lot, Wingra Oak Savanna*

Similar species Great Indian-plantain (*Arnoglossum reniforme*) and prairie or tuberous Indian-plantain (*A. plantagineum*) are even less common in Wisconsin. The former has stems that are green and conspicuously angled and grooved and leaves that are green on both sides. It is not found in the Arboretum. Prairie Indian-plantain produces relatively small rosettes of pale green leaves that are particularly distinctive (see the following species).

Prairie Indian-plantain
Tuberous Indian-plantain
Arnoglossum plantagineum
Synonym *Cacalia tuberosa*
Wisconsin threatened species

In addition to the broad, flat-topped cluster of subtly colored flower heads, this conspicuous plant also produces relatively small rosettes of pale green, parallel-veined leaves that are particularly distinctive. The numerous, rather large heads have only 5 involucral bracts and 5 florets. **Stout perennial, 2–6 feet tall, from a short, tuberous-thickened base.**

Leaves alternate, green on both sides, the basal and lowest ones long-stalked, the blades triangular-egg-shaped to commonly oval (obviously longer than wide), firm, the margin entire to wavy, with 5–7 prominent longitudinal veins; stem leaves few, scattered, and manifestly reduced upward, becoming stalkless or nearly so, sometimes slightly toothed. **Stem** tall and smooth, emerging from the center of the rosette in early summer. Leaves often perforated by flea beetles.

Flower heads in a short broad, flat-topped cluster at the top of the plant; involucre of equal bracts in 1 series, pale green, short-urn-shaped (pinched at the mouth), $1/4$–$1/2$ inch long, each bract with a wing-like ridge on the back; receptacle with a short conic projection in the center; ray florets absent; disk florets 5, the corolla pale yellowish or nearly white, the expanded tip equally 5-toothed. *Mid July–August*

Achenes columnar, crowned by abundant white, very fine and soft bristles. *Late August*

WI range and habitats Local in the southern two tiers of counties, most commonly in wet-medium to dry-medium prairies, less frequently on dry limy prairies and lower slopes of prairie bluffs; prefers rich moist sandy soils along rivers and streams or springy peaty ground; probably once common or abundant, now rare owing to the elimination of its deep-soil prairie habitat. *Greene Prairie*

In 1996 the Greene Prairie population of prairie Indian-plantain exploded, then decreased very rapidly in subsequent years. Although it can no longer be found, the species may still present, but local land stewards realize that until its population dynamics are understood, current management practices may fail to protect this species.

In their parallel venation, the leaves of prairie Indian-plantain resemble those of broad-leaved plantains (*Plantago* species). In these cases as well as in rattlesnake-master (*Eryngium yuccifolium*) the leaves constitute an evolutionary reduction to a midrib, followed by an expansion.

Asteraceae • Sunflower Family

FIELD SAGE-WORT
Field wormwood
Artemisia campestris
Synonyms *Artemisia caudata*,
A. campestris subspecies *caudata*

Young flower heads

Finely dissected leaves and elongated clusters of small nodding greenish heads make field sage-wort readily identifiable. Plants are often silvery-hairy when young but become nearly hairless when mature; they are neither densely woolly nor strongly aromatic like most of the artemisias. Biennial (rarely short-lived perennial), 1–4 feet tall, from a taproot.

Basal leaves 2 or 3 times pinnately compound or deeply cut (dissected), 1–8 inches long (including stalks), the linear or almost threadlike ultimate divisions less than $1/16$ inch wide. **Stem leaves** alternate, with axillary branchlets, smaller and less divided, the uppermost cut into 3s or simple, hairy on both surfaces to nearly hairless. First year leaves form a basal rosette. Second year **stems** solitary or 2–5 together.

Flower heads very numerous in elongate but narrow panicles 9–24 inches long (contracted and spike-like in small individuals), small and inconspicuous, composed entirely of disk florets; involucre $1/16$–$3/16$ inch high, its 12–16 bracts green with a red-brown midrib, thin, dry, and membranous, hairless; receptacle naked (without hairs); florets 21–41 per head, the outer female and fertile, the central ones sterile (ovary abortive), whitish, tubular. *Late July–mid September*

Achenes oblong, less than $1/16$ inch long, lacking a pappus. *Mid September–October*

WI range and habitats Very common throughout much of the state, in dry limy or sandy prairies, oak-pine barrens and open woodlands, inner beaches and dunes along the Great Lakes, sandbars, cliffs, and outcrops, often weedy on roadsides, pastures, abandoned fields, and even urban waste places. *West Knoll*

Field sage-wort is a complex species made up of a number of semi-distinct races. New World plants have been treated either as a separate species, namely *Artemisia caudata*, or as *A. campestris* subspecies *borealis* and *A. campestris* subspecies *caudata*, to which Wisconsin plants belong (Eurasian plants constitute subspecies *campestris*). The modern tendency is to recognize a single, wide-ranging species consisting of several geographical races (subspecies) but not to subdivide these into lesser varieties.

WHITE SAGE
Louisiana sage-wort, western mugwort
Artemisia ludoviciana

The manifestly whitened stems and leaves and colonial habit make white sage distinctive. It is further identifiable by its small nodding greenish heads, which are aromatic when crushed. **Low-growing, unbranched perennial, 1–3 feet tall, rhizomatous, often in extensive loose clones.**

Leaves alternate, lance- or lance-elliptic, 1¼–4 inches long, entire (mostly) or the lower ones sometimes irregularly and coarsely toothed or rarely even lobed, the upper reduced, linear-lance-shaped, entire. Herbage persistently grayish- or whitish-hairy on both sides or becoming irregularly hairless above. **Stems** single or 2 or more loosely clustered.

Flower heads numerous, inconspicuous, composed entirely of disk florets, arranged in small clusters, these together forming a narrow branching leafy panicle 3–17 inches long; involucre ⅛ inch long, its 10–15 bracts overlapping in several rows, membranous–margined, grayish- or whitish-woolly; receptacle not hairy; florets regular, the outer ones (6–9) female, the central ones (6–12) bisexual, fertile throughout the head, whitish, tubular, ¹⁄₁₆–⅛ inch long. *(Late) July–mid October*

Achenes ellipsoid-cylindrical, minute, hairless; pappus absent. *(Late) July–mid October*

WI range and habitats Throughout much of the state (absent far north-central), characteristic of dry-medium prairies as well as dry sandy, deep-soil medium and even moist prairies, somewhat weedy on roadsides and railroads. *Curtis Prairie (limestone knoll), West Knoll*

Indigenous peoples burned white sage to repel insects, and early European settlers burned it as incense. The name sage is used for western sagebrushes and several garden artemisias as well as salvias, which are placed in the Mint Family.

Similar species Wisconsin wormwoods and mugworts are a variable lot, half native and half introduced. The native prairie sage-wort (*Artemisia frigida*) has shorter, finely dissected leaves and heads with long-hairy receptacles and more flowers. Another native, saw-leaf mugwort (*A. serrata*), is much taller, with simple leaves that are dark green and hairless above and white beneath, and the lower ones regularly toothed.

NODDING BEGGAR-TICKS
Nodding bur-marigold
Bidens cernuus

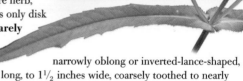

Nodding beggar-ticks is easily identified in the field by its opposite simple stalkless leaves that often grow together at the base, its nodding flower heads, and its habitat. Typically, yellow ray florets are present, imparting a handsome appearance to this widespread Northern Hemisphere herb, but rarely it lacks these and produces only disk florets. **Annual, 2–34 inches tall (rarely much taller), from fibrous roots.**

Leaves lance-linear to narrowly oblong or inverted-lance-shaped, mostly $1\frac{1}{2}$–7 inches long, to $1\frac{1}{2}$ inches wide, coarsely toothed to nearly entire. **Stems** usually somewhat bristly below.

Flower heads erect in bloom, nodding in fruit, $\frac{3}{8}$–2 inches broad; involucral bracts in 2 series, the 3–9 outer bracts leafy, unequal, much longer than the inner membranous ones; receptacle bracts narrow, yellow-tipped; ray florets 6–8 (rarely absent), neutral (with neither stamens nor pistil), $\frac{1}{4}$–$\frac{5}{8}$ inch long; disk florets bisexual, their corollas deep yellow and 5-toothed; anther tube purple-black. ***Mid August–early October***

Achenes purplish-brown, narrowly wedge-shaped, about $\frac{1}{4}$ inch long, ± flattened, with thickened pale margins, keels, and summit; pappus of 4 downwardly barbed awns. ***Early September–November***

WI range and habitats Throughout in damp to wet places, sometimes in shallow water: abundant on sandy, muddy, or boggy shores of streams, lakes, ponds, pools, and sloughs, common in swamps, bogs, marshes, sedge meadows, wet prairies, fens, and disturbed wet soils (e.g., pastures, ditches); native but weedy. ***Lower Greene Prairie***

Field biologists and duck hunters are well acquainted with the ubiquitous "beggar-ticks," the barbed, dart-like fruits that catch on fur, feathers, and clothing. Nodding beggar-ticks' insect visitors include bees, flies, and beetles, which suck nectar and sometimes collect or feed on pollen.

Highly variable in stature, leaf characteristics, head size, number of bracts, and number of florets per head, nodding beggar-ticks has been subdivided into many varieties, forms, and even segregate species of dubious merit. It is one of nine species of *Bidens* known to occur in Wisconsin, the majority of which have leaves that are cleft, divided, or compound. *Bidens* includes complexes of intergrading species that are separated on variable and technical characters, making them difficult to identify.

Pasture Thistle
Field thistle, prairie thistle
Cirsium discolor

Despite their ignoble reputation, thistles are quite attractive, their prickly leaves contrasting with the soft texture of the beautiful, brush-like flower heads. The leaves of our native pasture thistle are closely and persistently white-woolly on the lower surface, and the involucral bracts are spine-tipped. **Biennial (perhaps perennial), 3–7 feet high, from a taproot.**

Leaves deeply lobed and coarsely toothed, each tooth ending in a sharp spine, green and hairless above, white-felted beneath by a close covering of dense matted hairs. **Stems** without wings.

Flower heads solitary on each stalk-like branch end, about $1\frac{1}{2}$ inches high, the bell-shaped involucre of about 8 rows of strongly overlapping bracts, those of the outer rows terminated by abruptly bent spines; ray florets absent; disk florets numerous, the corollas pinkish-lavender, $1-1\frac{1}{4}$ inches long, tubular, with 5 long narrow lobes. *Mid July–early September*

Achenes light brown, $\frac{1}{8}$ inch long, equipped with a pappus of feathery bristles or "thistledown" for dispersal by wind. *Early August–September*

WI range and habitats Frequent south and west, sporadic northwest-central and northeast, in wet to dry-medium prairies and degraded remnants, less often in weedy pastures, sunny open woods, sedge meadows, and lakeshores, occasionally weedy in roadside ditches and waste places. *Curtis Prairie, Greene Prairie, Juniper Knoll, Sinaiko Overlook Prairie, Teal Pond, Visitor Center parking lot*

Common thistles are generally regarded as noxious, especially by farmers who battle to eradicate pernicious species in fields and pastures. Not all thistles are Eurasian invaders; about half of Wisconsin's *Cirsium* species are native Americans. Being good nectar producers, the flowers attract many bees, butterflies, and hummingbirds.

Similar species The leaves of another handsome native species, tall thistle (*Cirsium altissimum*), are lighter green and usually unlobed. The stout bull thistle (*C. vulgare*) and pernicious Canada thistle (*C. arvense*) are extremely spiny and very weedy. The former differs from pasture thistle by its spiny-winged stems and branches and rough (upper surface) leaves. Canada thistle is easily distinguished by its colonial habit and small heads with bracts tipped with weak prickles.

Asteraceae • Sunflower Family

Swamp Thistle
Cirsium muticum

Thick stems, spiny leaves, and large, lavender-flowered heads resemble those of other thistles, but its spineless involucral bracts and adoption of moist low habitats immediately identify our only swamp-loving native species. **Biennial, forming a rosette of leaves the first year and a flowering stem 2–6 feet high the second, from a weak taproot.**

Leaves pinnately lobed and toothed, weakly spiny, white-woolly when young, becoming green or greenish on both surfaces or grayish beneath by loosely matted hairs. **Stems** hollow, not winged.

Flower heads 1–3 at each branch end; involucre $5/8$–1 inch high, its bracts overlapping in about 8 rows, the outer ones egg-shaped, tipped with a very short, somewhat thickened prickle, portions cobwebby and sticky; florets bisexual, the corollas tubular, $5/8$–$1 3/8$ inches long, with 5 long narrow lobes. *Mid July–mid September*

Achenes dark brown with a yellow band at the apex, $1/8$–$1/4$ inch long; pappus bristles grayish-white, feathery-hairy, joined at the base and falling together. *Mid August–early October*

WI range and habitats Throughout but sporadic southwest, in the south characteristically in wet prairies and fens or less often in wet-medium to medium prairies, also in sedge meadows and tamarack swamps, in northern Wisconsin often in moist conifer or aspen-birch woods, rarely a roadside weed in burned or second-growth woods. *Curtis Prairie, Greene Prairie*

Because none of the involucral bracts is spine-tipped, their discouraging effect is thus muffled (hence the species epithet "mute"). Thistles are very successful at attracting bees, butterflies, and seed-eating birds. The American goldfinch loves the fruits of thistles and also uses thistledown as favorite nest-building material. Because thistles bloom relatively late, goldfinches wait until the second half of summer to build their nests.

Similar species Small individuals may resemble thistle relatives in our flora, especially spotted knapweed (*Centaurea stoebe*) and plumeless thistle (*Carduus acanthoides*), both introduced and ecologically invasive. Spotted knapweed lacks spines, although the involucral bracts, outlined in black and edged with fingerlike hairs, may at first appear to have spines. The pappus bristles in *Carduus* are minutely barbed rather than feathery-hairy.

Asteraceae • Sunflower Family

CANADIAN HORSEWEED
Fleabane, hogweed
Conyza canadensis
Synonym *Erigeron canadensis*

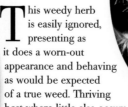

This weedy herb is easily ignored, presenting as it does a worn-out appearance and behaving as would be expected of a true weed. Thriving best where little else occurs, it varies greatly in height depending on the soil in which it grows. Healthy individuals are easily identified by their slender straight stems, numerous leaves, and numerous tiny heads with cream-colored corollas. **Annual, 3–60 inches tall, taprooted.**

Basal leaves surprisingly large but soon withering. **Stem leaves** alternate, linear to narrowly inverse-lance-shaped, tapering at the base, entire or the lower ones often toothed, progressively reduced, those of the upper half of the stem $1\frac{1}{2}$–4 inches long, $\frac{1}{8}$–$\frac{1}{4}$ inch wide. **Stems** unbranched (except within the inflorescence), coarsely spreading-hairy.

Flower heads numerous in an ample panicle, the convex disk hardly more than $\frac{1}{8}$ inch wide; involucre $\frac{1}{8}$ inch long, its bracts strongly overlapping, scarcely green or leafy, very narrow, nearly equal; ray florets about 25–40, female, fertile, with a threadlike corolla tube and extremely minute, flattened end ("ray"); disk florets about 5, bisexual, fertile. *Mid July–mid October*

Achenes flattened and 2-ribbed, bristly, with a pappus of abundant, dull white, soft bristles. *Mid July–early November*

WI range and habitats Almost ubiquitous, on roadsides, railroads, fields, prairies, edges of woods, sandy or rocky shores, also banks, bluffs, rocky flats, and sand plains, characteristic of cleared, burned, or otherwise disturbed ground (paths, gravel pits, dumps, lawns, gardens); surviving best in bare soil. *Curtis Prairie, Marion Dunn Prairie, Sinaiko Overlook Prairie, West Knoll*

At one time conyzas were considered species of *Erigeron*. This species may be called horseweed either on the supposition that it resembles a horse's tail, or because it is irritating to the nostrils of horses, or as a result of its use by Native Americans and settlers to treat strangury and other diseases in horses. A volatile oil, obtained from the leaves and tops by distillation, was once produced commercially, albeit on a limited scale, in Midwestern states. It was used as a flavoring for candies and soft drinks.

Asteraceae • Sunflower Family

PRAIRIE TICKSEED
Finger tickseed, prairie coreopsis, stiff tickseed
Coreopsis palmata

The tickseeds (*Coreopsis* species) comprise another large genus of attractive composites, and a number of species are grown in flower gardens. The low stems of prairie coreopsis produce distinctive leaves with a "birdfoot" appearance and in summer and autumn, bright golden-yellow heads with plastic-like involucral bracts. **Short-lived perennial, 16–35 inches tall, rhizomatous.**

Leaves opposite, 5 or more pairs, palmately 3–5-parted into lance-oblong segments, to $2\,3/4$ inches long and 2 inches wide, the base appearing as an indistinct winged stalk, scarcely or not at all reduced upward. **Stems** rigid, lined or grooved.

Flower heads solitary or few, on short stalks in loose panicles, the disk $3/8$–$5/8$ inch across; involucral bracts in 2 series of 8 each, the outer green, narrowly oblong, about equaling the inner, united at the very base, the inner yellow-brown, with margins free to the base; receptacle convex, with scaly bracts subtending the disk florets; ray florets 7–8, neutral, the ray $5/8$–1 inch long, 3-toothed, withering and falling before the achenes; disk florets bisexual and fertile, the corolla yellowish, 5-lobed. *Mid June–early August*

Achenes dark brown, inversely oblong-egg-shaped, nearly $1/4$ inch long, flat or becoming incurved, with narrow smooth marginal wings and a pappus of 2 firm teeth above the shoulders of the achene (or pappus greatly reduced or absent). *August–November*

WI range and habitats Very common southwest, in medium to dry prairies, these days especially dry lime and sand prairie relicts, also open sandy jack pine, scrub oak, or bur oak savannas, formerly (before indiscriminate herbicide spraying and mowing) along roadsides and railroads; well-drained soil and full sun. *Curtis Prairie, Greene Prairie, Juniper Knoll, Marion Dunn Prairie, Sinaiko Overlook Prairie, West Knoll*

The genus name is derived from two Greek words and means bedbug-like, alluding to the fruit.

Similar species Lance-leaf tickseed, also called sand coreopsis (*Coreopsis lanceolata*), carries solitary or few large heads on tall, somewhat leafless stalks. The leaves are mostly crowded near the base, and the few pairs of stem leaves are usually reduced and unlobed. It is locally common to sporadic in the southeastern half of Wisconsin but does not grow in the Arboretum.

Asteraceae • Sunflower Family

Purple Coneflowers

The species of *Echinacea* are all very showy plants that often occur in large numbers and put on an impressive display in bloom. The stout stems rise in patches on the prairies, and for two or more months of summer bear large heads with long, rose-purple rays and prickly, purple-brown centers. **Coarse perennials. Pale purple coneflower: 1–3 feet tall, from a thick taproot. Purple coneflower: 2–4 feet tall, from a woody rhizome or tough crown.**

Leaves alternate, simple, 3–8 inches long, the lower ones long-stalked, the rest with short winged stalks, commonly toothed. Pale purple coneflower has oblong-lance-shaped leaves with gradually tapering bases, entire margins, and 3 strong veins, whereas purple coneflower has egg- to broadly lance-shaped leaves with rounded bases, toothed margins, and 5 veins. Herbage is rough with short stiff hairs in both species, varying to hairless in purple coneflower.

Flower heads solitary on long stalks terminating the stems and few branches, with disk florets in the center and a whorl of ray florets around the edge; involucral bracts overlapping, leafy, nearly equal; disk convex, becoming egg-shaped, bearing persistent bracts narrowed into a rigid toothlike spine longer than the disk florets; ray florets sterile, the rays $1\frac{1}{4}$–$3\frac{1}{2}$ inches long; disk florets bisexual and fertile, the corolla purplish, 5-toothed. The rays in pale purple coneflower are light purplish-pink, $\frac{1}{8}$ inch wide, drooping, and papery in texture,

Pale Purple Coneflower
Prairie coneflower
Echinacea pallida
Wisconsin threatened species

whereas in purple coneflower they are magenta-purple, $3/16$–$3/8$ inch wide, spreading, and almost leathery. Pollen white in pale purple coneflower, yellow in purple coneflower. Pale purple coneflower: *Late June–August.* Purple coneflower: *mid August–mid September*

Achenes broadly wedge-shaped, 4-sided, $3/16$ inch long, hairless, surmounted by a low-toothed crown (pappus) and the withered corolla. *August onward (persisting into winter)*

WI range and habitats

Pale purple coneflower: Now rather rare in four southern counties, in medium to dry prairie remnants and along rights-of-way, less frequently on dry gravelly slopes and in gravel pits. *Curtis Prairie, Greene Prairie, Sinaiko Overlook Prairie, Visitor Center parking lot*

Purple coneflower: Adventive or escaped in the southern two tiers of counties, also Taylor County, in fields, roadsides, railroads, and vacant lots. Planted, not native to Wisconsin. *Curtis Prairie, Marion Dunn Prairie, Sinaiko Overlook Prairie*

The genus name is from the Greek *echinos,* hedgehog or sea-urchin, and refers to the sharp-pointed bracts of the receptacle. Showy, easy to grow, long-lived, and tough, these plants are popular for landscape use. Native farther south, purple coneflower is cultivated in preference to pale purple coneflower owing to its darker-colored flowers and greater ability to attract butterflies. The heads are also excellent for cutting. Pale purple coneflower and narrow-leaf purple coneflower (*Echinacea angustifolia*) were used by plains Native Americans to treat more ailments—everything from snakebite, burns, and stings to systemic infections—than any other group of plants. Modern herbalists endorse the drug echinacea as an immuno-stimulant and general-purpose wound healer.

PURPLE CONEFLOWER
Broad-leaved purple coneflower, eastern purple coneflower
Echinacea purpurea

Annual Fleabane
Eastern daisy fleabane
Erigeron annuus

Like daisy fleabane, annual fleabane is a common field weed throughout most of the U.S. and southern Canada. It has white- or rarely pinkish-rayed heads that are only $1/2$ inch wide and middle and upper leaves that are commonly toothed in the middle and fringed with coarse hairs. **Somewhat coarse annual, $1/2$–5 feet tall, with fibrous roots.**

Basal leaves broadly egg-shaped and unexpectedly large, to 4 inches long, contracted into a long, stalk-like base, coarsely toothed. **Stem leaves** egg- to lance-shaped, $3/8$–$1 1/4$ inches wide, tapering at the base, with a few large teeth or almost entire. **Stems** spreading-hairy, the hairs tending to be confined to the angles and ascending above.

Flower heads several to numerous, borne on somewhat naked stalks, the open cluster flat-topped or convex; involucre hemispheric, $1/8$–$3/16$ inch high, the bracts nearly equal, narrow, finely glandular and sparsely hairy; ray florets about 80–125, female, to $3/8$ inch long; disk florets numerous, bisexual and fertile, the corolla yellowish. *Late June–early October (later on resprouts)*

Achenes straw-colored, small, flattened, hairy; pappus of ray florets a single ring of several very short, slender scales (visible under 20× magnification), that of disk florets a double series of 10–15 long, hair-like bristles plus the very short scales. *Late June–early October (undoubtedly later)*

WI range and habitats Common throughout, in fields (cultivated or abandoned), pastures, woodlands (especially in clearings, old roads, trails), savannas, disturbed prairies, sedge meadows, wet thickets, swamps, roadsides, railroads, occasionally fencerows, farmyards, and gardens, also waste ground (e.g., gravel pits, dumps). *Curtis Prairie, Greene Prairie, East Knoll, Juniper Knoll, Marion Dunn Prairie, Sinaiko Overlook Prairie, Wingra Oak Savanna*

The fleabanes are clearly related to the dwarf fleabanes (*Conyza*) and asters (*Symphyotrichum*) but bloom in spring and early summer. *Conyza* and *Erigeron* usually have only 1 row of involucral bracts that scarcely overlap. Their ray flowers are usually more numerous and narrower (our species) than those of asters, and the disks remain yellow.

Similar species Annual fleabane is routinely misidentified as daisy fleabane (*Erigeron strigosus*), especially if it has narrow leaves. The species boundaries seem poorly defined; transitional plants with intermediate or recombined characters exist.

Robin's-plantain
Erigeron pulchellus

Having mostly basal leaves and only a few, relatively large heads, this spring bloomer at first sight looks like an out-of-season blue aster. Its heads resemble those of the more widespread fleabanes, but the ray florets are wider and the disk florets longer, and the leaves are relatively broad and almost velvety. **Biennial or short-lived perennial, 6–18 inches tall, producing a few superficial offsets and runners, forming patches.**

Basal leaves inverted-lance-shaped to almost round, $3/4$–5 inches long, shallowly toothed above the middle to nearly entire, 3- or 5-veined. **Stem leaves** few, lance-shaped to oblong or egg-shaped, stalkless and rounded or flanged at the usually clasping base, entire or toothed. Plants densely spreading-hairy.

Flower heads 1–4, 1–$1 1/2$ inches across (including rays), the open cluster flat-topped or convex; involucre hemispheric, $1/4$ inch tall, the bracts nearly equal, narrow, tapering to a narrow tip, ± hairy and somewhat sticky; ray florets about 50–80 (100), female, the rays violet to pinkish, linear-oblong, $1/4$–$3/8$ inch long; disk florets numerous, bisexual, yellow. *Mid May–mid June*

Achenes flattened, 2- or 4-veined, the pappus a single whorl of 30–35 soft bristles, alike in ray and disk florets. *Mid June–July*

WI range and habitats Chiefly south, also far west and Green Bay area, in dry-medium to medium, deciduous woods, oak savannas, and cedar glades (especially in grassy openings and cleared woodlands), locally common on grassy or brushy hillsides, partly shaded ravines, banks, dry prairies, and prairie pastures, occasionally in sandy fields and old quarries. ***Curtis Prairie, Upper Greene Prairie, Juniper Knoll***

The genus name comes from the Greek and means "old man in spring," possibly because some species are hoary. It is one of the earliest-blooming members of the Sunflower Family. Clones bloom mostly at the periphery, behaving like other prairie composites that have been shown to release toxins into the soil so as to inhibit the growth and development of other plants in the vicinity.

Although most familiar today from dry prairies, Robin's-plantain was probably one of the species typical of oak savannas during prehistoric times.

Asteraceae • Sunflower Family

Daisy Fleabane
Prairie fleabane, rough fleabane
Erigeron strigosus

The loosely branched or flat-topped summit of this fleabane produces a few new heads each day over a month-long period. Each head is on a naked stalk and has very numerous, linear white rays (sometimes tinged pink or blue). **Annual (rarely biennial), mostly 1–4 feet tall, with fibrous roots.**

Basal leaves elliptic to spoon-shaped. **Stem leaves** few, linear to lance- or inverted-lance-shaped, rarely more than $3/8$ inch wide, tapering at the base, entire or the lower shallowly toothed. **Stem** grayish with short, ascending or appressed hairs on both the sides and angles.

Flower heads several to numerous; involucre hemispheric, $1/16$–$3/16$ inch high, the bracts nearly equal, narrow; disk less than $2/5$ inch across; ray florets female, fertile, about 50–100, rays to $1/4$ inch long; disk florets perfect and fertile, yellow. *Early June–early October*

Achenes flattened, with a pappus of very short, bristle-like scales only (rays) or of several short outer scales and 10–15 longer inner bristles (disk florets). (Use 20✕ magnification to see details.) *Late July–late October*

WI range and habitats Very common throughout, in a variety of habitats, particularly dry to medium prairies, fields, sand barrens, roadsides, and railroads, also dry woods (pine, oak, juniper), sandy, gravelly, or rocky hillsides, beaches, flats, and quarries, and disturbed open areas generally. ***Curtis Prairie, Grady Tract knolls, Upper Greene Prairie, Juniper Knoll, Sinaiko Overlook Prairie***

The common name "fleabane" was first applied to species of *Erigeron* in Europe, where bundles of dried plants hung in a house or stuffed into mattresses were believed to repel fleas.

Daisy fleabane is the commonest *Erigeron* in northeastern North America, including Wisconsin. It typifies weedy composites like horseweed and dwarf fleabanes (*Conyza* species), hawk's-beards (*Crepis* species), and American burn-weed (*Erechtites hieracifolius*), all annuals that survive the winter as seeds and produce an abundance of fruits in summer.

Similar species Daisy fleabane and annual fleabane (*Erigeron annuus*) are closely related and sometimes difficult to distinguish. The former is usually a shorter slenderer plant with sparse foliage and short hairs pressed flat against the stems, whereas annual fleabane has more, often broader, toothed leaves and longer spreading hairs.

Boneset

Common boneset, thoroughwort
Eupatorium perfoliatum

Boneset often grows alongside its cousin, spotted Joe-Pye-weed, in wet meadows. The leaf bases are not only broad and stalkless but nearly always grown together around the stem, which appears to pass through the leaf. Rayless whitish heads in branched, shallowly dome-shaped clusters and abundant crinkly hairs further distinguish this easily recognized species. **Rather coarse perennial, 1–3 feet tall, from a weakly spreading rhizome.**

Leaves opposite, simple, lance-shaped, 4–8 inches long, the pairs fused together at the base around the stem, gradually tapering into a narrow tip, the margins weakly rounded-toothed their length, veiny, resinous-glandular beneath. **Stems** usually solitary.

Flower heads consist of (9) 10–16 (23) bisexual disk florets only; involucre $3/16$–$1/4$ inch long, its bracts in about 3 overlapping ranks, the outer less than half as long as the inner, gradually tapering to a point; corollas dull white, $3/8$ inch long, tubular, regular, with long style branches protruding far out of the corolla tube. ***Late June–September***

Achenes 5-angled and 5-veined, resinous-glandular, with a well-developed pappus of hair-like bristles. ***Late July–early October***

WI range and habitats Throughout, common in sedge meadows, low prairies, fens, shrub carrs, marshes, swamps, lowland forests (deciduous and/or coniferous), also lakeshores, streamsides, and wet cliffs, rarely in drier habitats but often weedy in heavily grazed pastures; open moist ground. ***Curtis Prairie, Lower Greene Prairie***

The genus is named for Mithridates Eupator (132–63 B.C.), an ancient king of Pontus (now part of Turkey), who was not only a successful general but also a skilled medicine man. When poisoned, he recovered by taking an antidote made from some species of the genus now bearing his name. Boneset was once sprinkled on the skin over broken bones in the belief that this would speed healing. Boneset tea, made from the leaves and flowering tops, was once administered not only as a tonic but also for treatment of old-fashioned illnesses called bone-break fever, a catarrhal disease like grippe, and ague or malarial fever.

Grass-leaved Goldenrod
Common flat-topped goldenrod
Euthamia graminifolia
Synonym *Solidago graminifolia*

As indicated by its common name and species epithet, this species has almost grasslike leaves and numerous, tiny yellow heads in a broad, flat-topped, terminal inflorescence. **Perennial, 1–3 feet tall, from well-developed, creeping rhizomes, forming patches or extensive colonies.**

Leaves uniform, stalkless, the lower ones soon falling, linear to linear-lance-shaped, 1½–5 inches long, ⅛–½ inch wide, entire, with 3 or 5 longitudinal veins evident without magnification, sparsely to moderately dotted with tiny resinous glands (under magnification). **Stems** freely branching in the upper half, the entire top of the plant flat. Plant rather densely leafy, hairless to densely short-hairy or rough to the touch.

Flower heads stalkless or nearly so in small compact clusters that together form a flat-topped inflorescence; involucre ⅛–3/16 inch high, its bracts overlapping in 3–5 series of different lengths, straw-colored, oblong, slightly sticky; receptacle rough but not chaffy; ray florets 15–25 (35), female and fertile, yellow, very small; disk florets fewer than the rays, 4–13 (mostly 5–10) per head, bisexual and fertile, yellow. *(Late July) early August–October*

Achenes with short straight hairs pressed closely against the surface, several-veined; pappus of numerous whitish, hair-like bristles. *Late August–October*

WI range and habitats Common throughout except for the southwest corner, in moist to dry, especially degraded prairies and sedge meadows, fallow fields, and edges of marshes, fens, bogs, and moist woods, somewhat weedy along fencerows, roadsides, and railroads; often in moist sandy, clayey, or peaty areas where soil has been removed. *Curtis Prairie, Greene Prairie, Juniper Knoll, Marion Dunn Prairie, West Knoll*

Euthamia has often been included with the "true" goldenrods, *Solidago,* from which it differs in gross aspect, resinous-dotted leaves, and disk florets fewer than the number of ray florets. Both grass-leaved goldenrod and slender goldentop (*Euthamia tenuifolia*) occur in several different forms and are imperfectly understood species.

Similar species Viscid grass-leaved goldenrod (*Euthamia gymnospermoides*) and slender goldentop (*E. tenuifolia*) are much less common in Wisconsin. They differ from grass-leaved goldenrod in having the principal leaves narrower (less than 3/16 inch wide, with only 1 strong or sometimes 3 evident veins) and the uppermost leaves and involucres somewhat sticky. Neither is found in the Arboretum.

Asteraceae • Sunflower Family

SPOTTED JOE-PYE-WEED
Eutrochium maculatum
Synonym *Eupatorium maculatum*

The various species of Joe-Pye-weed produce numerous heads in flat- or round-topped clusters on tall sturdy stems. Each head contains only tubular florets; there are no rays. In spotted Joe-Pye-weed the stems are red-purple or speckled with purple, the leaves are regularly whorled, and the inflorescence or its divisions are flattish-topped. **Perennial, 2–6 feet tall.**

Leaves (3) 4–5 in each whorl, simple, lance- to lance-elliptic or lance-egg-shaped, narrowed to a short stalk, coarsely toothed, with orange resin dots beneath (use 10✕ magnification). **Stems** solid, without a pale waxy bloom.

Flower heads in dense clusters, each of 9–16 (22) bisexual disk florets; involucre oblong, $1/4$–$3/8$ inch long, its bracts successively shorter in several overlapping series, blunt, often 3–5-veined; florets 9–16 (24), rose-purple, regular, with long style branches protruding far out of the corolla tube. *Early July–September*

Achenes 5-angled and -veined, to $3/16$ inch long, with orange glands and a well-developed pappus of short grayish, hair-like bristles. *Late July into October*

WI range and habitats Very common throughout, in wet prairies, sedge meadows, fens, marshes, alder thickets, wet woods, and along streams and lakes, less common in bogs and mucky pastures or drier habitats; wet soil and full sun. ***Curtis Prairie, Greene Prairie***

Various native groups used Joe-Pye-weeds to cure everything from arthritis to fevers to kidney disorders. It is popularly believed that the plants were named for a noted medicine man whose name was anglicized to Joe Pye. He supposedly cured early members of the Massachusetts Bay Colony when they contracted typhus, using a Joe-Pye-weed to induce sweating and thereby break fevers. An alternate hypothesis is that the name comes from a Native American word for typhoid fever, *jopi*, which became anglicized to Joe Pye.

Similar species Purple Joe-Pye-weed (*Eutrochium purpureum*) has very similar foliage, but its stems are green throughout (or purple only at the nodes) and faintly whitened with a waxy bloom, and the pale purple heads are in dome-shaped inflorescences and only 4–7-flowered. It grows in dry to medium woods and clearings.

Common Sneezeweed
Helenium autumnale

Common sneezeweed is distinguished, even when not in bloom, from everything else in the marsh by its winged stems. Contributing to the wonderful colors of early autumn, its clusters of heads are likewise distinctive; both the large spherical disk and wedge-shaped, drooping rays are bright yellow. **Perennial, 2–5 feet tall, from basal offshoots from a somewhat rhizomatous rootstock.**

Leaves alternate, numerous, lance-shaped to elliptic or narrowly egg-shaped, 1 1/4–6 inches long, tapering at both ends, the bases extending down the stem to form wings, distantly toothed, both surfaces dotted with microscopic resinous glands. **Stems** simple below, branched above.

Flower heads solitary on each stalk, several to many in an open leafy cluster, with a spherical naked receptacle; involucre leafy, its bracts in 2 series of about 8 each, the outer somewhat longer than the inner, gradually tapering to a slender tip; ray florets (10) 13–20, female and fertile, the ray 3/8–7/8 inch long, 3-toothed at the apex; disk 3/8–3/4 inch across, its florets very numerous, bisexual and fertile, the 5 corolla lobes glandular-hairy. *Mid July–mid October*

Achenes yellow or grayish-brown, top-shaped, 1/16 inch long, 4–5-angled and hairy-ribbed; pappus of 7 or 8 lance-shaped scales, these long-pointed, translucent, irregularly toothed. *August–October*

WI range and habitats Common in all but northernmost Wisconsin, in sunny or shady, moist to wet soil: wet prairies, sedge meadows, marshes, swales, floodplain forests, low open woods, sand and gravel bars of rivers and shores of lakes; usually in calcareous meadows, often where grazed. *Lower Greene Prairie, Wingra Oak Savanna*

In Native American languages the name for this plant means sneezing or inhalant. The dried, nearly mature, pulverized heads were used by the Menominee as a snuff to induce sneezing to loosen head colds and rid the body of evil spirits. A tea made from the florets was used by the Meskwaki to treat stomach problems. Livestock have been poisoned from eating large quantities of the flowering tops.

Sunflowers (*Helianthus* species) and coneflowers (*Echinacea, Rudbeckia,* and *Ratibida* species) differ from sneezeweed in having the ray petals entire or shallowly notched at the apex and the scaly-bracted disk flat to convex or conic, or if globular, then brown or greyish.

Saw-tooth Sunflower
Helianthus grosseserratus

Tall and aggressive, sawtooth sunflower is relatively easy to identify. Key traits are the smooth stems, evident leaf stalks, and characteristically drooping leaves with pale or whitened lower surfaces covered with appressed white hairs that are finer and shorter than in our other sunflowers. **Stout erect perennial, 3–7 (9) feet tall, from a short thick rhizome with thickened roots, forming patches.**

Basal leaves absent or inconspicuous. **Stem leaves** alternate (or opposite below), on a definite stalk to $1^{1}/_{2}$ inches long, lance- to oblong-egg-shaped, 4–8 (12) inches long, with low sharp, forward-pointing teeth or almost entire. **Stem** light red or sometimes green, hairless (except sometimes the uppermost parts), covered by a pale waxy bloom.

Flower heads usually many on stalks terminating the stems and branches, those on the lower stalks opening their flowers first; involucral bracts in several series, narrowly lance- to almost awl-shaped, loose and spreading, hairless or with short scattered hairs; receptacle with scaly bracts; ray florets 10–20, sterile, the ray yellow, $1-1^{1}/_{2}$ inches long; disk $^{3}/_{8}-^{3}/_{4}$ inch broad, its florets bisexual, fertile, yellow. *Late July–early October* (peaking in early September)

Achenes pale brown, $^{3}/_{16}$ inch long, moderately flattened, the pappus of 2 scale-like awns that soon fall. *Late August–October*

WI range and habitats Very common south, in wet-medium to dry-medium prairies (absent from dry or sandy prairies), including degraded prairies, sedge meadows, and thickets, frequent along streamsides, roadsides, ditches, and fencerows. *Curtis Prairie, Greene Prairie, Juniper Knoll, Marion Dunn Prairie, Sinaiko Overlook Prairie*

Not only do the flower heads look like symbols of the sun, but also, if the popular story is believable, they turn their faces ever toward the sun, tracking that great orb as it crosses the sky. Like other sunflowers, sawtooth sunflower attracts songbirds and butterflies.

Similar species Giant sunflower (*Helianthus giganteus*) can also be stout, tall, and branched above, but it has reddish-purple stems that are roughened with stiff white, sometimes scattered hairs; essentially stalkless leaves that are wedge-shaped at the base and more harshly hairy beneath; and hairy involucral bracts. Although very common throughout Wisconsin, this species is scarce in the Arboretum.

Asteraceae • Sunflower Family

SUNFLOWERS

The several *Helianthus* species in Wisconsin can be rather difficult to distinguish, especially these two semi-weedy species, which abound in dry open woodlands, edges of upland prairies, and disturbed areas. **Coarse perennials, 2–6 feet tall, from short, much-branched rhizomes, forming patches.**

Basal leaves absent or inconspicuous. **Stem leaves** opposite or those subtending inflorescence branches alternate, the larger ones lance- to lance-egg-shaped, mostly 3–8 inches long, often more than 2 inches wide, abruptly contracted (straight across to rounded or gradually rounded at the base) into a short, unwinged or slightly winged stalk, shallowly toothed to entire, very rough to stiff-hairy above, thick, 3-veined. Hairy sunflower has stems with minute stiff hairs nearly or quite throughout or if becoming sparsely hairy to hairless below, marked by blister-like bases of worn-off hairs. Rough-leaved sunflower has hairless stems or is somewhat rough to the touch upward only within or immediately below the heads.

Flower heads 1–several on stalks ³⁄₈–2 inches long terminating the stems and branches, collectively forming an open cluster in which the outer heads flower first; involucral bracts in several series, green throughout, broadly linear-lance-shaped, very loose, curved backward;

HAIRY SUNFLOWER
Rough Sunflower, oblong sunflower
Helianthus hirsutus

disk ¾–1¼ inches across, with dry scaly bracts; ray florets 10–15, neutral, yellow, the strap-shaped ray ⅝–1 inch long, falling off in due course; disk florets bisexual and fertile, yellow. *Late July–September*

Achenes dark, broadly inverse-egg-shaped, ⅛ inch long, hairless; pappus of 2 broadly lance-shaped awns that fall off early (without other scales).

WI range and habitats Very common throughout, especially south, abundant in oak woods, oak savannas, edges of woods, thickets, sand prairies, medium to moist prairies, old fields, fencerows, bluffs and ledges, roadsides and railroads. *Throughout the Arboretum* (both species)

Rough-leaved sunflower, hairy sunflower, and pale sunflower (*Helianthus decapetalus*)—all common wild sunflowers—form a polyploid complex. Each is highly variable, and they intergrade with each other and in the case of pale sunflower especially, hybridize with other sunflowers. It is extremely difficult to distinguish the woodland and hairy sunflowers from each other, and a strong case could be made for considering them members of a single variable species. It is equally difficult to separate forms of hairy sunflower with extremely short-stalked leaves from divaricate woodland sunflower (*H. divaricatus*), rare in eastern and southern Wisconsin, and woodland sunflower from Jerusalem-artichoke (*H. tuberosus*), common in central and southern Wisconsin. Divaricate woodland sunflower is a diploid species that probably contributed a set of chromosomes to either or both rough-leaved sunflower and hairy sunflower. The extensive intergradation between rough-leaved sunflower and Jerusalem-artichoke is not unexpected, given that they hybridize relatively frequently.

Rough-leaved Sunflower
Pale-leaved woodland sunflower
Helianthus strumosus

Western Sunflower
**Few-leaved sunflower,
naked-stemmed sunflower**
Helianthus occidentalis

This low-growing species of dry prairies is one of the most distinctive of the sunflowers. The mostly basal, relatively large leaves give the appearance of a rosette and are evident throughout the growing season. They are followed in late summer by a few, bright yellow, small heads held aloft on a nearly leafless stem. **Slender perennial, 1–4 feet tall, spreading slowly by slender shallow rhizomes to form colonies.**

Basal leaves conspicuous. **Stem leaves** 3–5 pairs mostly near the base (only 1 or 2 pairs of reduced leaves above), egg- to oblong-lance-shaped, 2–6 (8) inches long, $^5/_8$–$2^3/_4$ inches wide, becoming very small upward, stalkless, rough to the touch and resinous-glandular on both surfaces. **Stems** usually solitary, hairy at the base, less so to nearly hairless above.

Flower heads solitary or few, on long stalks terminating the stems and branches; involucral bracts in about 3 series, egg- to lance-shaped, very unequal, tightly pressed against and slightly shorter than the disk; receptacle convex, to $^5/_8$ inch wide, with scaly bracts; rays florets 10–15, neutral, deep yellow, $^3/_8$–$^3/_4$ (1) inch long; disk about $^1/_2$ inch across, its florets bisexual and fertile, yellowish. *Mid July–mid September (early October)*

Achenes 4-sided and laterally flattened, hairy along 1 margin and at the apex, becoming hairless; pappus of 2 dry thin, early-falling scales on the principal angles. *August–October*

WI range and habitats Widespread throughout (lacking from the Northern Highlands), in medium to extremely dry prairies, characteristic of sand prairies and oak and pine barrens, also on calcareous hill prairies and sandy fields, roadsides, and railroads; well-drained soil and full sun. *Curtis Prairie, Greene Prairie, West Knoll*

Clones of the prairie sunflowers (*Helianthus pauciflorus* and *H. occidentalis*) exhibit antibiotic and autotoxic effects, their dead plant parts emitting poisonous chemicals into the soil that inhibit the growth and development of other plants in the vicinity, including individuals of the same species.

Seeds of various sunflower species, including western sunflower, are relished by songbirds.

Asteraceae • Sunflower Family

STIFF SUNFLOWER
Few-leaved sunflower, showy sunflower
Helianthus pauciflorus

The short broad involucral bracts and red-brown or red-purple disk readily separate this late-blooming sunflower from all others in our flora. **Lanky perennial, 2–5 feet tall, spreading rapidly by stout rhizomes.**

Basal leaves frequently absent at flowering time. **Stem leaves** 6–15 pairs (or rarely a few of the upper alternate), gray to light green, lance-oblong to lance-egg- or lance-diamond-shaped, 3–10 inches long, $1/2$–2 inches wide, tapering into short winged stalks or stalkless, gradually reduced upward, very firm-textured and rough on both sides.

Flower heads few on long stalks terminating the stems and branches; involucral bracts in 3 or 4 series, appressed, oblong to narrowly egg-shaped, very unequal, broadly sharp-pointed to blunt, shorter than the disk; receptacle with scaly bracts; ray florets 10–21, neutral, the rays deep yellow, $5/8$–$1 3/8$ inches long; disk $5/8$–1 inch across, the florets bisexual and fertile, the corolla lobes reddish-brown. *(Early) late July–mid September*

Achenes dark-colored, inverted-oblong-egg-shaped, $1/4$ inch long; pappus of 2 broadly lance-shaped, principal scales, each with accessory awns at each side, none persistent. *[Early–late fall]*

WI range and habitats Frequent southwest, sporadic northward, in wet to dry (usually in medium and dry-medium) prairie remnants, sand prairies, sand barrens, dry steep calcareous prairies, roadsides, and railroads; well-drained soil and full sun. *Curtis Prairie, Greene Prairie, Juniper Knoll, Sinaiko Overlook Prairie, Visitor Center parking lot, West Knoll*

Sunflower blooms are often mistaken for single large flowers but are actually comprised of many small florets crowded into a head at the tip of a common stalk (receptacle). The tubular florets that comprise the round flat central disk produce the plentiful quantities of fruits (achenes, usually thought of as "seeds") that birds love. In the related silphiums, fruits are in a ring around the disk.

Stiff sunflower is variable in plant height and leaf number, shape, and size. It occasionally hybridizes with Jerusalem-artichoke (*Helianthus tuberosus*), yielding hybrid prairie sunflower (*H.* ×*laetiflorus*), distinguished by its yellow disk florets and occasional production of tubers.

Ox-eye
False sunflower, sunflower-everlasting
Heliopsis helianthoides

Ox-eye bears pairs of simple stalked leaves and produces a profusion of rather large, yellow-flowered heads starting in midsummer, signaling the golden months to come when all the yellow composites come into bloom. **Coarse, loosely branched perennial, 3–5 feet tall, from a tough crown.**

Leaves opposite, triangular to oblong-egg-shaped, (2) 3–5 inches long, ± straight (as if cut off) to heart-shaped at the base, coarsely toothed, 3-veined. Herbage from nearly hairless to rough-hairy.

Flower heads few to several (rarely solitary), $1\frac{1}{2}$–$2\frac{1}{2}$ inches across, on leafless stalks terminating the stem and branches; involucral bracts overlapping in 2 or 3 series, the outer ones somewhat leafy at the blunt to somewhat sharp tips, densely hairy; ray florets 10–16, female and fertile, their large corollas persistent on the achenes; disk $\frac{5}{8}$–1 inch broad, the florets bisexual and fertile, partly enclosed by the scaly bracts of the receptacle, brownish-yellow or purplish (corolla yellow, anthers dark).
June–early October

Achenes brown, conical at the base but 3-angled (rays) or 4-angled (disk florets) at the apex, $\frac{3}{16}$ inch long; pappus an obscure 2- or 3-toothed crown or sometimes obsolete or absent.
July–October

WI range and habitats Common throughout, in wet to dry prairies, edges of open woods and thickets, on floodplains, lakeshores, low or abandoned fields, roadsides, and railroads; in the north along wooded roadsides, lakes and streams; full sun or partial shade. *Curtis Prairie, Greene Prairie, Juniper Knoll, Marion Dunn Prairie, Sinaiko Overlook Prairie, Visitor Center parking lot, Wingra Oak Savanna*

The name *Heliopsis* is from the Greek *helios,* sun, and *opsis,* appearance, from the likeness of these plants to the sun or sunflower. The epithet *helianthoides* means the same thing, i.e., like *Helianthus,* sunflower.

Ox-eye is very shade tolerant and may be more of a savanna than a prairie species.

Similar species Flower heads of ox-eye are similar to those of rosinweeds (*Silphium* species), tickseeds (*Coreopsis* species), and especially true sunflowers (*Helianthus* species), but its ray florets persist on the achenes and fall with them, whereas in the other groups the withered ray florets fall before the achenes.

Asteraceae • Sunflower Family

LONG-HAIRED HAWKWEED
Hairy hawkweed, long-beard hawkweed, prairie hawkweed
Hieracium longipilum

The cylindrical, raceme-like inflorescence and conspicuous long hairs on the lower stem make this beautiful sand prairie dweller readily recognizable. Its small, dandelion-like heads close at the end of each day and reopen the next until seeds are set. **Perennial with milky juice, 16–60 inches tall, from a short, rhizome-like crown.**

Basal and lower stem leaves often crowded, pale green beneath, inverted-lance-shaped to narrowly elliptic, $3\frac{1}{2}$–12 inches long, densely long-hairy, entire. Stem hairs rusty to white, mostly $\frac{3}{8}$ (to $\frac{3}{4}$) inch long.

Flower heads few to numerous; involucre $\frac{1}{4}$–$\frac{3}{8}$ inch high, its bracts in 2 classes, a single series of nearly equal principal bracts subtended by small outer bracts, covered with a mixture of small, star-shaped and stiff blackish, mostly gland-tipped hairs; florets 40–60 (90), all bisexual and fertile, the corollas yellow, strap-shaped, 5-toothed.
Early July–mid August

Achenes ± prism-shaped, $\frac{1}{8}$–$\frac{3}{16}$ inch long, slightly tapered toward the base and apex, strongly ribbed and grooved; pappus bristles to $\frac{1}{4}$ inch long. *Late July–late September*

WI range and habitats Frequent southwest, especially along major rivers like the Chippewa, Black, and Wisconsin, characteristic of dry sandy prairies and sand barrens, occasionally on steep dry prairies and in jack pine woods, rarely weedy in sterile abandoned fields and roadsides; equally at home on disturbed and undisturbed soils. *Curtis Prairie, Upper Greene Prairie, Grady Tract knolls*

The names hawkweed and *Hieracium* are derived from the Greek *hierax*, hawk. According to legend, hawks would rend these plants and wet their eyes with the juice to sharpen their eyesight.

Hieracium is a huge genus represented in Wisconsin by naturalized weeds and American natives, 10 species and five hybrids in all. A native diploid (i.e., having the normal two chromosome-complements), long-haired hawkweed is easily classified, but overall the systematics of hawkweeds are complex and controversial.

Similar species Long-haired hawkweed is easily distinguished from such naturalized European immigrants as orange hawkweed (*Hieracium aurantiacum*) and king devil (*H. caespitosum*), which have leafless stems arising from a basal rosette of numerous, narrowly elliptic to inverted-lance-shaped leaves.

Asteraceae • Sunflower Family

FLAX-LEAVED ASTER
Stiff aster
Ionactis linariifolia
Synonym *Aster linariifolius*

Adapted to warm days and cool nights, the vivid violet heads of this sturdy plant outlast the complementary golden yellow of the goldenrods with which it is naturally associated. The late-season floral display and compact habit, when coupled with more technical characters like the double pappus and copiously long-hairy achenes, make flax-leaved aster readily identifiable. Its numerous narrow stiff leaves are unlike any other aster and confer both its common name and species epithet. **Low stiff perennial, 4–24 (mostly 8–16) inches tall, from a short tough crown.**

Basal rosettes absent. **Stem leaves** abundant, stalkless but not at all clasping, dark green, linear, 1–1½ inches long, to ⅛ inch wide, entire, rigid, with but 1 prominent vein (the midrib), hairless but the margins rough to the touch. **Stems** tufted, branched above, slender, minutely hairy (hairs hardly visible).

Flower heads 1 inch across, few to numerous (rarely more than 30), solitary at the tips of stiffly ascending stalks that are provided with bracts; involucre to ⅜ inch long, the bracts strongly overlapping in several series, appressed, unequal and in part leaf-like, linear-lance-

shaped, with fringed margins and greenish to purplish-tinged (inner ones) tips; ray florets 6–18, violet, the flattened ray 3/16–7/16 inch long; disk florets 25 or more. *Late August–mid October*

Achenes brown, spreading-hairy, the pappus bristles in 2 series, the inner of long, firm bristles, the outer of extremely short bristles (use a hand lens to see these). *Mid October–November*

WI range and habitats Confined to about 13 south-central counties, in dry open, sandy or rocky places, such as outcrops and bluff tops, sandy and gravelly prairies, and black oak barrens, occasionally on lakeshores, roadside banks, abandoned fields, and old quarries; acidic soils. *Upper Greene Prairie, West Knoll*

Flax-leaved aster is still locally common, but the number of extant sites for it has been declining during recent decades.

Similar species In Wisconsin this species might be confused only with aromatic aster (*Symphyotrichum oblongifolium*), which is similar in habit and foliage shape and texture. The latter is distinguished from flax-leaved aster by its glandular herbage, smaller involucre, greater number of ray florets per head, and simple pappus.

Asteraceae • Sunflower Family

FALSE-DANDELION
Two-flowered Cynthia
Krigia biflora

Pale, soft-looking, hairless foliage and few, medium-sized, long-stalked heads distinguish false-dandelion. **Slender perennial with milky juice, 6–28 inches tall, fibrous rooted.**

Basal leaves inverse-lance-shaped to oblong-elliptic, to 10 inches long and 2 inches wide, entire, toothed, or sometimes lobed (never prickly). **Stem leaves** 1 (occasionally 2 or 3, then appearing almost opposite because of the very short internodes), bract-like, stalkless and clasping, entire. **Stems** hairless and somewhat waxy whitened.

Flower heads 1–5 (6), erect, to $1^1/_8$ inch in diameter, solitary on long naked, branch-like stalks arising from upper axils; involucre bell-shaped, $1/_4$–$1/_2$ inch high, the 9–18 bracts in 1 series, essentially equal; florets all bisexual and fertile, the corolla yellow-orange. Inflorescence branches and involucre vary from completely hairless to sparsely to densely glandular-hairy. *Late May–mid July (August)*

Achenes dark brown, slightly tapering upward but not beaked, obscurely ribbed; pappus double, of an outer whorl of about 10 minute thin translucent scales plus an inner whorl of 20–35 unequal fragile bristles. *June–August (October)*

WI range and habitats Frequent throughout but northward only in pine barrens; mostly in wet prairies and open, dry to medium forests, pine and/or scrub oak woods, and thickets, less common in alder or cedar swamps and lakeshores, weedy on sandy hillsides, open fields, and along roadsides and railroads. *Curtis Prairie, Greene Prairie, Juniper Knoll*

The numerous composites with dandelion-like heads belong to the same tribe, whether plants like false-dandelion with orange to yellow flowers or chicory and the white-lettuces with blue to pink or whitish flowers. All have stems with milky juice and heads composed solely of ray florets with flattened, 5-toothed corollas.

Similar species Virginia dwarf-dandelion (*Krigia virginica*) is a slender annual with numerous narrow basal leaves and solitary small heads ($3/_4$ inch in diameter when fresh). Local in southwestern and south-central Wisconsin, it once grew on a formerly open, now overgrown sandy slope at the far east end of Greene Prairie but has not been seen there by local naturalists in at least 10 years.

FALSE BONESET
Kuhnia eupatorioides
Synonym *Brickellia eupatorioides*

For practical purposes false boneset may be thought of as an alternate-leaved Joe-Pye-weed, though the heads are cream-colored instead of pink, purple, or white, and the fruits exhibit one or two technical differences. Like those of Joe-Pye-weeds (*Eutrochium* species), blazing-stars (*Liatris* species), and ironweeds (*Vernonia* species), the heads are composed entirely of disk florets. The leaves, involucre, and achenes are hairy. **Perennial, 1–4 feet tall, from a short, somewhat stout crown surmounting a taproot.**

Leaves alternate (or opposite toward the base), the lower falling early and no larger than the middle ones, narrowly lance- to lance-diamond-shaped, $3/4$–4 inches long, variable in width, entire or sparingly rounded-toothed, 3-veined. Entire plant minutely hairy and abundantly (under 10✕ magnification) resinous-dotted.

Flower heads several in each of a series of rather dense, flat-topped aggregations at the ends of branches (or a few solitary or in 2s); involucres oblong, $3/8$–$1/2$ inch high, the bracts in about 4 series, strongly marked by 4–6 longitudinal lines; disk florets 7–33 per head, bisexual and fertile, the corolla creamy or light yellowish, $3/16$ inch long, 5-lobed at the summit. *(July) August–October*

Achenes dark purplish, cylindrical or spindle-shaped, about $1/8$ inch long, 10-ribbed, minutely gray-hairy; pappus of 20–30 white to brownish, feathery bristles. *Late August–October*

WI range and habitats South and west, characteristic of dry to dry-medium prairies, rocky or sandy prairie relicts on steep bluffs, and sand terraces of the Mississippi River, rarely on sand dunes but absent from the level sands of the Central Plain, occasionally along roadsides or railroads. *Curtis Prairie (limestone knoll), Marion Dunn Prairie, Sinaiko Overlook Prairie, Visitor Center parking lot*

Linnaeus named this genus after Dr. Adam Kuhn of Philadelphia, who brought him the living plant. As the species resembles *Eupatorium*, Linnaeus gave it the species epithet *eupatorioides*.

Similar species False boneset is similar to tall boneset (*Eupatorium altissimum*), which has opposite, prominently 3- or 5-veined leaves and smaller heads with only 5 florets and downy, scarcely lined involucral bracts. Moreover, its achenes have 5 angles, and pappus bristles are not feathery. It is infrequent in southern Wisconsin but is not found in the Arboretum.

Wild Lettuce
Canada lettuce, tall lettuce, tall wild lettuce
Lactuca canadensis

This relative of the sow-thistles (*Sonchus* species) and hawkweeds (*Hieracium* species) occurs in fields and waste places as well as prairies. Its small heads are composed wholly of bisexual florets with flattened rays with 5 terminal teeth. The strongly flattened achenes have threadlike beaks, the pappus therefore appearing to be stalked. **Biennial, 1–6 feet tall, with a basal rosette, from a taproot.**

Leaves about equal in size but variable in shape, the basal and lower usually deeply pinnately cut and often sickle-shaped, the midstem and upper ones mostly wavy-cleft or nearly lobeless, to 8 inches long, clasping, often arrowhead-shaped; upper leaves usually narrowed to a non-clasping base, weakly toothed or entire.

Flower heads numerous in elongate open panicles, the involucres narrowly urn-shaped before flowering, $^5/_{16}$–$^1/_2$ inch high, the bracts in several overlapping sets of unequal lengths, the outermost short; florets 17–22 per head, the rays yellow, $^1/_8$ inch long. *Late June–September*

Achenes blackish (tinged with brown), ellipsoid-egg-shaped, $^3/_{16}$–$^1/_4$ inch long (including beak), flat, cross-wrinkled, 1-veined on each face; pappus white, of hair-like bristles, falling early. *July–September*

WI range and habitats
Common throughout in wet to medium prairies, sandy fields, pastures, borders and trails in thin maple, aspen, or river bottom woods, limestone bluffs, shaded ravines, roadsides, and railroads. *Curtis Prairie, Grady Tract knolls, Greene Prairie, Juniper Knoll, Sinaiko Overlook Prairie*

Wild lettuce is quite variable, especially in leaf shape. The genus name is shared with garden lettuce, *Lactuca sativa*, and was derived from *lac*, milk, alluding to the thick white juice that exudes from broken stems.

Similar species The European weed prickly lettuce (*Lactuca serriola*) differs in having the leaves prickly along the midrib beneath, and the achenes about 7-veined on each face. The leaves of tall blue lettuce (*L. biennis*) have wider segments, the corollas are pale blue to white, and the pappus is light brown or grayish.

Daisy

Common daisy, ox-eye daisy
Leucanthemum vulgare
Synonym *Chrysanthemum leucanthemum*
Invasive weed

This popular garden plant and naturalized weed demonstrates well the construction of a composite bloom. What is commonly called a "flower" consists of many stalkless florets in a dense head 1–2 inches in diameter. The marginal "petals" are ray florets, each with one large white corolla, or *ray*. The round yellow center (the eye of ox-eye) is made up of tubular disk florets. **Perennial, 8–30 inches tall, from a short thick rhizome.**

Basal leaves spoon-shaped or inverted-egg-shaped, 1–2 inches long, from nearly entire to round-toothed, notched, or toward the base irregularly pinnate-lobed. **Stem leaves** clasping, strongly reduced or lacking upward. Plants typically smelling of sage.

Flower heads solitary at the ends of long leafless branches; involucre broad and flat, its many overlapping bracts lance-shaped to narrowly oblong, with a narrow, dark brown band near the margins; disk $3/8$–$3/4$ inch broad; ray florets 15–30, female and fertile, $3/8$–$3/4$ or more inch long; disk florets male or female. ***June–August (October)***

Achenes dark brown or black, narrowly oblong, $3/8$ inch long, with 10 white longitudinal ridges; pappus none. ***July–October***

WI range and habitats Common throughout, a weed of disturbed ground, especially roadsides, railroads, abandoned fields, and pastures, also clearings, trails, shores, and waste ground. ***Curtis Prairie, Greene Prairie, Juniper Knoll, Marion Dunn Prairie***

Ox-eye daisy is a close relative of the Shasta daisy and chrysanthemums. The English name *daisy,* or "day's eye," comes from the Anglo-Saxon *doeges-sege* ("dayeseye" of Chaucer), meaning "eye of the day," because the yellow disk resembles the sun. In France it is called Marguerite, because the English Queen Margaret of Anjou, who came from France, displayed the daisy in her coat of arms.

Predators such as assassin bugs and flower spiders like to conceal themselves beneath the flower heads of ox-eye daisy while waiting to capture unsuspecting bees and butterflies.

Similar species Two abundant weeds throughout Wisconsin, dog-fennel or stinking chamomile (*Anthemis cotula*; see Other Asteraceae) and pineapple-weed (*Matricaria discoidea*), also have white, daisy-like heads, but they are otherwise very different. Their leaves are delicately cut into fine linear segments, and their stems bear many small heads with conic receptacles.

Rough Blazing-star
Tall gay-feather
Liatris aspera

The blazing-stars and gay-feathers are distinguished by their solitary unbranched stems, narrow leaves in a close spiral, and rose-purple (rarely white) heads in spikes. Rough blazing-star is inconspicuous until late summer, when its spikes of bell-shaped heads suddenly blaze with bright pinkish-lavender, tubular florets. Its interesting involucral bracts have broad, often purplish, irregularly toothed edges. **Perennial, 16–40 inches tall, from a ± spherical, fleshy, bulblike crown (corm) $3/4$–2 inches thick.**

Leaves alternate, simple, the basal ones slightly diamond shaped, 2–6 (lowermost to 16) inches long, the upper ones gradually reduced and becoming shorter than the heads they subtend, linear to narrowly lance-shaped, entire.

Flower heads 10 or more, relatively large, all about the same size, $3/8$–$5/8$ inch high, in spikes (heads stalkless) or racemes (occasionally on stalks); involucre $5/16$–$5/8$ inch high, the middle and inner bracts spatula-shaped to narrowly oblong, petal-like at the spreading, rounded tips, the dry papery margins white to pink, wavy or toothed but scarcely cut at the apex; ray florets absent; disk florets 14–25 (30) per head. *Mid July–mid September*

Achenes 10-ribbed, not angled; pappus of numerous, straw-colored or creamy bristles about $1/4$ inch long, finely barbed but not feathery. *(Late July) August–September (October)*

WI range and habitats Frequent throughout although rare in the Northern Highlands except for the northeastern and northwestern pine barrens, in dry to medium prairie remnants, less common in black oak or jack pine savannas and other dry open places, especially open woods in sandy or rocky areas (e.g., bluffs, riverbanks); well-drained, usually acidic soils, in full sun. *Curtis Prairie, Grady Tract knolls, Greene Prairie, Juniper Knoll, Sinaiko Overlook Prairie*

Long intermingled styles, matching the corollas in color, protrude well beyond the mass of florets, contributing to the showiness of the heads and at the same time conferring a touch of grace to these bold plants.

Cylindrical Blazing-star
Few-headed blazing-star
Liatris cylindracea

Despite being the most diminutive of our *Liatris* species, cylindrical blazing-star is nonetheless attractive owing to its dapper habit, which makes it superior for small gardens. The few-to-several (or as few as one), pinkish-purple, oblong heads have hard shiny involucral bracts that are quite different from those of our other species. **Slender perennial, 10–23 inches tall, from a spherical, bulblike underground stem (corm).**

Leaves alternate, simple, linear, the lowest small, the principal ones $3\frac{1}{2}$–9 inches long and $\frac{1}{8}$–$\frac{1}{4}$ inch wide, the upper much smaller, entire, rigid, weakly dotted with translucent glands. **Stems** hairless or very sparsely and minutely hairy.

Flower heads (1) 4–20 in a raceme-like arrangement, each subtended by a leafy bract, of disk florets only; involucre oblong, $\frac{3}{8}$–$\frac{3}{4}$ inch high, the bracts in several series of different lengths, purple to green, lying flat against one another, mostly blunt or rounded except for the short sharp tip, leathery, hairless; receptacle without scaly bracts; ray florets absent; florets 10–35 per head, the corolla tube cylindrical, the inner surface of the 5 equal lobes hairy. *Mid July–early September*

Achenes somewhat cylindrical (not angled) but pointed at the base, 10-ribbed, the pappus bristles feathery, that is, with numerous fine lateral barbs 15 or more times the diameter of the bristle. *August–October*

WI range and habitats Infrequent southwest, most common on dry prairies, also limestone bluffs and sandy riverbanks, rarely on roadsides or railroads; dry, often limy soils. *Upper Greene Prairie*

Liatris is a name of unknown derivation. Moreover, there is no apparent reason for calling the species of *Liatris* "blazing-stars." The heads have neither the shape nor color conventionally ascribed to stars although the late-season color they provide is sufficiently brilliant to merit being termed "blazing."

Because flower heads are borne on long stems and are long-lasting, blazing-stars make excellent cut flowers. This distinctive species is unparalleled as a native plant for rock gardens.

Asteraceae • Sunflower Family

Northern Plains Blazing-star
Showy blazing-star
Liatris ligulistylis

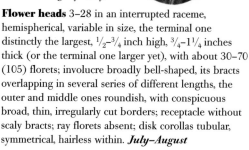

The brilliant, rose-purple flowers of the Northern Plains blazing-star are crowded thickly together in close heads within sheathing bells (involucres) of flat, red or purple bracts. The heads are stalked (occasionally nearly stalkless) and on average larger than those of our other *Liatris* species. They are attached to stiff stems at wider intervals than are those of rough blazing-star. **Somewhat coarse perennial, 1–4 feet tall, from an enlarged underground stem.**

Leaves alternate, simple, very numerous, the lower ones broadly lance-shaped, 3–5 (10) inches long and $1/4$–$1\,1/2$ inches wide, the upper much smaller, entire, dotted with translucent glands. **Stems** hairless or sometimes lightly hairy.

Flower heads 3–28 in an interrupted raceme, hemispherical, variable in size, the terminal one distinctly the largest, $1/2$–$3/4$ inch high, $3/4$–$1\,1/4$ inches thick (or the terminal one larger yet), with about 30–70 (105) florets; involucre broadly bell-shaped, its bracts overlapping in several series of different lengths, the outer and middle ones roundish, with conspicuous broad, thin, irregularly cut borders; receptacle without scaly bracts; ray florets absent; disk corollas tubular, symmetrical, hairless within. *July–August*

Achenes 10-ribbed, not angled, the pappus of numerous stout bristles, merely finely barbed (not feathery). *August–early September*

WI range and habitats Sporadic, mostly southeast and west-central, in medium prairies, especially deep-soil railroad and roadside prairies, less common in low prairies, lakeshores, edges of swamps, and dry sandy places; in northern Wisconsin adventive on sandy roadsides and railroad embankments. *Greene Prairie*

The Northern Plains blazing-star is unsurpassed as a natural nectar source for monarch butterflies, and after the flowers are gone, the seed heads are relished by goldfinches.

Similar species Northern Plains blazing-star is superficially similar to and often grows near or with rough blazing-star (*Liatris aspera*), and hybrids or introgressants (hybrid derivatives resulting from backcrossing) between the two appear to be relatively frequent. Northern Plains blazing-star differs from rough blazing-star by its thicker heads, longer pappus, and hairless corolla tube. Its involucral bracts are sometimes minutely fringed with marginal hairs, whereas in rough blazing-star the margins are fringeless.

Thick-spike Gay-feather
Prairie blazing-star,
thick-spike blazing-star
Liatris pycnostachya

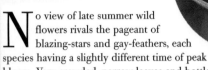

No view of late summer wild flowers rivals the pageant of blazing-stars and gay-feathers, each species having a slightly different time of peak bloom. Very crowded, narrow leaves and bottlebrush-like wands of closely crowded, small stalkless magenta heads are trademarks of this species and the next, which are among the few members of the Sunflower Family to produce heads in a spike. **Stiff tall perennial, 2–5 feet tall, from an enlarged woody underground stem.**

Leaves alternate, simple, linear-lance-shaped, mostly 4–24 inches long and to $3/8$ inch wide, gradually reduced upward and passing into bracts, entire, dotted with translucent glands. **Stems** usually few together, coarsely hairy throughout or only above.

Flower heads $3/8$ inch high, $1/4$–$1/3$ inch across, in crowded cylindrical spikes 6–18 inches long; axis of spike hairy; involucre oblong to narrowly top-shaped, $3/8$ inch high, the bracts overlapping in several series of different lengths, green or reddish-purple, tapering to sharp-pointed tips that are bent outward, hairy; receptacle without scaly bracts; ray florets absent; disk florets 3–7 (12), the corollas tubular, $1/4$–$3/8$ inch long, the tube lacking (or with very few) hairs within. *July–early September*

Achenes 10-ribbed, not angled, the pappus of numerous, minutely barbed bristles $3/16$–$1/4$ inch long. *(Mid July) August–September*

WI range and habitats Locally common southwest, on medium to wet-medium prairie remnants and fens, sometimes in peat marshes, bogs, and wet prairie relicts along roads and railroads. *Curtis Prairie, Greene Prairie, Juniper Knoll, Sinaiko Overlook Prairie*

Although we think of these conspicuous, sometimes ill-defined species as belonging to the prairie, others grow in the northeastern and southern U.S. Some, including thick-spike and marsh gay-feather, are grown as garden ornamentals.

Unlike most spike-like inflorescences, which start blooming at the bottom, those of all members of *Liatris* start blooming at the top. Thus bees, which usually fly to the bottom of a spike and work their way upward, when visiting blazing-stars come to younger flowers before older ones, versus other plant groups in which bees visit older flowers first.

Similar species See marsh gay-feather (*Liatris spicata*).

Asteraceae • Sunflower Family

Marsh Gay-Feather
Dense gay-feather, marsh blazing-star,
sessile blazing-star, sessile-headed blazing-star
Liatris spicata
Wisconsin special concern

The tallest member of this colorful genus, marsh gay-feather produces a bold magenta spike to 13 inches long of strictly stalkless, closely crowded heads. An eastern species, it is replaced on the great majority of Wisconsin prairies by its close cousin, thick-spike gay-feather (*Liatris pycnostachya*). **Perennial, 2–6 feet tall, from a short, bulblike, underground stem.**

Leaves alternate, numerous, the lower ones narrowly elliptic or narrowly oblong, 4–16 inches long, $3/16$–$3/4$ inch wide, the others linear, reduced upward, somewhat dotted with translucent glands, entire. **Stems** hairless or nearly so.

Flower heads $1/6$–$1/3$ inch across, of 5–10 (14) blue-purple (white) disk florets (only), in dense spikes; axis of spike hairless; involucre oblong, $1/4$–$7/16$ inch high, of several series of overlapping bracts of different lengths, green or purplish, the tips lying flat, prevailingly rounded or blunt (outer ones sometimes ± sharp-pointed); florets 5–13 per head, their corollas tubular, hairless (or very nearly so) within at the base. *Late July–early August*

Achenes somewhat cylindrical but basally pointed, 10-ribbed, the pappus of numerous, minutely barbed bristles. *August–early September*

WI range and habitats Once locally abundant in extreme southeastern Wisconsin, in sandy calcareous, moist to medium prairie remnants, calcareous springy sites, and limy sand flats along the Lake Michigan shore. Planted in *Curtis, Greene, and Marion Dunn prairies* but not native to the Arboretum

Marsh gay-feather thrives in full sun and ordinary soil and makes an excellent garden plant. It is much in demand by florists, who use it in bouquets. Like its confreres, it attracts nectar-feeding butterflies and seed-seeking birds. The root contains a volatile oil and a resin and once saw limited use as a diuretic and uterine tonic.

Similar species Marsh gay-feather co-occurs with thick-spike gay-feather (*Liatris pycnostachya*), and occasional specimens are suggestive of hybridization between them. Both have stalkless, densely crowded heads, but the involucral bracts of the former have appressed blunt tips, whereas those of thick-spike gay-feather terminate in a sharp or gradual point that spreads outward or bends backward. The axis of the inflorescence in marsh gay-feather is hairless; that of thick-spike gay-feather is usually beset with spreading crinkly hairs.

Ragworts, Groundsels

Ragworts are pretty, albeit small herbs that look very much like daisies with yellow ray and disk florets. Unlike most prairie composites, they bloom early in the season and are easily overlooked owing to their short blooming period and low stature. They are lightly woolly with long white soft hairs when young but soon become nearly hairless except in the axils. **Short-lived, sparsely leafy perennials, 8–24 inches tall, from a short persistent, over-wintering crown.**

Basal leaves long-stalked, coarsely toothed. **Stem leaves** few, distant, and much reduced, lance-shaped to oblong, ± deeply cut into pinnate lobes. Northern Ragwort has chiefly inverted-lance-shaped to elliptic basal leaves $3/8$–$2 1/2$ inches long with wedge-shaped bases and smaller, rounded to sharp teeth. It sometimes produces short slender runners. Western golden ragwort has egg- to broadly lance-shaped basal leaves $3/4$–$1 1/2$ inches long or sometimes smaller, slightly heart-shaped to straight across at the base and bluntly toothed. It reproduces via basal offshoots (never runners).

Flower heads 2–12 in loose, flat-topped clusters (terminal head blooming first), small, the principal involucral bracts in 1 series, subtended by a few short accessory bracts at the base, $1/4$ inch long or less; receptacle naked (not chaffy); ray florets 5–10, fertile, without stamens, their corollas $1/4$ or less inch long; disk florets tubular, 5-toothed. Northern ragwort: *mid May–July;* western golden ragwort: *mid May–June*

Achenes columnar, 5–10-ribbed, hairless; pappus of copious white soft bristles. *June–July* (northern ragwort to mid September in the far north)

Northern Ragwort
Balsam groundsel, balsam ragwort, northern meadow groundsel
Packera paupercula
Synonym *Senecio pauperculus*

WI range and habitats

Northern ragwort: Fairly common throughout, in medium prairies, fens, bracken grasslands, meadows, savannas, open aspen or oak woods, and streambanks; generally in moist calcareous sites. *Greene Prairie*

Western golden ragwort: Rare southeast, in moist prairies, fens, and sunny springy sites. *Upper Greene Prairie*

Packera is segregated from *Senecio,* which is one of the largest genera of plants when broadly interpreted. The genus name honors John G. Packer, a distinguished North American botanist and authority on the flora of Alberta. The species epithets *paupercula* and *pseudaurea* are Latin for "poor little" and "false aurea," respectively.

Northern ragwort is a complex species, encompassing many morphological and cytological races, four in Wisconsin.

Similar species These two species are not easy to distinguish, especially if the basal leaves have withered. Furthermore, they are very similar to prairie ragwort (*Packera plattensis*), plants of which, however, are darker green, have more persistent tangled woolly hairs, broader-lobed basal leaves, more deeply cut stem leaves, and grow mostly on dry bluffs and gravel terraces. There is no habitat in the Arboretum for prairie ragwort or for golden ragwort (*P. aurea*), which differs in having large, ± strongly heart-shaped basal leaves and grows in cool swampy places.

WESTERN GOLDEN RAGWORT
Heart-leaved groundsel, western heart-leaved groundsel
Packera pseudaurea
Synonym *Senecio pseudaureus*

Wild Quinine
American feverfew, eastern feverfew, eastern parthenium
Parthenium integrifolium
Wisconsin threatened species

Wild quinine blooms for a long time, providing a wonderful mainstay through the summer for the continuously changing floral display of the medium prairies. The large coarse leaves and button-like heads of purest white, clustered at the top of a tall stem, are utterly distinctive. **Aromatic perennial, 20–40 inches tall, from a tuber-like root that often branches to form underground runners.**

Basal leaves long-stalked, to 8 inches long. **Stem leaves** alternate, egg- to broadly lance-shaped, the lower ones 4–8 inches long, sometimes with a winged, stalk-like base or (especially the smaller upper ones) stalkless and clasping, scalloped to toothed, roughened with stiffish hairs on both surfaces. **Stems** single or clustered, lacking latex.

Flower heads numerous in flat-topped arrangements; involucre $1/4$ inch tall and $1/2$ inch wide, the bracts in 2 series of 5 each, broadly oblong (outer) to somewhat circular (inner ones), nearly equal; receptacle small with scaly bracts; ray florets 5, female and fertile, tiny, the "mouse-ear" rays short and thick, entire or 2-lobed at the tip; disk florets functionally male (pistil sterile), funnel-form, the corolla 5-toothed; stigmas brownish-black. *Late June–early September*

Achenes blackish-brown, inverted-egg-shaped, 3-sided, fused to and falling with their attached scale and contiguous pair of disk florets, bearing a pappus of 2–3 awns on the summit. *Mid August–October* (notorious for poor fruit set)

WI range and habitats Once probably common in the southern two tiers of counties, now rare in medium (black-soil) and wet-medium (level sandy) prairie remnants and bur oak openings. *Curtis Prairie, Greene Prairie, Sinaiko Overlook Prairie*

The genus name comes from the Greek word *parthenos,* virgin, and might refer to the fact that the disk florets are infertile, that, is, with a sterile, seed-bearing organ (pistil). Common names indicate medicinal usage; for example, the heads, when dried and crushed, yield an astringent.

Glaucous White-lettuce
Purple rattlesnake-root
Prenanthes racemosa

Handsome white-lettuces or rattlesnake-roots (*Prenanthes* species) have nearly unbranched stems, erect and leafy except in the vicinity of the heads, and whitish, creamy, or in the case of glaucous white-lettuce, pink or purplish florets with strap-shaped corollas. Nodding when in bud, the heads become ascending as they open. **Coarse perennial, 1–5 feet tall, with milky juice and tuberous-thickened, fibrous roots.**

Leaves alternate, variable, the lower persistent, stalked, inverted-egg- or inverted-lance-shaped to elliptic, 3–16 inches long overall, the upper ones reduced, becoming stalkless and weakly clasping. **Stem** and leaves blue-green, hairless (except in the inflorescence).

Flower heads in a narrow elongate panicle made up of a series of very short, raceme-like branches; florets all bisexual and fertile; involucre of 8–13 principal bracts (inner ones in a double row) and several smaller outer ones, purplish or blackish, $3/8–1/2$ inch long, conspicuously hairy; receptacle small; florets 14–26 per head, the corollas flattened, 5-toothed at the apex. *(End of August) early September–October*

Achenes reddish- or yellowish-brown, short-cylindrical, indistinctly 5-ribbed, beakless, hairless; pappus of numerous buff-colored bristles that fall off. *Mid September–mid October*

WI range and habitats At one time prevalent in deep-soil prairies in the southeast and sporadic in the Driftless Area, now fairly rare, restricted to dry to wet-medium prairie remnants along railroads and rocky or gravelly hill prairies. *Greene Prairie*

> The origin of the name rattlesnake-root has not been satisfactorily explained, but the name implies that some member or members of this group were used to ward off rattlesnakes or treat their bites. Linnaeus took the Greek words *prenes*, drooping, and *anthos*, flower, and turned them into the genus name.
>
> **Similar species** Except for lion's-foot (*Prenanthes alba*), the four white-lettuces in Wisconsin have become rare through habitat destruction. Lion's-foot has distinctive lower leaves, hairless involucres, and a cinnamon-brown pappus. It occurs in the Arboretum, but neither rough white-lettuce (*P. aspera*) nor Midwestern white-lettuce (*P. crepidinea*) is found there. The former has hairy stems and leaves and cream-colored florets. The latter has broadly triangular-egg- or arrowhead-shaped leaves and a more ample inflorescence of slightly larger heads than the other species.

Sweet Everlasting
Fragrant cudweed, cat's-foot,
old-field-balsam, old-field cudweed,
rabbit-tobacco
Pseudognaphalium obtusifolium
Synonym *Gnaphalium obtusifolium*

The loose woolly covering of the stem and leaves, coupled with the small white, tightly compacted heads, help identify this native weed, which is further distinguished by a rather characteristic scent (something like maple syrup or brown sugar). **Stiff annual (or biennial?), (4) 10–32 inches tall, from a taproot.**

Leaves alternate, lance-shaped to linear, to 4 inches long, the upper surface green (often drying brown), the short, gland-tipped hairs beneath concealed by the ± persistent covering of dense matted hairs, the margins entire but often wavy.

Flower heads in round-topped clusters standing above the leaves, the inflorescence small to ample depending on plant size; involucre bell-shaped, off-white, soon tan or yellow-tinged, of numerous papery, scale-like bracts; ray florets absent; disk florets about 75–125, the outer ones female and fertile, a few (about 5) in the center of the head bisexual, their corollas yellow and minutely 5-toothed. *August–early October*

Achenes olive-green to yellowish-brown, narrowly egg-shaped, small, wrinkled but veinless and hairless; pappus of about 20 roughened, hair-like bristles. *September–October*

WI range and habitats Locally common throughout except far northwest, in sandy fields, sand barrens, degraded dry prairies, pastures, roadsides, and occasionally open dry woods; especially common in burned and cut-over areas; sandy or rocky soils. *West Knoll*

Gnaphalium means "locks of wool" in Greek. The common name "cudweed" was applied to the softly furry European species, which were once used to prevent chafing, such as on the backs of beasts of burden, and relieve irritation, apparently including that endured by external parts of the female genitals during childbirth.

Sweet everlasting and the related pearly everlasting (*Anaphalis margaritacea*), if picked while the heads are young, will dry and appear as if fresh for up to a year or more. These are also related to pussy-toes (*Antennaria* species), which differ in producing elongate creeping stems and unisexual, pure white (or pinkish) heads.

Similar species Of the 6 cudweed species in Wisconsin, sweet everlasting is most similar to cliff cudweed (*Pseudognaphalium saxicola*), a state-threatened species endemic to cliffs and ravines in south-central Wisconsin. It is not found in the Arboretum.

Yellow Coneflower
Globular coneflower
Ratibida pinnata

Showy drooping rays make this species more recognizable than any other yellow coneflower. The disk, hemispherical or somewhat spherical and gray when young, becomes twice as tall as thick and turns brown at maturity. **Perennial, $1\frac{1}{2}$–4 feet tall, from a woody rhizome or short crown.**

Leaves basal and alternate, at least the lower ones deeply pinnate-divided, the 3–7 segments linear to lance-shaped or larger ones lance-egg-shaped, $\frac{1}{8}$–$\frac{3}{4}$ inch wide, entire or coarsely toothed, 3-veined, with short stiff hairs and translucent glandular dots on both surfaces. **Stems** light or grayish green, with dark green lines and appressed straight hairs.

Flower heads few to several, each solitary at the top of a naked branch; involucral bracts 10–14 in 2 series, leafy, narrowly oblong, the outer spreading or bent backward; disk $\frac{3}{8}$–1 inch high, the scaly bracts hood-shaped, velvety at the apex, subtending the ray as well as the disk florets; ray florets 6–10 (13), the ray 1–$1\frac{3}{4}$ inches long, obscurely 3-toothed; disk florets bisexual and fertile, the corolla yellowish-green, the lobes becoming purple-brown. *July–early October*

Achenes dull brown, inverted-lance-shaped, tiny, flattened with wing-like edges, the scales of the receptacle ± clasping the achenes and falling with them; pappus none. *Late September–October*

WI range and habitats Very common in the region of limestones south and west, in dry, medium, and wet prairies, often in degraded dry prairies and along roadsides and railroads, reappearing in the north in scattered colonies along forest edges; full sun. *Curtis Prairie, Greene Prairie, Juniper Knoll, Marion Dunn Prairie, Sinaiko Overlook Prairie, Wingra Oak Savanna*

This is one of the most popular plants for use in wildflower meadows and restorations. Like round-headed bush-clover and black-eyed Susan, it is an early-successional species, growing well in a variety of soils and never failing to produce flowers in profusion.

Similar species Long-headed coneflower (*Ratibida columnifera*) is much less common in Wisconsin and does not occur in the Arboretum. It has narrowly oblong disks, relatively short rays, and a pappus of 1 or 2 teeth.

BLACK-EYED SUSAN
Rudbeckia hirta

This abundant and familiar prairie flower has bright heads with leafy involucral bracts and orangish-yellow, down-curved rays (sometimes with a brownish base) that contrast beautifully with the purplish-black, firm-textured disk (the "black eye"). **Coarse biennial or short-lived perennial, 1–3 feet high, from a cluster of fibrous roots.**

Leaves alternate, simple, the lower inverted-lance-shaped to elliptic, on winged stalks or nearly stalkless, the middle ones inverted-lance-shaped to oblong or lance-linear, 2–5 inches long, entire or weakly toothed, often 3-veined. Plant covered all over with gray bristly hairs.

Flower heads 2–3 inches across, terminating the stems and branches; involucral bracts in 3 series, bristly-hairy; receptacle hemispheric or egg-shaped, with sharp-pointed (but not spine-tipped) scaly bracts; rays (8) 10–15 (21), lacking pistil and stamens, $^3/_4$–$1^1/_2$ inches long; disk turning dull brown, the numerous florets bisexual and fertile, their corollas tubular, 5-toothed. *Mid June–September (October)*

Achenes purplish or black, $^1/_{16}$ inch long, 4-angled, hairless but with fine lines; pappus none or a tiny, crown-like border. *Late July–October*

WI range and habitats Very common throughout, in wet to dry-medium prairies, savannas, fields, pastures, fencerows, along roadsides and railroads, and in other open habitats, including waste places; somewhat weedy. *Curtis Prairie, East Knoll, Greene Prairie, Juniper Knoll, Marion Dunn Prairie, Sinaiko Overlook Prairie, Wingra Oak Savanna*

Black-eyed Susan is a complex species, consisting of several intergrading varieties. It is commonly grown as a garden flower, and many horticultural varieties exist, including the 'Gloriosa Daisy.' Like echinaceas and ratibidas, rudbeckias are remarkably easy to cultivate and will thrive in situations that are too demanding for less hardy species; as a result, all three genera of coneflowers are included in wildflower seed mixes.

Similar species Distinguish rudbeckias from sunflowers (*Helianthus* species) on the basis of receptacle shape (flat to convex in true sunflowers) and leaf arrangement (never opposite in black- and brown-eyed Susans), and from yellow coneflowers (*Ratibida* species) by the distinctly 4-sided achenes. Their unusually well-developed receptacles or "cones" are even more reminiscent of those of purple coneflowers (*Echinacea* species), which, however, have purple to white rays and spine-tipped bracts on the receptacle.

Sweet Coneflower
Sweet black-eyed Susan
Rudbeckia subtomentosa

This close cousin of black-eyed Susan produces firmer foliage on taller stems and an abundance of yellow, sunflower-like heads with dull reddish- or purplish-brown disks. **Coarse perennial, $1\frac{1}{2}$–$4\frac{1}{2}$ feet tall, from a stout rhizome, producing leafy basal offshoots.**

Leaves alternate, distinctly stalked, egg-shaped or less often lance-elliptic, simple or some of the larger ones 3-lobed and the terminal lobe egg-shaped, 3–5 inches long, minutely toothed, resinous-glandular. **Stems** (above) and leaves (especially beneath) gray-downy.

Flower heads several, terminating the stems and branches, on stalks 2–8 inches long; involucral bracts in 3 series, narrow, leafy, spreading or bent backward, hairy toward the blunt to sharp tips; receptacle hemispheric or egg-shaped, its bracts scaly with short whitish sticky hairs at the blunt or rounded apex; rays 12–20, infertile, each $\frac{5}{8}$–$1\frac{1}{4}$ inches long, obscurely 3-toothed at the apex; disk $\frac{3}{8}$–$\frac{1}{2}$ inch broad, its florets very numerous, bisexual and fertile, the corolla lobes dark purple or brown. *Late July–early September*

Achenes dark brown, 4-angled, barely $\frac{1}{8}$ inch long, hairless, the pappus a minute crown-like border. *[Mature about one month after bloom]*

WI range and habitats Rather rare across the southern 2–3 tiers of counties, in medium or wet-medium prairies, marshes, river bottoms, and low habitats along roadsides and railroads, most common in relatively undisturbed, bottomland habitats (lowland savannas, edges of woods, thickets, marshes, prairies) along the lower Wisconsin and Sugar rivers; drought resistant. *Curtis Prairie, Greene Prairie, Juniper Knoll, Marion Dunn Prairie, Sinaiko Overlook Prairie, Lower Wingra Oak Savanna*

The name sweet coneflower comes from the anise- or vanilla-like scent emitted by the leaves and receptacle when bruised or crushed.

As in all composites, the rays (if present) open first; then the perimeter disk florets of the central cone open, forming a blossoming circle that contracts day by day as it creeps inward toward the top.

Similar species See brown-eyed Susan (*Rudbeckia triloba*). The sunflower-like wild golden-glow (*R. laciniata*) differs from sweet coneflower and black-eyed Susan (*R. hirta*) in having hairless stems, larger leaves, the principal ones 5–7-lobed, and only 6–10 rays surrounding the greenish-yellow central cone (disk). It blooms during July and August in wet thickets.

Brown-eyed Susan
Three-lobed coneflower
Rudbeckia triloba

This plant resembles its cousin the black-eyed Susan, but the thinner leaves often with 3 broad lobes, smaller heads with fewer rays, and presence of a pappus make it readily distinguishable. **Short-lived perennial, 1–5 feet tall, from a stout rhizome.**

Leaves variable in shape and size, the basal ones broadly egg-shaped and long-stalked, the principal stem leaves narrower and short- or unstalked, tending to be 3-lobed, often more than 2 inches long. **Stems** much-branched, green or dark-colored, with scattered long hairs to nearly hairless.

Flower heads several, terminating the stems and branches, only 2 inches broad; involucral bracts about 8, green and leafy, spreading or bent backward, hairy toward the blunt-to-sharp tips; receptacular bracts hairless, their tips abruptly contracted into a short awl-shaped tooth; disk dark purple-brown, hemispheric or egg-shaped, not elongating, $3/8$–$1/2$ inch broad; ray florets 6–13, infertile, the ray wholly yellow or orange at the base, $3/4$–$1 1/2$ inches long, obscurely 3-toothed apically; disk florets very numerous, bisexual and fertile, funnelform, the corolla lobes dark purple, narrowed into a tubular base, 5-toothed. *August–mid October*

Achenes a little less than $1/8$ inch long, 4-angled, hairless, with a minute crown and a pappus of 2–4 short teeth or these obsolete. *October through winter* (achenes persist until the following April)

WI range and habitats Locally frequent south, in second-growth woods and thickets along riverbottoms, weedy prairies, old fields, and borders of marshes and fens, spreading and/or escaping along roadsides, fencerows, railroads, and waste places, including dumps, vacant lots, alleys, and unkempt yards; native although often inhabiting severely disturbed communities. *West Curtis Prairie, Upper Greene Prairie, Marion Dunn Prairie*

The flowers of any species of *Rudbeckia* can be used to make a superior golden dye. As in all composites, the anthers form a tube and open inward at maturity. The pollen is pushed out through the anther tube by the growing style.

Similar species See Black-eyed Susan (*Rudbeckia hirta*).

Rosinweed
Prairie rosinweed, whole-leaf rosinweed
Silphium integrifolium

Although dissimilar vegetatively, silphiums share many inflorescence characters. Rosinweed's leaves are uniform in shape, size, and distribution and are very rough to the touch on both sides. Its medium-sized, bright yellow heads have many ray florets surrounding a darker yellow mass of disk florets. **Coarse erect perennial, 2–6 feet tall, from a short tough crown or short rhizome.**

Leaves opposite, egg-shaped to lance-elliptic, the larger ones 3–6 inches long, stalkless and often clasping, entire or slightly toothed, rigid. **Stems** and leaf surfaces variously velvety to minutely roughened or nearly hairless.

Flower heads 2–3 inches across, several together in short open clusters; involucre $3/8$–$3/4$ inch high, the bracts leafy and somewhat overlapping, glandular-hairy in the middle of the back and fringed with marginal hairs; receptacle with lance-shaped bracts subtending ray as well as disk florets; ray florets 11–22 in 2–3 series, fertile, the ray $3/4$–2 inches long; disk $5/8$–1 inch across, its florets bisexual but sterile. **Late June–early September**

achene

buds

Achenes inverted-egg-shaped or somewhat circular, to $3/8$ inch long and wide, conspicuously flattened, broadly winged and deeply notched, the wings continuous above with the 2 sharp teeth of the pappus. **September–mid October**

WI range and habitats Once common south in dry to wet prairies, rarely adventive (or escaped) farther north, still frequent in prairies and along railroad rights-of-way and roadsides; prefers medium-dry to medium-wet soils in full sun. *Curtis Prairie, Greene Prairie, Juniper Knoll, Marion Dunn Prairie, Sinaiko Overlook Prairie, Lower Wingra Oak Savanna*

Rosinweed is an excellent plant for wildflower meadows and prairie restorations. The pollen and nectar attract long- and short-tongued bees, various flies, and some butterflies, including sulphurs and painted ladies. The seeds are a favorite food for birds such as sparrows, goldfinches, and redpolls. Like other *Silphium* species, its sap contains a resin.

Hybrids between rosinweed and cup-plant (*Silphium perfoliatum*), variously intermediate between the parental species, have been collected on Curtis Prairie and on a railroad prairie relict west of Brodhead, Green County.

COMPASS-PLANT
Silphium laciniatum

This classic prairie plant produces deeply cut, stiff leaves, the lower very large, and tall rough resinous stems that bear several large, sunflower-like heads in a raceme-like arrangement. **Coarse perennial, 3–7 feet tall, from a tremendous taproot.**

Basal leaves commonly 1 or more feet in length, long stalked, deeply pinnate-divided or -lobed into lance-shaped or linear segments, these entire or often further divided. **Midstem leaves** similar but smaller, becoming strongly reduced and merely lobed or toothed upward, the bases stalkless or clasping. Herbage coarsely white-hairy.

Flower heads 2–5 inches across, stalkless or short-stalked; involucre shallowly bell-shaped, $3/4$–$1\frac{1}{2}$ inches high, its bracts more than $3/16$ inch wide, the middle ones with long-pointed, leafy tips; disk flat; ray florets 15–25 (34) in 2–3 series, female and fertile, ray petals $3/4$–2 inches long; disk florets apparently bisexual but sterile, the yellow corollas obscured by the long scaly bracts. *July–August*

Achenes oval to somewhat circular, barely more than $1/4$ inch long and wide, conspicuously flattened and broadly winged, notched at the apex; pappus of 2 teeth confluent with the wings. *Late July–early October*

WI range and habitats

Formerly common south, now fairly rare, in wet to dry but especially medium prairies in full sun; until recent decades persisting along roadsides and railroads, but nowadays even these lonely reminders of very rare habitats are vanishing. *Curtis Prairie, Greene Prairie, Juniper Knoll, Marion Dunn Prairie, Visitor Center parking lot*

The basal leaves stand vertically and tend to align in a north-south direction (hence the name compass-plant), the surfaces thus avoiding the full force of mid-day radiation. The large nutritious seeds of this and other silphiums are relished by songbirds. The stems produce a copious resinous juice that served as chewing gum for Native American children.

This species hybridizes with prairie dock (*Silphium terebinthinaceum*), producing plants that are exactly intermediate between the parental species.

Similar species *Silphium* heads resemble those of sunflowers (*Helianthus* species), tickseeds (*Coreopsis* species), and ox-eye (*Heliopsis* species), but among other characters are readily distinguished by fruits only in a ring around the disk. A little experience will have even casual botanists at once recognizing the unique "gestalt" of a *Silphium* bloom.

Cup-plant
Silphium perfoliatum

Splendidly sculpted leaves identify this grand plant. Opposite and large, the bases of each upper pair are grown together around the thick square stem, forming prominent cups at the nodes. Sunflower-like heads with large yellow rays enhance its exceptional character. **Coarse perennial, 3–8 feet high, from a stout, rhizome-like base.**

Leaves mostly opposite, egg- to triangular-egg-shaped, 6–12 (14) inches long, the lower ones narrowed to broadly winged stalks, the midstem and upper ones fused at the base, the stem appearing to pass through the leaves (perfoliate), grossly toothed, rough-hairy.

Flower heads 2–3 inches across, several to many in a panicle; involucre $3/8$–1 inch high, the bracts in 2 series, elliptic, blunt, fringed with marginal hairs; receptacle flat with scaly bracts; ray florets in 2–3 series, fertile, ray petal $3/4$–1 inch long; disk $5/8$–1 inch wide, the florets about (15) 18–26, sterile. *July–September*

Achenes inverted-egg-shaped, $1/4$ inch long and wide, flattened, prominently winged and deeply notched at the apex; pappus lacking or often of 2 teeth confluent with the wings. *September–October*

WI range and habitats Chiefly southwest, frequent at the edges of low woods, river and stream banks through wet forests, and low prairies, common along railroads and roadsides and in fields; in the north locally established along wooded roadsides; medium to moist soils, open or shaded. *Curtis, Greene, and Marion Dunn Prairies, Juniper Knoll, Sinaiko Overlook Prairie, Visitor Center parking lot, Wingra Oak Savanna*

Cup-plant is persisting and spreading slightly from introductions on the Bad River, Lac du Flambeau, and Menominee reservations. The Winnebago attached supernatural powers to the plant, and the Ojibwe made a medicinal extract from the roots. Rainwater caught in the leaf cups attracts thirsty birds and butterflies, and later in the season goldfinches come to gorge on the fruits.

Prairie Dock
Basal-leaved rosinweed, prairie rosinweed
Silphium terebinthinaceum

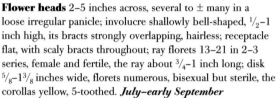

Although most notable for its "elephant-ear" basal leaves, the rather numerous, large heads and tall, nearly leafless stems help make prairie dock one of the most striking of all prairie plants. The floral structures are very similar to those of the other silphiums. **Coarse perennial, 3–9 feet tall, from a large woody taproot.**

Basal leaves 4–20 inches in length, long stalked, egg- to oblong-egg-shaped, heart-shaped, rounded, or nearly straight across at the base, sharply toothed, hairless or rough to the touch. **Stem leaves** few and strongly reduced, resembling large bracts. **Stems** hairless or becoming so, exuding a resinous juice.

Flower heads 2–5 inches across, several to ± many in a loose irregular panicle; involucre shallowly bell-shaped, $1/2$–1 inch high, its bracts strongly overlapping, hairless; receptacle flat, with scaly bracts throughout; ray florets 13–21 in 2–3 series, female and fertile, the ray about $3/4$–1 inch long; disk $5/8$–$1 3/8$ inches wide, florets numerous, bisexual but sterile, the corollas yellow, 5-toothed. *July–early September*

Achenes strongly flattened and firmly winged, notched at the top; pappus of 2 insignificant teeth confluent with the winged margins. *Late August–September*

WI range and habitats Southeast (barely entering the Driftless Area in Adams and Juneau counties), locally common on wet to dry-medium prairie remnants, rarely persisting along fencerows, railroads, and roadsides, surviving severe degradation of its prairie habitat by means of its large taproot; deep, alkaline or neutral soils. *Curtis Prairie, Greene Prairie, Juniper Knoll, Marion Dunn Prairie, Visitor Center parking lot, Wingra Oak Savanna*

The chief function of the large leaves is to manufacture food, but they also may shade competing plants and the moist ground in which the plant grows, slowing evaporation. Aesthetically, they add a wonderful, season-long contrast in texture to the changing mosaic of thin-stemmed wildflowers and prairie grasses. Like those of compass-plant, the basal leaves tend to align their edges in a north-south direction so that the blades face east-west. Clasp a prairie dock leaf between your hands on a hot summer day, and you will discover that the leaf is several degrees cooler than the surrounding air owing to evapotranspiration.

Canadian Goldenrod
Common goldenrod
Solidago canadensis
Synonym *Solidago altissima*

Our commonest goldenrod is considered a weed, not only because it forms large patches but also because it is erroneously blamed for causing hay fever. It has finely hairy stems, narrow, 3-veined leaves, and a handsome, plume-shaped inflorescence. **Perennial, 1–5 (6) feet high, from short to long, horizontally creeping rhizomes.**

Leaves numerous (basal leaves none, lower stem leaves soon withering), lance-linear to -elliptic, (1) 2–5 (6) inches long, stalkless, not particularly reduced upward, with 2 prominent lateral veins semi-parallel with the midvein, sharply toothed margins, and finely hairy (at least on main veins) lower surfaces.

Flower heads few-flowered and small, $1/12$–$1/8$ inch long, on spreading or backward-curved, 1-sided, raceme-like branches that together form a large, pyramid-shaped panicle; involucral bracts somewhat overlapping in several graduated series, yellowish, sharp-pointed; receptacle small; ray florets (8) 10–18, female and fertile, the rays yellow, $1/8$ inch or less long; disk florets 2–8, bisexual and fertile, yellow. *(Mid) late July–late September (early October)*

Achenes short-hairy, the pappus of white, hair-like bristles. *August–early November*

WI range and habitats Abundant throughout in a wide range of habitats, particularly fallow fields, disturbed prairies, old pastures, roadsides, railroads, fencerows, and edges of marshes, swamps, and thickets, occasionally in open woods and pine-oak barrens; moist to dry, open or partly shaded, often degraded habitats; our weediest goldenrod. *Throughout the Arboretum*

This species is extremely variable in plant height, leaf characters, inflorescence and head size, and floret number. Stems often have conspicuous galls—swellings caused by the attacks of small flies or moths that lay their eggs in the tissue. Goldenrods have the heavy sticky pollen of insect-pollinated plants, whereas the much less obvious ragweeds, which bloom at the same time, produce tremendous amounts of wind-borne pollen and are the major source of late-season allergies.

Similar species Missouri goldenrod (*Solidago missouriensis*) and late goldenrod (*S. gigantea*) also have 3-veined leaves, but these plants have stems that are hairless below the inflorescence (and even in the inflorescence in the former). Stems of late goldenrod may have a waxy bloom on the surface that is lacking in Canadian goldenrod.

Late Goldenrod
Giant goldenrod, smooth goldenrod
Solidago gigantea

Many kinds of goldenrods flood late summer and autumn fields and roadsides with gold, especially late goldenrod and its close cousin Canadian goldenrod. Both grow fairly tall and have lance-shaped, 3-veined leaves and tiny flowers (florets) in small close heads. The mid-stem leaves are usually the largest, and lower stem leaves fall off before flowering time. Despite its common name, late goldenrod does not bloom later than the others. **Perennial, 2–4 (5) feet tall, from extensive horizontal rhizomes.**

Leaves (basal leaves none) not particularly reduced in size upward, narrowly elliptic to lance-shaped, the larger ones $2\frac{1}{2}$–7 inches long, 3-veined (with 2 lower lateral veins parallel to the midvein), toothed, hairless (or very slightly hairy on the undersurface veins). **Stems** hairless (rather sparsely hairy among the heads) and often waxy-whitened below the inflorescence.

Flower heads $\frac{1}{12}$–$\frac{1}{8}$ inch long, borne on spreading or backward-curved, 1-sided, raceme-like branches that together form a fairly dense panicle; involucre $\frac{1}{8}$ inch high, its bracts overlapping in several series, of different lengths, blunt to sharp-tipped; receptacle small; ray florets mostly 8–15 (17), female, yellow, the ray $\frac{3}{16}$ inch long; disk florets 6–10 (12), bisexual, yellow. *Late July–September*

Achenes with sparse hairs closely pressed against the surface; pappus of hair-like bristles $\frac{1}{8}$ inch long. *August–October*

WI range and habitats Abundant throughout, in sedge meadows, fens, edges of bogs, streams, and ponds, and moist beaches, less common in dry to wet prairies, fallow fields, railroad embankments, and brushy roadsides; rich moist soils, in shade or sun. *Throughout the Arboretum*

The name *Solidago* is derived from the Latin *solidare,* to strengthen or make whole, in other words to cure. It was given to this genus because the European species *S. virgaurea* was reputedly used to heal wounds at the time of the Crusades.

Similar species Smooth goldenrod strongly resembles Canadian goldenrod (*Solidago canadensis*) but is minutely hairy only in the inflorescence; the stem is hairless and covered with a waxy bloom, and the leaves are often hairless beneath. Moreover, its involucral bracts are greener, firmer, and less sharply pointed than those of Canadian goldenrod.

Asteraceae • Sunflower Family   115

MISSOURI GOLDENROD
Solidago missouriensis

Panicles of Missouri goldenrod are dense and about as thick as long, and if not nodding at the top, at least have the lower branches backward-curved and 1-sided. When it goes to seed and dries out, the inflorescence turns fluffy, resembling a mass of feathers. **Perennial, 1–3 feet tall, from elongate rhizomes, forming colonies.**

Basal and lower leaves linear-lance-shaped, almost always less than 1 (rarely more than $5/8$) inch wide, absent at flowering time, ± strongly 3-veined. **Stem leaves** reduced upward, becoming lance-shaped to linear, often subtending axillary tufts of small leaves, slightly yellowish, firm. **Stems** hairless.

Flower heads small, oriented along the upper sides of spreading or backward-curved branchlets; involucre $1/8$–$3/16$ inch high, its bracts overlapping in several series of different lengths; receptacle small, with involucre-like bracts near the margin among the florets; ray florets 7–13, yellow; disk florets 8–14, yellow. Inflorescence branches and head stalks hairless or essentially so. *(Late July) early August–September*

Achenes hairless to sparsely hairy. *Mid August–October*

WI range and habitats
Frequent in southwest prairie areas, in dry to medium, often disturbed prairies and dry open slopes, river terraces, roadsides, and railroad rights-of-way, sometimes on sandy prairies, blowout dunes, and steep hillsides. *Curtis Prairie, East Knoll, Greene Prairie, Sinaiko Overlook Prairie*

This species is very similar to and intergrades with the appropriately named early goldenrod (*Solidago juncea*), which begins to bloom before other goldenrods. Early goldenrod has a short hard crown as the main over-wintering organ although it reportedly may produce short or even creeping rhizomes. Compared to Missouri goldenrod, its leaves are greener, less firm, and less harshly roughened along the margins. Its basal and lower leaves are often more than 1 inch wide, often present at flowering time, and 1- or faintly 3-veined. The uppermost leaves and bracts are characteristically very small but conspicuous. Formerly present on Curtis Prairie but not reported since the 1970s, early goldenrod is more likely to be found in fields, embankments, fencerows, and edges of open woods than on prairies.

Similar species See Canadian goldenrod (*S. canadensis*) and late goldenrod (*S. gigantea*).

Old-field Goldenrod
Dyer's-weed goldenrod,
gray goldenrod
Solidago nemoralis

Ash-green stems and leaves help distinguish gray goldenrod, a small plant with gracefully ascending stems and chiefly basal, narrow leaves. The inflorescence often assumes a wand-like or club-like shape and nods at the summit but varies to more ample, in either case occupying the upper $1/3$ of the plant and having at least the lower branches recurved and/or 1-sided. **Short-lived perennial, $1/2$–2 feet high, forming clumps from a somewhat rhizomatous crown at or just beneath the surface of the ground.**

Leaves long-tapering into a winged stalk, inverted-lance-shaped, 2–6 (10) inches long, weakly or scarcely 3-veined, rounded-toothed or the upper ones essentially entire; middle and upper leaves not especially numerous, from not much to conspicuously reduced in size up the stem, becoming short-stalked to stalkless, often with axillary tufts, 1-veined or pinnate-veined. Herbage densely and minutely spreading-hairy throughout.

Flower heads few-flowered and small, oriented along the upper sides of short, spreading or backward-curved, raceme-like branchlets, forming a rather small but dense, usually 1-sided to definitely pyramid-shaped panicle; involucre $3/16$–$1/4$ inch high, its bracts overlapping in several series of graduated lengths; receptacle not chaffy; ray florets 5–9, female and fertile, yellow, short; disk florets 3-6, bisexual and fertile, yellow. *Mid August–September (October)*

Achenes $1/16$ inch long, whitish and somewhat silky or less often brownish and weakly hairy with coarser hairs; pappus of hair-like bristles $1/8$ inch long. *Early September–October*

WI range and habitats Very common throughout, on dry prairie remnants, old fields, pastures, black oak savannas, jack pine woodlands, edges of oak woods, railroad embankments, and sand blowouts and dunes (including near Lake Michigan); dry, sandy, rocky, or clayey soils and full sun. ***Curtis Prairie, Grady Tract knolls, Greene Prairie, Juniper Knoll, Sinaiko Overlook Prairie***

This species is variable in ploidy level, leaf shape and size, inflorescence density, head size, and achene hairiness. Robust forms possess reduced upper leaves and open inflorescence branches and tend to occupy the deciduous forest rather than the Great Plains portion of the overall species range.

Asteraceae • Sunflower Family

Riddell's Goldenrod
Solidago riddellii

This unmistakable species has numerous, sickle-shaped leaves—narrow, firm, many of them longitudinally folded and curved—and bright yellow heads disposed in a terminal, flat-topped cluster. Perennial, 1–3 feet tall, from a short thick crown just beneath the surface of the ground, sometimes also rhizomatous.

Basal and lower leaves well-developed and persistent, tapering into long winged stalks, 6–15 times longer than wide, with gradually tapering, sharp tips, entire, tending to have 3 or more longitudinal veins at the base. **Middle and upper leaves** similar but gradually reduced, linear-lance-shaped, long-tapering into a stalkless clasping or ± sheathing base. **Stems** and leaves hairless (except for a little minute hairiness in the inflorescence).

Flower heads numerous and crowded in a flat- or round-topped inflorescence in which the inner head-bearing stalks are progressively shorter; involucre $^3/_{16}$–$^1/_4$ inch long, its bracts overlapping in several series of different lengths, blunt or rounded and ± lined; receptacle small, not chaffy; ray florets 7–9, female, yellow, small; disk florets 10–20, bisexual, yellow. *Early August–late September*

Achenes 5–7-veined and hairless or nearly so; pappus of many equal, hair-like bristles. *September–October*

WI range and habitats Occasional southeast, in wet calcareous prairies, sedge meadows, and fens, also in medium prairies and edges of marshes, rarely in moist roadside ditches. ***Curtis Prairie, Greene Prairie***

Solidago is one of our larger and more taxonomically difficult genera. Some species such as Riddell's goldenrod are distinctive, but many others are superficially similar and sometimes intergrading, making identification especially challenging.

Similar species In Wisconsin this species might be confused only with Ohio goldenrod (*Solidago ohioensis*), a rare inhabitant of wet prairies and fens in southeastern Wisconsin and interdunal depressions in Door County. It has flat leaves with blunt or somewhat sharp tips and hairless inflorescence branches and head stalks. It does not occur in the Arboretum.

Stiff Goldenrod
Rigid goldenrod
Solidago rigida

This handsome goldenrod presents its relatively large heads in a large, flat-topped inflorescence atop a rigidly erect stem. Its leaves alone—short and broad, minutely rough-hairy, and tinted grayish-yellow—distinguish stiff goldenrod from all other prairie wildflowers. **Relatively robust perennial, 1–4 feet high, in clumps from a short stout crown.**

Leaves from egg- or lance-shaped to elliptic or oblong-elliptic, the basal and lower persistent, long-stalked, to 10 inches long and 4 inches wide, the midstem and upper ones greatly reduced, egg-shaped, 1–2 inches long, becoming stalkless and ± clasping, entire, stiffish, 1-veined. Leaves and stems densely short-hairy.

Flower heads in congested, flat-topped clusters 2–10 inches across; involucre $^{3}/_{16}$–$^{3}/_{8}$ inch high, its bracts very blunt, firm, lined with several veins in a median strip (visible under magnification); disk $^{3}/_{16}$–$^{3}/_{8}$ inch wide; ray florets 7–14, female, yellow; disk florets 17–31 (35), bisexual, yellow. *Early August–late September*

Achenes 10–15-veined, hairless or nearly so; pappus of hair-like bristles $^{3}/_{16}$ inch long. *Early September–October*

WI range and habitats Very common southwest, in prairie remnants, often with weedy species in degraded dry prairies, overgrazed pastures, old fields, and restored prairies and wildflower meadows, spreading somewhat beyond the limits of the prairie areas into northeastern and northwestern parts of the state along roadsides, railroads, and sandy lakeshores; in a variety of soils, from moist clays to dry sands. *Curtis Prairie, Greene Prairie, Marion Dunn Prairie, Sinaiko Overlook Prairie, West Knoll*

Goldenrods are very easy to transplant and under cultivation may produce extraordinarily large inflorescences and an abundance of heads. They are especially effective when planted alongside the asters with which they associate in the wild. Unnaturally oversized, cultivated plants of these genera can be viewed in the native plant garden alongside the Visitor Center.

Similar species Ohio goldenrod (*Solidago ohioensis*), Riddell's goldenrod (*S. riddellii*), and grass-leaved goldenrod (*Euthamia graminifolia*) also have dense, flat-topped inflorescences, but they are otherwise very different. Leaf shape alone will easily distinguish these species, which in contrast to stiff goldenrod grow in moist places.

Asteraceae • Sunflower Family

Showy Goldenrod
Solidago speciosa

Resplendent for a goldenrod, this colorful species is capable of producing an abundance of blazing yellow heads in a dense to loose, large and somewhat raceme-like or narrowly panicled inflorescence to 10 inches long. Like all goldenrods, its fruits are equipped with a small parachute for dispersal by wind. **Perennial, 1–4 feet tall, from a ± vertical, hard crown at or just beneath the soil surface.**

Leaves basal as well as on the stem, the lower ones persistent (and less than 7 times as long as wide) and the others gradually decreasing in size upward, or the lower falling and all remaining ones smaller and uniform, scarcely sheathing at the base; stem leaves several to numerous (rarely fewer than 25), those of the midstem usually elliptic or narrowly so, firm, entire. **Stems** solitary or (usually) clumped, hairless except for inflorescence branchlets and stalks bearing heads.

Flower heads in a large, generally dense (occasionally open), somewhat raceme-like or narrowly pyramid-shaped inflorescence, commonly with ascending branches (unbranched in small plants), these neither nodding nor 1-sided; involucre $1/8$–$3/16$ inch long, its bracts yellowish, blunt to rounded, sometimes sticky; receptacle small, not chaffy; ray florets (5) 6–8, female, yellow; disk florets 7–9, bisexual, yellow. *(Mid) late July–early October*

Achenes $1/16$ inch long, essentially hairless; pappus of hair-like bristles. *Mid August–late October*

WI range and habitats Locally common southwest, spreading slightly beyond the limits of the prairie areas into sandy parts of the northeast and northwest, in remnant medium to dry prairies, fields, black oak savannas, open jack pine stands, bur oak openings, open ridges, outcrops, and river terraces, spreading along sandy roadsides and railroad rights-of-way, and on beaches and dunes along Lake Michigan; increases in frequency after fire. *Curtis Prairie, Grady Tract knolls, Greene Prairie, Sinaiko Overlook Prairie*

This species is quite variable in plant size, leafiness (number, persistence, and size of leaves), and inflorescence branching.

Heath Aster
White prairie aster
Symphyotrichum ericoides
Synonym *Aster ericoides*

The many asters native to North America bloom from August to October and together with the goldenrods reign supreme on the prairies and along roadsides and railroads. Commonest among the white asters are frost aster and heath aster. The latter is distinctive in its numerous small, grayish-green leaves and profusion of crowded small heads. **Low compact perennial, 12–30 (40) inches tall, colonial by creeping rhizomes.**

Leaves oblong, $1/2$–2 inches long, stiff, entire, rough-hairy, the lower and middle ones falling before flowering time, those of the branches numerous, tiny. **Stems** much branched near the top.

Flower heads densely crowded, ± 1-sided on numerous short arching branchlets; involucre $1/8$–$3/16$ inch long, its bracts strongly overlapping, unequal, with blunt (outer ones) to abruptly ± bristle-pointed, spreading tips, short-fringed on the margin and short-hairy on the back; ray florets 8–20, the rays to $1/4$ inch long; disk florets 8–12 (15), yellowish, turning purple. *Late August–late October*

Achenes purplish-brown, 7–9-ribbed, almost silky with appressed straight hairs; pappus bristles copious. *Early September–October*

WI range and habitats Common south, sporadic northeastward and absent northwest, in dry-medium to wet-medium prairies, open oak and pine woods, and sandy riverbanks and lakeshores, surviving in old cemeteries, orchards, and fencerows, weedy on roadsides, railroads, sandy fields, pine plantations, and quarries; most abundant in dry soil in full sun.

Curtis Prairie, Greene Prairie, Sinaiko Overlook Prairie, Visitor Center parking lot

This large, temperate-zone genus is abundant in North America, including Wisconsin, where 23 species and 40-some intragenus hybrids occur. Although the many species are variable, non-botanists can learn to identify the most obvious asters, but the process is challenging and much judgment is needed when making identifications.

Similar species See frost aster (*Symphyotrichum pilosum*), which has larger heads with more numerous florets and involucral bracts with awl-shaped tips. Heath aster occasionally hybridizes with New England aster (*S. novae-angliae*), producing amethyst aster (*S. ×amethystinum*). Its involucres are hairy but never glandular, and its achenes are hairy.

Shining Aster

Shiny-leaved aster
Symphyotrichum firmum
Synonyms *Aster firmus, A. lucidulus,
A. puniceus* var. *firmus*

This attractive tall aster is identified by the compact leafy appearance of the top of the plant, the almost uniform leaves with ± strongly lobed, clasping bases, and the large involucre of gradually tapering, narrow-tipped bracts. The rays are a beautiful, pale lavender color, ranging to whitish. **Stout perennial, 2–5 feet tall, from elongate horizontal rhizomes, forming showy colonies.**

Leaves oblong-lance-shaped, 3–6 inches long, to $1\frac{1}{2}$ inches wide, clasping at the base and partly surrounding the stem, entire, hairless, those toward the inflorescence conspicuously crowded and not much reduced. **Stems** simple to much-branched above, hairless to sparsely hairy especially in lines extending downward from the leaf bases.

Flower heads several to many in diffuse panicles, on short stalks, not 1-sided; involucre $\frac{1}{4}$–$\frac{1}{2}$ inch tall, the outer bracts about equaling or even longer than the inner ones, whitish with a pale green, barely dilated central line, linear or oblong, gradually tapering into a long, narrow tip, hairless on the back; ray florets 30–40, the rays $\frac{3}{8}$–$\frac{1}{2}$ inch long; disk florets yellow. ***End of August–early October***

Achenes hairy or nearly hairless, veined, the pappus of numerous hair-like bristles $\frac{1}{4}$–$\frac{5}{16}$ inch long. ***Late September–October***

WI range and habitats Fairly common east and southeast, all but absent northwest, in sedge meadows, wet to wet-medium prairies, fens, tamarack bogs, swampy woods and thickets, occasionally in wet fields, low pastures, ditches, and moist depressions in upland woods. ***East Curtis Prairie, Lower Greene Prairie***

Similar species Shining aster hybridizes with several related species, especially panicled aster (*Symphyotrichum lanceolatum*). More often, it intergrades with swamp aster (*S. puniceum*), and some taxonomists think that it should be treated as a subspecies or variety of the latter. Swamp aster has rough hairs uniformly distributed around the red stems, at least under the heads. The leaves are not crowded and are often distantly toothed; the involucral bracts are often more gradually tapered to an even more slender tip; and the rays are darker (dark lavender, lilac, or rose-purple, rarely white). Swamp aster grows in sun or shade and is less common in wet prairies and sedge meadows than shining aster.

Smooth Aster
Smooth blue aster
Symphyotrichum laeve
Synonym *Aster laevis*

This adaptable floral treasure is similar in form to sky-blue aster, both featuring relatively large heads with long, blue-violet rays and yellow centers in an open panicle. The waxy, bluish-green color and firm leathery texture of the hairless foliage make it one of our more readily identifiable asters. **Perennial, 1–4 feet tall, from a short rhizome or short crown.**

Basal and lowest stem leaves narrowed into a winged stalk. **Middle and upper leaves** egg- to lance-shaped or oblong, the main ones more than $3/8$ (often more than 1) inch wide, lobed or somewhat heart-shaped at the base, at least some clasping the stem; uppermost leaves reduced to clasping bracts.

Flower heads several to many on stalks with greatly reduced bracteal leaves; involucre $3/16$–$1/4$ inch high, its bracts overlapping and closely appressed, unequal, blunt to sharp-pointed, with short green, narrowly diamond-shaped tips; ray florets 15–25, $5/16$–$5/8$ inch long; disk florets about 25. *(Mid) late July–early October*

Achenes $1/8$ inch long, hairless, the pappus bristles copious, usually reddish. *Late September–mid November*

WI range and habitats Common throughout most parts (absent from the Northern Highlands except for barrens areas), in dry to medium open woods, barrens, and prairies, on banks and clay bluffs along Lake Michigan, roadsides, railroads, and fencerows, occasionally on lakeshores, pastures, quarries, gravel pits, and fallow gardens; various soil types, in full sun or partial shade. *Curtis Prairie, Greene Prairie, Sinaiko Overlook Prairie, West Knoll, Wingra Oak Savanna*

Asters, fleabanes, and goldenrods are similar to one another. Moreover, each genus contains a multitude of species that can be difficult to distinguish. In *Symphyotrichum* the involucral bracts are in graduated series and are partly or wholly green at the tips, whereas in *Erigeron* (fleabanes) they are nearly equal and not green at the tips. In *Solidago* (goldenrods) the ray florets are almost always yellow.

Similar species See sky-blue aster (*Symphyotrichum oolentangiense*). Short's aster (*S. shortii*) differs from smooth aster and sky-blue aster in having the leaves uniformly long-triangular and all but the uppermost heart-shaped at the base. The panicles are leafier, and the rays are typically pale violet instead of deep blue.

Panicled Aster
White panicle aster
Symphyotrichum lanceolatum
Synonyms *Aster lanceolatus, Aster paniculatus, A. simplex*

This bushy wetland aster covers itself with an abundance of narrow thin hairless leaves and numerous small, white-flowered heads with shallowly lobed disk corollas. **Rather tall perennial, 1–6 feet tall, colonial by long rhizomes.**

Basal and lower stem leaves often withered or fallen by flowering time. **Principal leaves** lance-shaped to linear, stalkless or tapering to a stalk-like base, mainly $1\frac{1}{2}$–6 inches long and $\frac{1}{4}$–$\frac{3}{4}$ inch wide (sometimes broader or narrower), sharply toothed to entire. **Stems** hairy in lines above.

Flower heads numerous in a diffuse or elongate, leafy panicle, on short stalks, sometimes very small and mostly or somewhat 1-sided on the raceme-like branches, or larger and not 1-sided; involucral bracts unequal, narrow and sharp-pointed, hairless or sometimes fringed with hairs on the margin, whitish but with a green midvein extending to the tip; ray florets 20–40, white, $\frac{3}{4}$–$\frac{1}{2}$ inch long; disk florets (15) 20–30, the corolla lobes turning purplish after flowering. *Early August–mid October*

Achenes gray, somewhat compressed, thinly hairy, 4- or 5-ribbed; pappus of numerous tawny slender bristles. *Late October*

WI range and habitats Common throughout, in marshes, sedge meadows, prairies, shrub carrs, fens, open bottomland woods and thickets, edges of swamps, bogs, and low woods, shores of lakes, rivers, and streams, sometimes locally common in old fields, pastures, rights-of-way, fencerows, and ditches; moist or swampy, open or shaded ground. *Curtis Prairie, Greene Prairie, Juniper Knoll, Marion Dunn Prairie, West Knoll*

This is probably our most common and variable aster. It intergrades and hybridizes with practically every other Wisconsin species having the same base chromosome number, for example, northern bog aster (*Symphyotrichum boreale*), smooth aster (*S. laeve*), calico aster (*S. lateriflorum*), and swamp aster (*S. puniceum*)—12 species in all.

Similar species Panicled aster is similar to calico aster (*S. lateriflorum*) in many respects but differs in having long creeping rhizomes, more numerous, often longer rays on larger heads, and less deeply lobed disk corollas.

Calico Aster
Goblet aster, side-flowering aster
Symphyotrichum lateriflorum
Synonym *Aster lateriflorus*

The color and disposition of the tiny heads give this abundant aster its common names, but its growth habit also helps distinguish it. Low-growing and sprawling or bushy, it often forms a floral hedge along woodland edges. **Clumped perennial, 1–3 (4) feet tall, from a persistent short crown.**

Basal and lower stem leaves generally falling. **Principal leaves** narrowly oblong and entire to lance- or lance-elliptic and toothed, 2–4 (6) inches long, to $3/4$ (usually less) inch wide, hairy only along the midrib beneath. **Stems** spreading to arching, rather slender.

Flower heads ± numerous but not crowded, on 1-sided, raceme-like branches, these forming a diffuse panicle; involucre nearly $1/4$ inch high, its bracts in 3 or 4 series, blunt or sharp-pointed, the dilated green or purplish portion of the central line small and the green tip flat; ray florets 9–15, white or faintly purplish, $1/4$ or less inch long; disk florets 10–15 (20), cream-colored, turning magenta before withering. *Late July–early October*

Achenes gray, somewhat spreading-hairy, with 3-5 obscure ribs, the pappus of dull white soft, hair-like bristles in 1 series. *Late October*

WI range and habitats Abundant throughout, in dry to moist forests, woodlands, and thickets, especially along borders, trails, and clearings, on floodplains, fens, and clay bluffs along the Great Lakes, often weedy in grazed woods and semi-shady gardens; prefers moist soil and partial shade. *Curtis Prairie (edges), Gallistel Woods, Juniper Knoll, Lower Greene Prairie, Marion Dunn Prairie, Noe Woods, Wingra Oak Savanna, Wingra Woods*

Calico aster, one of our most common and variable asters, is particularly complicated, consisting of 3 poorly defined varieties and hybridizing in Wisconsin with 8 other species.

Similar species Heath aster (*Symphyotrichum ericoides*) and frost aster (*S. pilosum*) are distinguished from calico aster by having involucral bracts with almost spine-like, short sharp tips. Panicled aster (*S. lanceolatum*) and Ontario aster (*S. ontarionis*) differ by their creeping rhizomes. The former has disk corollas less deeply lobed and leaves hairless beneath; the latter has corollas similar to those of side-flowering aster but leaves that are hairy across the surface beneath.

Asteraceae • Sunflower Family

New England Aster
Symphyotrichum novae-angliae
Synonym *Aster novae-angliae*

Superb clusters of rose-purple heads win New England aster a prize for beauty in the autumn floral pageant. Its leaves are more numerous and more closely bunched than in other asters, and each has two basal, earlike lobes that half-clasp the hairy sticky stem. **Fast-growing perennial, 1–5 feet tall, from a tough branching crown.**

Leaves alternate, lance-shaped to oblong, $3/4$–4 inches long, $3/8$–$3/4$ inch wide, entire, hairy on both surfaces. **Stems** often clustered, stoutish, branching above, quite leafy to the top. Whole plant coarsely hairy and somewhat glandular-sticky.

Flower heads 1–2 inches across, mostly 30–50 in a panicle; involucre $1/4$–$3/8$ inch tall, its bracts loosely overlapping in 2 or 3 series, linear and awl-tipped, densely glandular and scantily hairy, usually flushed with purple; ray florets about 45–100, pink to red- or blue-violet (rarely white), $1/2$–$3/4$ inch long; disk corollas red-violet. *Late August–early October*

Achenes with appressed straight hairs; pappus bristles numerous, about $3/16$ inch long. *Late October–November*

WI range and habitats Abundant southeast, more local westward and absent in much of the north, in medium to wet-medium prairies, sedge meadows, fens, and swales, moist woods and thickets, streambanks and lakeshores, often in disturbed, sometimes dry, fields, pastures, fencerows, and ditches, somewhat weedy along roadsides, railroads, and quarries; usually in open, moist to wet places. *Curtis Prairie, Greene Prairie, Juniper Knoll, Marion Dunn Prairie, Sinaiko Overlook Prairie, Visitor Center parking lot, Wingra Oak Savanna*

The species epithet was given by Linnaeus, who first described this aster from specimens taken in New England. This common, hardy, easy-to-grow species blooms well into late fall and makes an excellent cut flower. New England aster is the parent of many cultivated varieties having relatively large ray florets in a range of colors. Native Americans used the leaves to alleviate skin rashes, including poison-ivy. The Shakers used them in cosmetics to keep the complexion clear. The Ojibwe of the Midwest smoked the leaves of this and other asters in pipes to create an odor attractive to deer.

Aromatic Aster
Symphyotrichum oblongifolium
Synonym *Aster oblongifolius*

The compact form, small stiff leaves, and late-season floral display of this aster would make it a fine species for enhancing any dry prairie or dry wildflower planting or rock garden. One of the last asters to bloom, the entire plant is covered with relatively large, purplish flower heads. **Perennial, 4–28 inches tall, rhizomatous, forming patches.**

Leaves alternate, the lowermost soon falling, oblong, mostly $3/8$–$1 1/2$ inches long and $1/8$–$1/4$ inch wide, rounded but not lobed and barely to slightly clasping at the base, entire. **Stems** variously bristly hairy and glandular (especially upward), giving the plant a sticky feel.

Flower heads few to many at the ends of the branchlets, not crowded, forming a (sometimes flat-topped) panicle; involucre hemispheric, $5/16$–$1/2$ inch broad, the bracts weakly overlapping in 3 or 4 series, with long green, abruptly spreading tips, glandular-hairy on the back; ray florets about 30, the corollas bluish-purple, $3/8$ inch long. *Early September–early October*

Achenes silky hairy or sparsely so, the pappus bristles numerous, yellowish-white. *Late October–November*

WI range and habitats Infrequent south-central and west, on dry, rocky, gravelly, or sandy prairies, rarely in scrub oak-jack pine woods or grassy brushy road banks; often on bluffs, outcrops, boulders, or talus or in spots with sparse vegetation. ***Curtis Prairie (limestone knoll), West Knoll***

Like so many prairie species, aromatic aster is now fairly rare in Wisconsin.

Asters have heads composed of two types of flowers: a dense mass of bisexual tubular florets on a central disk, and an outer circle of larger female florets, each with one strap-shaped ray, which give the heads a star-like appearance. The involucres are even more star-like during and after the shedding of the fruits, when their tiny, straw-like bracts spread radially.

Sky-blue Aster
Azure aster, prairie heart-leaved aster
Symphyotrichum oolentangiense
Synonym *Aster azureus, A. oolentangiensis*

This attractive aster's bright blue rays reflect the late summer sky, together with those of smooth aster creating an azure haze among the profuse grass stems. The leaves alone—deep green to almost bluish, rough like sandpaper above and rather stiff-hairy beneath—are enough to distinguish it. **Low perennial, 1–3 feet tall, from a short, somewhat rhizomatous crown.**

Basal and lowermost stem leaves egg- to lance-shaped, to 6 inches long, with somewhat heart-shaped or abruptly straight (as if cut off) bases and winged stalks. **Middle and upper stem leaves** few, egg-lance- to linear-lance-shaped, somewhat stalked but tapering (not necessarily heart-shaped) at the base, entire or some with an occasional tooth.

Flower heads several to many in a narrow open panicle, solitary at the tips of stiffly ascending stalks, these provided with minute, awl-shaped bracteal leaves; involucre $1/4$ inch high, its bracts overlapping in 3–5 series, linear-lance-shaped, white toward the base, with a broadly diamond-shaped green tip; ray florets 13–18 (20), with blue to blue-violet rays $1/4$ inch long; disk florets 15–30, with yellow corollas that turn purple after flowering. *Mid August–early October*

Achenes dull purple or purple-streaked, hairless or nearly so, with 4 or 5 weak ribs; pappus of cream- or rose-tinged, hair-like bristles. *Late September–October*

WI range and habitats Locally common east, south, and west (absent from the acid bedrock of the Northern Highlands), in dry to dry-medium prairies, occasionally medium to wet-medium prairies, and open upland woods and savannas (pine, oak, and/or aspen), on calcareous bluffs, cliffs, ledges, limestone flats, and Lake Michigan dunes, also along roadsides and railroad rights-of-way; dry, sandy, loamy, gravelly, or rocky soils. *Curtis Prairie, East Knoll, Greene Prairie, Juniper Knoll*

Similar species See smooth aster (*Symphyotrichum laeve*).

Frost Aster

Awl aster, hairy aster,
white old-field aster
Symphyotrichum pilosum
Synonym *Aster pilosus*

This abundant fall aster is identified by short leaves on stiff arching branches, straight spreading hairs on the stems and leaves, medium-sized heads with white rays surrounding a yellow disk, and involucral bracts with distinctively shaped, minutely spiny tips. A very hard frost is required to kill the flowers, making this the last aster remaining in bloom and giving it one of its common names. **Spindly, bushy-branched perennial, 1–5 feet tall, from a short tough crown.**

Basal and lower leaves falling off or the basal persistent. **Stem leaves** linear-lance-shaped to inversely lance-elliptic and 1–3 (4) inches long, those above and on branches numerous and reduced, becoming awl-shaped, stalkless but not clasping. Leaves and stems sometimes nearly hairless.

Flower heads often numerous, 1-sided on spreading leafy branches that form a diffuse panicle; involucre bell-shaped, about $1/4$ inch high, the bracts in 4 or 5 series, ascending, some of them gradually tapering into marginally inrolled, awl-shaped (almost spine-like), translucent tips (best seen under magnification); ray florets 16–20 (30), the corolla flat, $3/16$–$3/8$ inch long; disk florets about 30–40, light yellow, turning lavender or red-violet. *Late August–mid October*

Achenes whitish or gray, minutely silky, obscurely 2–4-ribbed, the pappus of numerous, dull white bristles. *October–November*

WI range and habitats Very common south and central, reappearing northward on stony shores in Door County and glacial lake barrens in the northwest, in dry-medium prairies and borders of woods, sandy shores, and edges of marshes and low prairies, but now (almost all medium prairies having been destroyed) on disturbed grassy hillsides, fields, pastures, fencerows, roadsides, railroads, and waste ground (e.g., vacant lots, city streets, mine tailings); quite weedy, mostly in unshaded disturbed ground. *Curtis, Marion Dunn, and Sinaiko Overlook prairies, Juniper Knoll, Visitor Center parking lot, West Knoll, Wingra Oak Savanna*

Bees and butterflies, including the monarch during its fall migration, and other insects use the frost aster for food, as do birds during the winter, for it holds its fruiting heads above the snow.

Similar species Frost aster is much coarser than heath aster (*Symphyotrichum ericoides*) and has sparser longer hairs, longer leaves, larger heads, and gradually tapering, marginally inrolled involucral bracts.

Silky Aster
Western silvery aster
Symphyotrichum sericeum
Synonym *Aster sericeus*

Unique, silvery-silky foliage and large, reddish-violet flower heads, all somewhat clustered toward the ends of the few wiry branches, make silky aster not only absolutely distinctive but also one of the most attractive of all the asters. **Low-growing perennial, 8–28 inches tall, from a short branching woody crown.**

Basal and lower stem leaves soon withering and falling. **Middle and upper stem leaves** uniformly small (slightly reduced on the branches), stalkless, lance-egg-shaped to oblong-elliptic, to $1^1/_2$ (mostly $^3/_8$–$^3/_4$) inches long, blunt but with a short abrupt tip, entire, firm, silvery-silky on both surfaces. **Stems** clustered, wiry and brittle.

Flower heads several–many in a widely branched panicle (few per branch), the involucre $^1/_4$–$^3/_8$ inch high, its bracts weakly overlapping in few series, their tips loose or spreading; uppermost leaves encroaching on the head, forming a secondary involucre; rays 14–20 (25), deep violet to rose-purple, $^3/_8$–$^5/_8$ inch long; disk florets yellowish-brown or rose-tinged. *Mid August–early October*

Achenes purple (brownish when weathered), 7–10-ribbed, hairless, the pappus of copious white bristles, becoming tawny in age. *Late September–late October*

WI range and habitats Locally common south and west (absent from the forested northeast), in sandy, gravelly, or rocky dry prairies, also old sandy fields and roadsides, sometimes under oaks, jack pine, or red cedar but usually in the open; often in limy soils.

Curtis Prairie, Upper Greene Prairie

Like many of its prairie associates, silky aster is becoming increasingly rare, surviving mostly on steep rocky prairies unlikely to have been plowed or overgrazed. It excels in the rock garden and is becoming popular for more general landscape use.

IRONWEED
Common ironweed, smooth ironweed
Vernonia fasciculata

The terminal, flat-topped clusters of medium-sized, intense reddish-purple heads tower over almost all other wildflowers of the late-summer marsh, making rich spots of color and providing bees with nectar during a time when little else is in flower. The very numerous, willow-like leaves, spiny-toothed and conspicuously dark-dotted beneath, are quite distinctive. **Stout perennial, 3–6 feet tall, from a short tough crown.**

Basal leaves absent at flowering time. **Stem leaves** alternate, evenly distributed along the stem, dark green, spreading, linear to narrowly lance-shaped, the middle ones 3–6 inches long and to $1^3/_4$ inches wide, tapering to a slender tip, sharply toothed, the teeth almost bristly, strongly dotted beneath with small pits (visible when dry). **Stems** often red or purple.

Flower heads numerous, slenderly bell-shaped, $1/_2$ inch across, in dense, round-topped clusters $1^1/_2$–4 inches across; florets all tubular (ray florets absent); involucre $3/_{16}$–$3/_8$ inch long, its bracts numerous, with rounded to somewhat pointed tips; receptacle naked; florets 10–26, bisexual, tiny, the corollas tubular. *Mid July–September* (peaking in early August)

Achenes ribbed, hairless; pappus tawny to purplish, of numerous bristles in 2 series, the inner of long slender bristles, the outer series of very short, scale-like bristles. *Late July–September* (peaking in late August)

WI range and habitats Locally frequent southwest, occasional farther north, especially in central and northwestern parts of the state, in wet-medium prairies and open river bottom forests, swamps, and marshes, often prominent in low overgrazed pastures (cattle will not eat it); moist soil and full sun. *Curtis Prairie, Greene Prairie*

The genus is named for William Vernon, English botanist of the late 17th century. The common name may refer to either the tough stem or the rust-red fruits that form after the heads are through blooming. Various species of butterflies make ironweed one of their preferred nectar sources.

Asteraceae • Sunflower Family

OTHER ASTERACEAE

GIANT RAGWEED
Great ragweed, horse-cane
Ambrosia trifida
Native; potentially invasive erect annual forb, 2–9 feet.
Flowers August–September

STINKING CHAMOMILE
Dog-fennel, Mayweed
Anthemis cotula
Introduced, naturalized; erect annual forb, 4–24 inches.
Flowers May–October

BURDOCK
Arctium minus
Introduced, naturalized; ecologically invasive and really annoying erect biennial forb, 1½–5 feet.
Flowers July–October

Plumeless Thistle
Spiny plumeless thistle
Carduus acanthoides
Introduced, naturalized; ecologically invasive erect biennial forb, 1–4 feet.
Flowers July–October

Spotted Knapweed
Centaurea stoebe subspecies ***micranthos***
Synonyms *Centaurea biebersteinii, C. maculosa*
Introduced, naturalized; weedy erect biennial/perennial forb, 1–4 feet.
Flowers July–October

Chicory
Blue-sailors
Cichorium intybus
Introduced, naturalized; weedy erect perennial forb, 1–6 feet.
Flowers July–October

Canada Thistle
Creeping thistle, field thistle
Cirsium arvense
Introduced, naturalized; ecologically invasive and noxious weed; erect perennial forb, 1–4 feet.
Flowers June–October

Asteraceae • Sunflower Family

Bull Thistle
Cirsium vulgare
Introduced, naturalized; ecologically invasive erect perennial forb, 2–6 feet.
Flowers June–October

Flat-top Aster
Parasol aster, tall flat-topped white aster
Doellingeria umbellata
Synonym *Aster umbellatus*
Native; erect perennial forb, 1–5 feet.
Flowers July–September

Fireweed
American burn-weed
Erechtites hieracifolius
Native; erect perennial forb, 2–80 inches.
Flowers July–October

Field Hawkweed
Meadow hawkweed, yellow king-devil
Hieracium caespitosum
Introduced, naturalized; erect perennial forb, 15–30 inches.

Prickly Lettuce
Lactuca serriola
Synonym *Lactuca scariola*
Introduced, naturalized; erect annual/biennial forb, 1–5 feet.
Flowers July–September

Elm-leaved Goldenrod
Solidago ulmifolia
Native; erect perennial forb, 1½–4 feet.
Flowers July–October

Prickly Sow-thistle
Spiny sow-thistle
Sonchus asper
Introduced, naturalized; perennial forb, 12–40 inches.
Flowers July–October

Swamp Aster
Bristly aster, purple-stem aster
Symphyotrichum puniceum
Synonym *Aster puniceus*
Native; erect perennial forb, 1–6 feet.
Flowers August–October

Dandelion
Taraxacum officinale
Introduced, naturalized; erect perennial forb, 2–12 inches. A benevolent flower that produces early spring nectar and pollen for bees, yet is apomictic—setting seed before the bud ever opens.
Flowers April–November

Goat's-Beard
Fistulous goat's-beard, greater sand goat's-beard, yellow salsify
Tragopogon dubius
Introduced, naturalized; erect perennial forb, 1–3 feet.
Flowers May–July

Orange Jewelweed
Orange touch-me-not, spotted touch-me-not
Impatiens capensis
Synonym *Impatiens biflora*

Conspicuous in late summer, the curious, pouch-like flowers of this easily recognized herb hang from filamentous stalks like so many windsocks. Seedlings come up simultaneously in spring, often in great profusion. **Annual, 2–5 feet tall.**

Leaves alternate, egg-shaped or oval, 1–3½ inches long, soft, pale beneath, the rounded teeth with a short abrupt tip. **Stems** hollow, somewhat succulent or watery.

Flowers few per raceme, strongly irregular, normally orange with crimson spots (rarely whitish, lemon-yellow, or unspotted), $3/4$–$1\,1/4$ inches long; sepals 3, colored, the 2 lateral small, the lowest one large, forming a sac that extends backward into a slender spur $1/4$–$1/2$ inch long; petals apparently 3, 1 upper and 1 on each side, but the lateral ones each with a lobe and presumably representing 2 fused petals. Spur curved sharply forward and lying parallel to the sepal sac. *Mid June–mid September*

Fruit a fleshy green capsule $3/4$ inch long, at maturity exploding with startling suddenness when touched or shaken, scattering the several seeds; seeds mottled green to brown, with 4 corky ridges. *Late August–early October*

WI range and habitats: Abundant throughout, on stream banks, lakeshores, marshes, fens, swamps, wet thickets, moist spots in upland woods, around springs and seeps, ravines, and ditches; wet or damp soils. *East Curtis Prairie, Lower Greene Prairie, Sinaiko Overlook Prairie, Lower Wingra Oak Savanna, lowlands throughout the Arboretum*

The species epithet, *capensis,* is in error, its author believing that the plant came from the Cape of Good Hope. Bumblebees, honeybees, and the ruby-throated hummingbird appear to be effective pollinators, although bees and wasps also regularly bite holes in the spurs to rob the flowers of nectar without pollinating them.

Two types of flowers are produced, the familiar larger ones, which if successfully pollinated, produce 2–5 seeds, and minute flowers that are self-fertilized early in their development. Their sepals and petals never expand but are forced off by the growing capsule, which produces 1–3 seeds.

Similar species This species could be confused only with the less frequent, less weedy pale touch-me-not (*Impatiens pallida*), which has light yellow, sparingly spotted flowers with a shorter spur that is bent at a right angle.

Boraginaceae • Borage Family

Hoary Puccoon
Lithospermum canescens

A spring favorite, hoary puccoon produces flat-topped clusters of bright, golden-yellow to yellow-orange flowers during a long season from spring into early summer. A colony of these handsome plants in full bloom never fails to gladden the hearts of nature lovers, although the dropping of the corollas as the season progresses gives an untidy appearance to the scene.

Hoary perennial, 5–14 inches tall, from a vertical woody taproot.

Leaves numerous, the lower short, the well-developed ones lance-shaped to elliptic, $1/2$–2 inches long, $3/16$–$3/8$ inch wide. Stems, leaves, bracts, and calyx densely covered with fine straight hairs pressed closely against the surface, the hairs lacking a nipple-like base.

Flowers crowded toward the ends of 1–3 uncoiling stem and branch tips that, blooming sequentially, elongate and straighten to simulate leafy racemes; sepals 5, slightly fused at the base, $1/8$–$1/4$ (flower) to $1/4$–$5/16$ (fruit) inch long; petals 5, joined to make a slender tube $1/4$–$1/2$ inch long below an abruptly spreading, flat portion cleft into 5 broad rounded entire lobes; stamens 5. Corollas have appendage-like folds opposite each lobe that almost close the mouth of the tube. *Late April–late June*

Fruits deeply cleft into (1–3) 4 white (faintly yellowish), egg-shaped nutlets to $1/8$ inch long, very hard, smooth and shiny. *Early June–late July*

WI range and habitats Widespread but lacking from the Northern Highlands, common in undisturbed, dry to medium (or even seasonally moist) prairies, open woodlands such as jack pine and oak barrens, on bluffs and sometimes roadsides and railroads; well-drained, calcareous (seldom sandy) soils. *Curtis Prairie, Grady Tract knolls, Greene Prairie, Sinaiko Overlook Prairie*

The genus name is from the Greek and simply means stone seed, from the nutlets. In both hoary and hairy puccoon the flowers are of two forms. In short-styled flowers the stamens are inserted at the top of the corolla tube, and the thread-like style of the seed-bearing pistil extends only to the middle of the tube. In long-styled flowers the anthers are attached near the middle of the corolla tube, and the long style reaches to or beyond the top of the tube.

Hairy Puccoon
Carolina puccoon, plains puccoon
Lithospermum caroliniense

Hairy puccoon is similar to hoary puccoon except that it is covered with coarser, less dense, slightly nipple-based hairs. Its bright, yellow-orange corollas average only slightly larger, but the calyx lobes are definitely longer. Like other members of the Borage Family, the flowers all grow out of one side of the branches of a modified inflorescence that uncoils as they open successively along its length. **Perennial, 8–18 inches tall, from a thick deep, dye-stained taproot.**

Leaves crowded, those at the base of the stem scale-like or reduced, the fully developed ones numerous and uniform, lance-shaped to narrowly oblong, mostly 2 or less inches long.

Flowers mostly on one side of branches coiled in bud from the top downward, these elongating and straightening to form false racemes 4–8 inches or more long; sepals 5, nearly free, $3/8$ (in flower) to $3/8$–$5/8$ (in fruit) inch long; corolla $3/8$–$7/8$ inch long and about as broad, with a cylindrical tube and an abruptly spreading, shallowly saucer-shaped, expanded portion cleft into 5 broad rounded lobes; stamens 5. *Early May–July (August)*

Fruits (1–3) 4 white, egg-shaped nutlets $1/8$ inch long, smooth and shiny. *Late June–September* (nutlets may persist on the dead stems until the following spring)

WI range and habitats Mostly south, extending into sandy areas in the northeast and northwest, characteristic of sand barrens, oak and pine woodlands (especially black oak savannas), sandy prairies, old fields, and dunes (including along Lake Michigan), on bluffs and ridges and along roadsides and railroads; dry, especially sandy soil. *Grady Tract knolls*

"Puccoon" is a Native American name for the red dye derived from the sap, especially in the roots, of this and other members of the Borage Family.

Populations of this puccoon contain plants of three types: those with showy, yellow-orange, cross-pollinating flowers; those with inconspicuous, yellow-green, self-pollinating flowers; and those with flowers of both types. The cross-pollinating flowers are either long-styled or short-styled and are pollinated by bumblebees and by smaller bees and butterflies. The self-pollinating flowers, whether on the same plant as cross-pollinating flowers or not, develop late in the season and are uniformly small with short styles.

Other Boraginaceae

STICKSEED
Beggar's-lice, wild comfrey
Hackelia virginiana
Native; erect perennial forb, 1–3 feet. Well-named for its clothes- and fur-infesting fruits. Flowers July–September

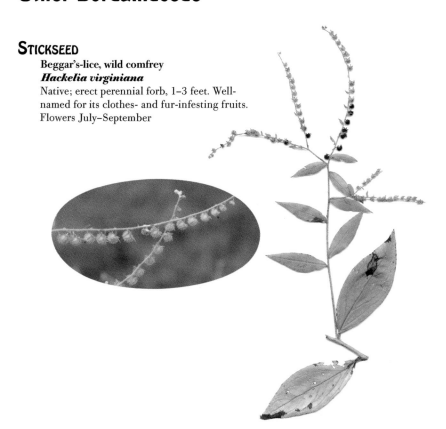

Sand Cress
Lyrate rock-cress
Arabis lyrata

Bearing a few small white flowers, this little plant is scarcely showy; nonetheless, it is one of the charming native representatives of this family. As in almost all crucifers, it has 4 sepals, 4 petals borne in opposite pairs, and 6 stamens arranged in a distinctive pattern, 4 longer than the remaining 2. **Low, wiry-stemmed biennial or perennial, 3–19 inches tall, with a dense basal rosette.**

Basal leaves pinnate-dissected or -lobed, $3/8$–$1 1/2$ inches long. **Stem leaves** simple, linear or inverse-lance-shaped, without lobes or segments, narrowed to a stalkless base that is never clasping. Stems often branched from the base.

Flowers in a terminal bractless raceme, white, on stalks that remain spreading to ascending; sepals 4, falling off after flowering; petals 4, arranged in the form of a cross, $1/8$–$1/4$ inch long; stamens 6, in 2 whorls, of which 4 are longer than the other 2. *April–September*

Fruits 2-chambered, cylindrical capsules (siliques) $3/8$–$1 3/8$ inches long, less than $1/16$ inch broad, erect on slender ascending stalks $1/4$–$5/8$ inch long, splitting lengthwise into two halves when ripe, the walls falling away and leaving a membranous partition; seeds in 1 row within each chamber, oblong, flat. *Mid May–September*

WI range and habitats Very common, particularly southwest, on beaches, dunes, and sand flats, dry prairies, dry open woods and cedar glades, cliffs and outcrops, weedy in pastures, gravel pits, and along railroads and roadsides; especially in open, sandy or rocky places. *Grady Tract knolls*

This family has the alternate name Cruciferae, meaning "cross-bearers." With few exceptions, the flower of any member of the Mustard Family is easily recognized by the numerical regularity of the parts and the cross-like pattern of the petals, which are diagonal to the sepals. Plants produce mustard oils.

Similar species Nine other species of rock-cress have been found in Wisconsin. Accurate identification of these plants requires careful attention to details of plant hairiness and leaf shape and especially characters of the fruits. Most are much larger plants than lyre-leaved rock cress and do not have divided basal leaves.

Yellow-rocket
Winter-cress
Barbarea vulgaris

If this familiar, spring-season weed has any charm, it is in producing an eye-catching display of color where growing in profusion. Its stems arise from the center of a prominent, overwintering basal rosette of dark green, shiny leaves, and bear bright yellow flowers in branched inflorescences at the top. **Biennial or perennial (rarely annual?), 12–30 inches tall.**

Leaves basal and alternate, the lower fully divided or pinnately lobed, with somewhat circular to egg-shaped or broadly oval divisions and a larger broader lobe at tip, the upper ones inversely egg-shaped to rounded and coarsely toothed or lobed, clasping by deeply cleft bases. **Stems** single, branched above; plants hairless.

Flowers in elongating racemes, not subtended by bracts, cross-shaped, about $^1/_4$ inch long, with 4 sepals, 4 petals, and 6 stamens, of which 4 are longer than the other 2. *Beginning of May–mid July*

Fruits slender, many-seeded capsules (siliques) $^3/_4$–$1^1/_2$ inches long, appressed or ascending on horizontally spreading stalks, the walls splitting from the bottom into 2 pieces and leaving a frame (to which the seeds are attached) and a thin partition; seeds gray, short-oblong to nearly square. *Mid May–September*

WI range and habitats Very common, especially south, frequently pernicious in fields and pastures, borders of and disturbed spots in woods, prairies, and sedge meadows, streambanks and lakeshores, along roadsides and railroads, and in waste places. *Throughout the Arboretum*

A native of Eurasia, yellow-rocket competes well in damp soil. The Mustard Family (Brassicaceae or Cruciferae) contains many plants that are successful weeds, whether European crucifers introduced into North America, or native American crucifers that have made themselves perfectly at home in Europe. The pungent watery juice of the leaves and the peppery taste of the seeds are ± characteristic of the entire family. The family is important economically, for besides many widespread weeds, it includes cabbages, mustards, radishes, turnips, watercress, and many ornamental plants.

Similar species Yellow-rocket is often mistaken for one of the mustards (*Brassica* species), which are annuals that bloom later. Their fruits are transversely segmented into two very unequal sections, the terminal one of which is a non-opening, flat or angled beak.

Other Brassicaceae

Hoary-Alyssum
Hoary false madwort
Berteroa incana
Introduced, naturalized;
erect annual/perennial forb, 10–28 inches.
Flowers May–October

Dame's Rocket
Hesperis matronalis
Introduced, naturalized; ecologically invasive erect biennial/perennial forb, 2–3 feet. Flowers May–September

Fieldcress
Field pepper-weed
Lepidium campestre
Introduced, naturalized; erect annual/biennial forb, 6–30 inches.
Flowers May–July

Small Pepper-Weed
Prairie pepper-grass
Lepidium densiflorum
Introduced, naturalized; annual/biennial forb, 3–24 inches.
Flowers May–July

Tall Tumble Mustard
Sisymbrium altissimum
Introduced, naturalized;
annual forb, 7–39 inches.
Flowers May–September

Eastern Prickly-pear
Opuntia humifusa
Synonyms *Opuntia compressa, O. rafinesquei*

Large broad stem segments and vivid yellow flowers make eastern prickly-pear an utterly distinctive member of our flora. Succulent leafless spiny shrub, usually only 1 or 2 stem segments tall, with mostly fibrous roots, prostrate or spreading, forming mats to 3–6 feet in diameter.

Leaves reduced to awl-shaped, fleshy scales, soon falling. **Stems** of 1 to several, circular to oblong or inverse-egg-shaped, flattened segments ("pads") 1–9 inches long and 1–4 inches wide, with circular cushions (areoles) of tan to brown wool and detachable barbed bristles (glochids); spines when present 0–3 per areole, $1/8$–$1^{3}/_{8}$ inches long, straight.

Flowers blooming singly along upper edges of pads of the previous year, radially symmetrical, stalkless, $1^{1}/_{2}$–3 inches wide; tepals numerous, all showy and petal-like, 1–$1^{1}/_{2}$ inches long; stamens numerous; ovary inferior. *Mid June–July*

Fruits berry-like, inverse-egg-shaped, green, becoming red to red-purple, $1^{1}/_{4}$–2 inches long, $1/2$–$3/4$ inch thick, fleshy, edible, with many tan seeds. *Mid July–October*

WI range and habitats Southwest, in sands and on bluffs and ledges, mainly along the Wisconsin and Black rivers, in sand barrens, sand prairies, open jack pine-scrub oak woods, and cedar glades. *West Knoll*

The pads do not readily break off, but once separated survive desiccation and root easily when transplanted. The small barbed bristles break loose when touched and penetrate clothing and skin; although irritating, they probably are not as injurious as the long rigid spines.

Water-conserving adaptations of cacti include water storage tissue, reduction or loss of leaves, and shallow root systems to absorb as much water as possible from brief showers.

Similar species Our other native cactus, brittle prickly-pear (*O. fragilis*), differs in having dark green, turgid pads (scarcely flattened) that are smaller and readily detaching. The cushions have white wool, and the fruits are dry and often spiny. It is rare in northwestern and central Wisconsin and does not occur in the Arboretum.

Partridge Pea
Golden cassia, locust-weed, sleeping-plant
Chamaecrista fasciculata
Synonym *Cassia fasciculata*

This attractive wildflower has rich green compound foliage that is complemented by showy, golden-yellow flowers starting in midsummer. The petals serve as a handsome setting for the anthers, 4 of which are yellow and the other 6 purple or reddish. The leaves are sensitive to being touched, and the leaflets will close together if disturbed. **Low annual, 7–31 inches tall, from a taproot.**

Leaves once pinnate-compound, with numerous (5–18 pairs) oblong leaflets $3/16$–$7/16$ inch long; leaf stalk with a depressed, saucer-shaped gland near the middle; basal paired appendages (stipules) persistent, conspicuously lined. **Stem** solitary, branched.

Flowers in reduced racemes, on stalks arising above the axil, slightly irregular (i.e., not radially symmetric), 1–$1 3/8$ inches across; sepals 5, free from one another; petals 5, separate, almost equal, overlapping; stamens 10, unequal, in 2 sets (9 + 1). Flower stalk with two little bracts. ***Early July–early October***

Fruits flat oblong pods $1 1/4$–$4 1/4$ inches long, containing 2–12 seeds. The two valves of the mature pods become elastic when dry, finally bursting and throwing the seeds up to several yards. ***Late August–early October***

WI range and habitats Locally common on alluvial sands of the Mississippi, St. Croix, and Wisconsin River bottoms, on sandy banks, slopes, and bases of sandstone bluffs, in dry-medium prairies or prairie openings in woods, spreading to old fields, grassy roadsides, and quarries; a pioneer of disturbed sandy soil. ***Lower Greene Prairie***

The flowers of partridge pea are distinguished from those of the Pea Family by having corollas that are only somewhat irregular (the upper petal inside of and smaller than the 2 lateral ones) and by having separate stamens (the filaments not joined into a tube), some of which may be imperfect or differently colored from the others. Plants produce mirror image flowers, that is, the anthers and style emerge from opposite sides in alternating flowers in an inflorescence. Flowers open for 1 day. They provide no nectar reward but fall into the "buzz" pollination type: large bees milk the anthers through vibration, causing the release of a cloud of pollen through terminal pores in the anthers.

Harebell
Bluebell bellflower
Campanula rotundifolia

The low stature, thin stems, and narrow leaves of this graceful plant are adaptations to withstanding the strong winds that blow across the open meadows, cliffs, and beaches where it grows. The delicate, blue-violet, nodding flowers are veritable "blue bells." **Perennial, 6–18 inches tall, with milky juice.**

Basal and lower leaves often clustered, from nearly round to inversely lance-shaped, soon withering. **Stem leaves** narrowly lance-shaped (toward the base) to linear (upward).

Flowers 1–several in a loose, simple or branched inflorescence, at least the terminal ones long-stalked at the ends of branches; calyx adherent to the ovary, its 5 lobes narrowly triangular, to $3/8$ inch long; corolla bell-shaped, $1/2$–1 inch long, shallowly 5-lobed; stamens 5, attached at the very base of the corolla, the filaments expanded at the base. *June–September*

Fruits nodding, inversely egg-shaped to ellipsoid capsules $1/8$–$1/4$ inch in diameter, opening by 3 pores near the base to release the numerous minute brown seeds. *June–September*

WI range and habitats Frequent throughout all but northeast and northwest-central, in sandy woods, dry to dry-medium prairies, cedar glades, sandstone or limestone bluffs, cliffs, and outcrops, also on shores and dunes, spreading into fields, embankments, and roadsides. *Curtis Prairie (limestone knoll)*

This is the plant celebrated in song and story as the bluebells of Scotland, where it was once also called witches-thimbles. Its association with witches may have given it the common name harebell: northern Europe legend says witches used it to turn themselves into hares that would bring bad luck to people.

The stamens mature before the stigma will accept pollen carried by insects from other flowers. If cross-pollination fails to occur, the stigma resorts to bending down and contacting its own pollen, dropped earlier into the base of the flower. This system encourages outcrossing but at the same time ensures that seeds will be produced, albeit by selfing.

Similar species Marsh bellflower (*Campanula aparinoides*) is a smaller species with weak, reclining or entangled stems that are roughened on the angles with stiff, backward-directed hairs. It has very small flowers (white to palest blue) and fruits at the ends of threadlike stalks and grows in wet habitats.

Deptford Pink
Dianthus armeria

The attractive Deptford pink, which originated in England and Europe, was an early introduction into the gardens of the colonists. Its vivid, reddish-purple, sweet-scented flowers illustrate a marked characteristic of the Pink Family: the disposition of flowers in 3-flowered clusters, in each of which the terminal flower blooms before the others in that cluster. In Deptford pink the flower branches are so short, the inflorescence is nearly an umbel or head. **Stiff slender annual or biennial, 6–18 inches tall, with a taproot.**

Leaves opposite, joined at the base into a short sheath, linear or lance-linear, almost grasslike, $3/4$–3 inches long, entire. **Stems** regularly forking above into pairs of branches.

Flowers about $3/8$ inch across, in small tight clusters at tips of branches (or the lower ones solitary), in which usually only one opens at a time; sepals united into a cylindrical tube $3/8$–$5/8$ inch long, 5-toothed, hairy, with 20–25 fine veins, accompanied at the base by appressed leafy bracts from over $1/2$ as long as, to surpassing, the calyx; petals 5, with a long narrow, stalk-like base about equaling the calyx and abruptly spreading, gnawed-looking or irregularly toothed blades $1/8$–$3/8$ inch long; stamens 10; styles 2. *Late June–beginning September*

Fruits 1-chambered, ellipsoid capsules $3/8$–$1/2$ inch long, slightly shorter than the persistent calyx, opening by 4 teeth at the apex; seeds numerous, dark brown, shield-shaped. *August–September*

WI range and habitats
Sporadic, mainly along roads, trails, edges of dry fields, pastures, marshes, and more severely disturbed areas (driveways, vacant lots, lawns). *Grady Tract knolls, Upper Greene Prairie*

Dianthus, the genus of pinks, carnation, and sweet William, was given its proud name by Theophrastus about 300 B.C. It is derived from the Greek *Dios,* Jupiter, and *anthos,* flower, in other words, Jove's own flower. A few of the true pinks are now casual weeds throughout much of eastern North America, where their presence along roadsides and in waste areas is not unwelcome.

Bladder Campion
White campion, white cockle
Silene latifolia
Synonyms *Lychnis alba, Silene pratensis*

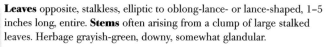

The large white fragrant flowers, which attract night-flying moths, are usually seen in closed or withered condition, for they open at dusk and remain so until about 9:00 a.m. the following morning. The stamens and pistils are in separate flowers on different plants (dioecious). The calyx tube of male flowers is ellipsoid, and that of female flowers inflated, forming a sort of bladder with a narrow mouth. **Biennial or short-lived perennial, 8–40 inches tall.**

Leaves opposite, stalkless, elliptic to oblong-lance- or lance-shaped, 1–5 inches long, entire. **Stems** often arising from a clump of large stalked leaves. Herbage grayish-green, downy, somewhat glandular.

Flowers $5/8$–$3/4$ inch wide, the clusters in an open forking inflorescence; male flowers producing 10 stamens only (functioning ovary absent); female flowers with a 1-chambered ovary with 5, 6, or more styles; calyx 5-toothed, in male flowers 10-veined, in female flowers 20-veined; petals 5, separate, with a sharply narrowed base and a 2-cleft blade with a pair of minute appendages at the base. *May–early October*

Fruits firm-textured, egg-shaped capsules, $1/2$–$5/8$ inch in diameter, opening at the tip into twice as many teeth as there are styles. Seeds grayish-black. *Late May–October*

WI range and habitats Abundant throughout, in fields, pastures, roadsides, embankments, edges of woods, swamps, marshes, and bogs, also shores; any open, disturbed, or waste habitat: railroad ballast, gravel pits, grounds of buildings, gardens; a naturalized weed. *Throughout the Arboretum*

Marked characteristics of the Pink Family include the swollen stem-nodes, opposite leaves connected by a line at the base, and flowers disposed in flattish clusters in which the central flowers bloom first.

Similar species This Eurasian species is most similar to sticky cockle (*Silene noctiflora*), another European immigrant with fragrant, white or pinkish flowers that open in the evening. The calyces of two other European natives, glaucous campion (*S. csereii*) and bladder campion (*S. vulgaris*), are also inflated. All three have bisexual flowers, 3 styles, and 3-chambered fruits that open by 6 teeth. None of these casual weeds is found in the Arboretum.

STARRY CAMPION
Widow's-frill
Silene stellata

This somewhat leggy native wildflower is recognizable at once by its whorled leaves and fringed petals, from which it gets both its common name and species epithet: its brilliant white flowers look like stars. It is less well-known than bladder campion (*Silene latifolia*) or the catchflies (*Lychnis* species, *Silene* species). **Erect, short-lived perennial, 1–4 feet tall, from a thickened taproot.**

Leaves stalkless, in whorls of 4 (or lowest or uppermost often opposite), lance-egg-shaped to lance-linear, $1 1/4$–4 inches long, $1/2$–$1 1/2$ inches wide, rounded at the base, entire. **Stems** several, swollen at the joints. Whole plant very finely downy.

Flowers nodding, several to many in loose elongate panicles 4–20 inches long; sepals united for $1/2$ or more their length into a bell-shaped cup, strongly inflated, 10-veined; petals 5, $3/8$–$3/4$ inch long, the gradually narrowed base not clearly differentiated from the horizontally divergent, flat blade, which is slashed or cut into finger-like lobes; stamens 10 and styles 3, all projecting beyond the petals. *Early June–mid August (September)*

Fruits 1-chambered, nearly round capsules about as long as and included in the calyx, opening at the top and forming 6 teeth; seeds purplish-brown, covered with small, pimple-like projections. *July–September*

WI range and habitats Locally frequent in the southeast and occasional in river valleys in the Driftless Area, in open upland oak woods, medium or sandy woods and woods borders, cedar glades, riverbanks and river terraces, sometimes in deep-soil prairies, prairie borders, and grassy ditches, occasionally along roadsides, railroads, and fencerows. *Curtis Prairie, Juniper Knoll, Wingra Oak Savanna*

The genus name *Silene* is derived from the Greek for saliva, in reference to the stickiness of the stem and calyx of some species.

Ten species of *Silene* occur Wisconsin, the majority introduced. All have 5 sepals joined to form a tube or cup, often bladdery, and 5 petals on narrow stalks with horizontally divergent, flat blades that are once-cut, 2-lobed, or notched, or in the case of starry campion, fringed with several deep cuts.

Other Caryophyllaceae

Mouse-ear Chickweed
Cerastium fontanum
Introduced, naturalized; biennial/perennial forb, 3–22 inches.
Flowers May–June

Bouncing-Bet
Soapwort, sweet Betty
Saponaria officinalis
Introduced, naturalized; erect perennial forb, 1–3 feet. Growing in patches.
Flowers July–October

Celastraceae • Bittersweet Family

Bittersweets

These woody twining vines climb or sprawl over shrubs, trees, fences, or rocks. Their greenish to greenish-yellow flowers are rather small and not very showy, unlike the berry-like, orange-yellow capsules. **Moderately slender, woody vines, climbing to 20–30 or more feet, spreading by root suckers.**

Leaves simple, alternate, on short stalks, finely blunt-toothed, smooth. Oriental bittersweet has nearly circular to broadly inverse-egg-shaped leaves mostly 1½–3 inches long and wide, ending in an abrupt short point, with minute glassy hairs along the main veins beneath (use a hand lens to see the hairs). American bittersweet has elliptic to oblong-egg- or egg-shaped leaves mostly 2–3 inches long and 1–2 inches wide, tapering gradually to a point, without any hairs.

Flowers with 5 free petals and 5 stamens, all inserted on the margin of a broad flat disk that lines the bottom of the calyx. Oriental bittersweet has 1–3 bisexual flowers in small axillary clusters. American bittersweet is quite or nearly unisexual: wholly male, mainly male with a few female flowers, or wholly female, or female with some bisexual flowers. It has 6–18 flowers in a terminal raceme or panicle 1–3 inches long. *Late May–June*

Fruits globular, 3-chambered capsules, splitting in autumn to disclose the seeds, which are completely covered with a bright scarlet, fleshy appendage (aril); seeds 1 or 2 per chamber, reddish-brown, ellipsoid. *Early August–midwinter*

Oriental Bittersweet
Asian bittersweet
Celastrus orbiculata
Invasive weed

WI range and habitats

Oriental bittersweet: Rare, locally naturalized at scattered sites in several counties, on edges of woods (including alluvial woods), thickets, fencerows, roadsides, and open disturbed areas. *Southwest Grady Savanna, West Knoll, Wingra Woods*

American bittersweet: Frequent throughout, in dry to moist woodland borders, savannas, thickets, fencerows, bluffs, cliffs, and outcrops, sandy, gravelly, or rocky hillsides, prairies, dunes along Lake Michigan, embankments, and roadsides. *East Knoll, Juniper Knoll, Sinaiko Overlook Prairie, Visitor Center parking lot, West Curtis Prairie, Lower Wingra Oak Savanna*

A native of East Asia, Oriental bittersweet is potentially one of the most devastating of the invasive species present in the eastern U.S. It is a rampant grower, capable of overtopping trees, and is able to invade natural areas and out-compete native species, degrading habitats and causing extensive ecological damage.

These two species can be distinguished from one another by the disposition of the flowers and the presence or absence of leaf hairs, and less reliably by leaf shape and growth habit. They are sometimes grown as ornamentals and are sold by florists or collected from the wild for use in dried decorative arrangements. Every fall from the East and South to the prairies, American homes are decorated with bouquets and wreaths of American bittersweet because of the salmon-colored fruits and bright seeds. Bluebirds, robins, hermit thrushes, and red-eyed vireos feast on the "berries." Even though birds readily eat them, the fruits are possibly poisonous to humans. Herbalists collected the bark of the root and stem of American bittersweet for medicinal purposes.

AMERICAN BITTERSWEET
Climbing bittersweet
Celastrus scandens

ROCK-ROSES

Our rock-roses are low, shrub-like inhabitants of dry sunny prairies and barrens. They are notable for producing flowers of two forms: the primary or early ones open only once, in sunshine, displaying large yellow petals that perish by the next day; the later flowers, much smaller and mostly lacking petals, remain closed and are self-pollinated. **Perennials, mostly 8–24 inches tall, clustered at the tips of a branching crown.**

Leaves on the main stem narrowly oblong, narrowly elliptic, or inverse-lance-shaped, commonly $3/4$–$1 3/4$ inches long (smaller on the branches), the upper surface not completely masked by the covering of star-shaped hairs. Bicknell's rock-rose has star-shaped hairs only, whereas young leaves of common rock-rose have scattered long simple hairs mixed with the star-shaped ones.

Flowers with 5 sepals, the 2 outer much narrower than and attached to the 3 inner ones, and 5 petals, these absent (rarely rudimentary) in self-fertilizing flowers; stamens 10–50 (outcrossing flowers) or 3–8 (self-fertilizing flowers). Bicknell's rock-rose has 3–several petal-bearing flowers in a loose terminal raceme, whereas common rock-rose has only 1 (rarely 2) petal-bearing flower that is soon overtopped by lateral branches. Later both species produce numerous, self-pollinated flowers crowded in dense small clusters on short axillary branches.

BICKNELL'S ROCK-ROSE
Hoary frostweed
Helianthemum bicknellii

stellate hairs

Bicknell's rock-rose: ***Mid June–July***, self-fertilizing flowers August–September; common rock-rose: ***Mid May–mid July***, self-fertilizing flowers mid July–August

Fruits 1-chambered, leathery capsules containing few to many, medium to dark brown seeds. Capsules produced by petal-less flowers of Bicknell's rock-rose are less than $1/_8$ inch in diameter, and the seed surface forms an indistinct fine network (use 10✕ magnification to see this). Capsules of petal-less flowers of common rock-rose are slightly larger, $1/_8$ inch or a little more in diameter, and the seed surface is minutely warty with nipple-shaped projections. Bicknell's rock-rose: ***Mid July–late October;*** common rock-rose: ***(June) July–mid October***

WI range and habitats Common south and central, following the Cambrian sandstone northeastward and reappearing northwest on outwash plains, in sand barrens, black oak and jack pine barrens, bur oak openings, prairie-like habitats on rocky slopes, bluffs, and outcrops, also sandy lakeshores. Bicknell's rock-rose: ***Grady Tract knolls;*** common rock-rose: ***Grady Tract knolls, Upper Greene Prairie***

The genus name is derived from the Greek words *helios*, sun, and *anthemon*, flower.

Swink and Wilhelm describe how "On a frosty late November morning, if you are lucky, you may see crystals of ice protruding from cracks in the lower part of the stem, hence [for both our rock-roses] another common name, Frostweed."

Similar species See intermediate pinweed (*Lechea intermedia*).

Common Rock-rose
Frostweed, long-branch frostweed
Helianthemum canadense

Intermediate Pinweed
Large-pod pinweed
Lechea intermedia

This slender inconspicuous herb is easily overlooked, but like the rock-roses (*Helianthemum* species), it or one of its fellow pinweeds is almost invariably present in southern Wisconsin's dry sandy prairies and barrens. The very numerous, tiny flowers, drab to brown and spherical, make pinweeds readily identifiable to genus. **Low perennial, 8–24 inches tall, producing basal offshoots with numerous crowded leaves late in the season.**

Leaves alternate, green, narrowly oblong, small, mostly $1/4$–$3/4$ inch long, appressed-hairy beneath only on the midrib and margins. Plants with simple hairs only (no star-shaped hairs). **Stems** solitary or few, with spreading branches.

Flowers in a slender leafy panicle $1/3$–$1/2$ the height of the plant; sepals 5, the 2 outer shorter and narrower than and attached to the 3 broad inner ones; petals 3, dark red, smaller than the sepals, withering but persisting; stamens irregular in number, the anthers minute; stigma dark red and very feathery.

Fruits ± spherical capsules, incompletely 3-chambered, enclosed by the persistent calyx; seeds 4–6, pale brown, shaped like the sections of an orange, at least partly enveloped by a thin, grayish-transparent membrane.

WI range and habitats Frequent south-central and central, following the Cambrian sandstone northeastward and reappearing northwest on outwash plains, in dry to medium, sandy or gravelly prairies, black oak and/or jack pine barrens, dry upland woods, sand barrens, sandy lakeshores, sandstone or quartzite bluffs, cliffs, and escarpments, and sandy grassy slopes and banks (including pastures). *Grady Tract knolls*

Similar species Six members of the Rock-rose Family are frequent to common and usually grow near one another in Wisconsin's dry to medium prairies, oak savannas, barrens, sand flats and dunes, abandoned fields, and sandy or rocky ground generally. Species of *Lechea* are difficult to tell apart. Their leaves and stems range from green and hairless to pale or gray because of a covering of fine close simple hairs, whereas our *Helianthemum* species always have foliage that is grayish or whitish from a dense covering of star-shaped hairs.

Common Spiderwort
Blue-jacket, smooth spiderwort
Tradescantia ohiensis

Like those of the Lily Family, the leaves of spiderworts have tubular sheaths and parallel veins, and the flowers are radially symmetrical with parts in 3s and 6s. Only a few flowers open at a time, their blue to rose or white petals showy but ephemeral. The filaments of the stamens are covered with long hairs, a very distinctive feature. **Somewhat succulent, tufted perennial, 16–24 inches tall, with fleshy roots.**

Leaves 3–8, linear-lance shaped, 4–13 inches long, the blade extended basally into a sheath that entirely surrounds the stem. Plants hairless (sheaths infrequently lightly hairy) and whitened with a waxy bloom.

Flowers numerous in umbel-like clusters in the axils of 2 leaf-like, folded bracts, on reflexed unequal stalks; sepals 3, separate, equal, boat-shaped, hairless or with a weakly developed tuft of non-glandular hairs near the tip; petals 3, separate, usually blue, equal, soon shriveling; stamens 6, the filaments with abundant long soft hairs. *Late May–early August*

Fruits dry capsules that open violently upon ripening to release 2–6 gray oblong seeds, these wrinkled and ridged. *Late June–September*

WI range and habitats Common southwest, in dry to moist prairies, barrens, open oak and pine woodlands, bluffs and dunes, old fields, fencerows, roadsides, and railroads. *Curtis Prairie, Grady Tract knolls, Greene Prairie, Juniper Knoll, Marion Dunn Prairie, Sinaiko Overlook Prairie*

The petals open in the morning and wilt around midday. They are noted for their propensity to shrivel and deliquesce into an aqueous jelly. John Tradescant (1570–1638), a Dutchman and head gardener to Charles I of England, lends his name to this genus. His son visited Virginia and returned with several new plants, among them the original *Tradescantia*.

Similar species Common spiderwort is superficially very similar to Wisconsin's other two wild spiderworts, long-bracted spiderwort (*Tradescantia bracteata*) and western spiderwort (*T. occidentalis*), both of which differ in having glandular-hairy flower stalks and sepals. Both are rare in upland prairies, and neither grows in the Arboretum.

Hedge Bindweed
Morning-glory
Calystegia sepium
Synonym *Convolvulus sepium*

Trailing or usually twining over other vegetation, fencerows, and banks, hedge bindweed winds its stems counterclockwise around available supports. Relatively small leaves, triangular with two symmetrical basal lobes, and the large, white to delicate pink flowers are distinctive. **Perennial, 3–7 feet long, from creeping rhizomes.**

Leaves alternate, long-stalked, triangular-, lance-, or oblong-egg-shaped, 1–4 inches long, heart- or arrowhead-shaped at the base, the basal lobes aligned with the blade or mostly angled outward, blunt to pointed or coarsely 2- or 3-toothed.

Flowers 1 or 2 from many axils, on stalks 1–5 inches long; bracts egg-shaped, enclosing and surrounding the calyx; sepals 5, overlapping, elliptic to lance-egg-shaped, thin; petals 5, united into a funnel-shaped corolla about 2 inches long and broad, very shallowly lobed; stamens 5, inserted on the base of the corolla; style 1, threadlike. *June–early October*

Fruits 1-chambered capsules $3/8$–$1/2$ inch in diameter, 1–4-seeded; seeds black, $3/16$ inch long, shaped like quarter sections of a sphere. *Early August–November* (seldom setting seed)

WI range and habitats Common except north-central, on dry to wet-medium prairies, edges of upland and bottomland woods and thickets, fencerows, fallow fields, pastures, lakeshores, streamsides, sedge meadows, shrub carrs, and such disturbed areas as roadsides, railroads, ditches, dikes, and boat landings. *Curtis Prairie, Greene Prairie, Visitor Center parking lot*

Many poorly defined varieties and subspecies and even segregate species have been recognized, some native to North America, others introduced from Eurasia. Several have been suspected to occur in Wisconsin, but in the absence of a careful study of our plants, we assume most are native.

The flowers are twisted shut at the top in bud and close when it is dark, overcast, or raining.

Similar species The more common field bindweed (*Convolvulus arvensis*) resembles hedge bindweed, but its leaves and flowers are smaller, its calyx is not enclosed by large bracts, and its capsule is 2-celled. Although native to Europe, it is now of worldwide distribution. It has spread to roadsides, railroads, fields, gardens, and waste places across the U.S. and southern Canada, where it has often become a troublesome weed, including in Wisconsin, which lists it as a **noxious weed.**

Gray Dogwood
Northern swamp dogwood, panicled dogwood
Cornus racemosa
Synonym *Cornus foemina* subspecies *racemosa*

This bushy shrub combines opposite entire leaves with parallel lateral veins, small flowers with parts in 4s, and an inferior ovary. Gray dogwood is noteworthy not only for its clusters of creamy white flowers and equally handsome white (rarely pale blue) fruits, which form continuously all summer, but also for its ecological importance. **Shrub, 3–6 feet high, forming thickets.**

Leaves lance-shaped to elliptic, mostly 1–3 inches long, tapering to a long point, both sides nearly hairless or with minute white, T-shaped hairs closely pressed against the surface, with 3 or 4 (5) lateral veins on each side. First-year branchlets 2-edged, with reddish or tan bark that eventually turns gray; pith brown (pale brown or white in 2nd-year branchlets, visible when cutting across a branchlet).

Flowers in terminal, convex to pyramid-shaped, panicle-like clusters 1–2 inches across; sepals forming a minutely 4-toothed, shallow cup; petals oblong-lance-shaped, $1/8$ inch long, spreading, inserted on the margin of a disk that surmounts the ovary. *May–June*

Fruits small stone-fruits (drupes) $1/4$ inch in diameter, containing 1 seed, tipped with minute calyx lobes. Branchlets and stalks of the fruit cluster bright red. *July–September*

WI range and habitats Abundant (absent from a few north-central counties), in dry to moist forests and woodlands, woodland and field borders, thickets, fencerows, medium to wet prairies, marshes, swamps, and along streams, rivers, and lakes, also roadsides; dry to wet, open or shaded places. ***Throughout the Arboretum***

Gray dogwood occurs in a wider range of habitats than any other native shrub and increases with the degree of openness and disturbance, its spread aided by its habit of proliferating from roots and undoubtedly also by birds.

Similar species Silky dogwood (*Cornus amomum*) also has brown pith, but the inflorescence is flat-topped or convex, the fruits are dark blue, the calyx lobes are longer, and the hairs on the lower leaf surface are longer and spreading, curled, or wavy. Red osier dogwood (*C. stolonifera*) is a larger shrub with long branches capable of rooting where they touch the ground. It is distinguished by bright red, 1st-year stems, leaves with 5 or more pairs of lateral veins, and white pith.

Other Cornaceae

Silky Dogwood
Blue-fruited dogwood
Cornus amomum
Native; shrub of wetlands, 3–9 feet.
Flowers June–September

pith

Red Osier Dogwood
Cornus stolonifera
Native; shrub of wetlands, 6–10 feet.
Flowers May–August

About Sedges

Sedges may be frequent or common in prairies, but they are generally overlooked. At least 7 genera and 38 species have been identified on the Arboretum's prairies; the following pages include sedges easily seen from visitor paths. *Carex,* taxonomically one of the most perplexing of all genera, is by far the largest genus in Wisconsin with 156 species, many difficult to tell apart.

Sedges resemble grasses and are often mistaken for them, but the two families are readily distinguished by certain superficial characteristics. The **stems** of sedges are not jointed, often contain solid pith, and for the most part are *triangular* in cross-section. The **leaves** when present are in 3 vertical rows or ranks and have **sheaths** with united edges ("closed"); **ligules** are rarely present except in *Carex*.

The tiny, wind-pollinated **flowers** lack ordinary sepals and petals and are borne in compact units called ***spikes.*** The spikes are the basic unit of the inflorescence, and an understanding of their structure is essential for success in identifying these plants. The flowers are bisexual or unisexual, each one solitary within the axil of a single, scale-like bract (***scale***). *Carex* flowers are unisexual, with males and females either borne on different parts of the same spike, or in different spikes on the same plant (monoecious); very occasionally the sexes are borne on different plants (dioecious). The **fruit** is a lens-shaped or 3-angled ***achene,*** which in *Carex* is enclosed in a special sac-like bract called the ***perigynium.***

If your specimen has stems that are round in cross-section and flowers subtended by few to many overlapping scale-like parts, consult the section on the Grass Family (Poaceae) or Rush Family (Juncaceae).

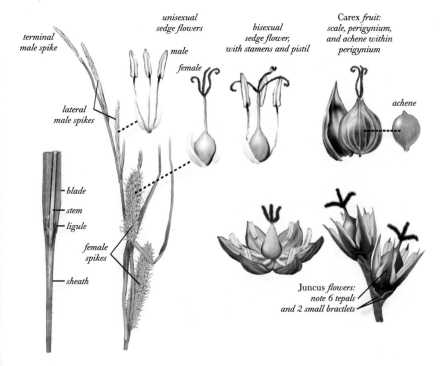

Bicknell's Sedge
Bicknell's oval sedge
Carex bicknellii

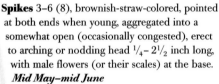

This tall slender sedge produces spherical to slightly oblong spikes that are all alike, rounded and somewhat prickly (from the ascending beaks) at the apex, rounded or usually narrowed to the base. The large translucent perigynia, $1/8$ inch broad with relatively short beaks and broad wings, are quite distinctive. **Tall clumped perennial, 16–40 inches tall, from short black tough rhizomes.**

Leaves 3–4 per stem, pale green, long, about $1/8$ inch wide, flat, thin, rough along the margins. Sheaths tight about the culm, green- and white-spotted on the back, white-transparent on the inner surface. Leafy sterile shoots few.

Spikes 3–6 (8), brownish-straw-colored, pointed at both ends when young, aggregated into a somewhat open (occasionally congested), erect to arching or nodding head $1/4$–$2 1/2$ inch long, with male flowers (or their scales) at the base. *Mid May–mid June*

Perigynia broadly elliptic to nearly circular, $3/16$–$1/4$ inch long, very thin and membranous, the wings turning yellowish- to orangish-brown and of uneven width, with fine distinct parallel veins on both faces. Female scales copper-colored to light brown with a green midstripe, lance-egg-shaped, blunt or slightly pointed, reaching to about the base of the beak. Achene visible through the mature perigynium wall. Stigmas 2, light reddish-brown, short. *Mid June–September*

WI range and habitats Occasional to locally common south, in medium to wet-medium prairies, dry to moist sandy prairies, rock outcrops, and prairie-like habitats, such as old fields, grassy roadsides, and quarries. *Curtis Prairie, Greene Prairie, Juniper Knoll, low prairie south of Teal Pond, Sinaiko Overlook Prairie*

Similar species Bicknell's sedge resembles fescue sedge (*Carex brevior*) and field oval sedge (*C. molesta*). Both have smaller firmer perigynia, those of the former with nearly circular bodies that are veinless on the inner face and erect-ascending beaks, those of the latter egg-shaped in outline, with distinct veins on the inner face and spreading beaks.

Buxbaum's Sedge
Carex buxbaumii

The whitish to pale blue-green perigynia of this wet-meadow sedge contrast beautifully with their purplish-black to -brown, very narrow scales. The bisexual terminal spike usually has female flowers at the tip and male flowers below. **Loosely clumped perennial, 8–30 inches tall; rhizomes cord-like, not far-creeping.**

Leaves 2–4 on the lower $\frac{1}{2}$ of the stem, ascending, blue-green, covered with a pale waxy bloom, $\frac{1}{4}$–20 inches long (to 32 inches on sterile shoots), $\frac{1}{16}$–$\frac{1}{8}$ inch wide, keeled. Basal sheaths purplish-red, bladeless; ligule longer than broad. **Stems** sharply triangular (below middle) to winged (above).

Terminal spike sometimes wholly male but usually composed of both male and female flowers with male flowers at the apex and base or only at the base. **Lateral spikes** 2–4, female, egg- or oblong-egg-shaped, $\frac{1}{4}$–$\frac{3}{4}$ inch long, erect or ascending, very short-stalked. Bracts sheathless or barely sheathing, with purple auricles and short blades, the lowest from shorter than to exceeding the inflorescence. *May*

Perigynia elliptic-egg-shaped, $\frac{1}{8}$ inch long, convex and faintly many-veined on both sides, rounded at both ends, somewhat leathery, granular, beakless, shallowly 2-toothed. Female scales lance-shaped with a pale green midrib and prolonged, sharp-pointed or awned tip,

exceeding the perigynia (varying to pointed and shorter than the perigynia). Stigmas 3, reddish-brown, rather short, breaking off of the mature triangular achene. *Late May–early August*

WI range and habitats Common southeast, rare and scattered in the far northwest; characteristic of low prairies, sedge meadows, calcareous marshes, fens, peaty shores, and old bogs, less frequently in wet openings in woods and swamps, also in wet sandy swales, cobble beaches, and rocky shores along the Great Lakes. *Greene Prairie*

Several other Wisconsin carices have reddish bases with pinnate fibrils, including among wetland species common tussock sedge (*Carex stricta*), common lake sedge (*C. lacustris*), and broad-leaved woolly sedge (*C. pellita*). All are otherwise very different from one another.

Hayden's Sedge
Long-scaled tussock sedge
Carex haydenii

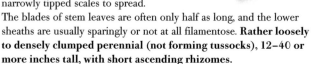

This sedge looks much like common tussock sedge, but the perygynia are broadest at or slightly above the middle and at maturity become inflated, causing the gradually tapering, narrowly tipped scales to spread. The blades of stem leaves are often only half as long, and the lower sheaths are usually sparingly or not at all filamentose. **Rather loosely to densely clumped perennial (not forming tussocks), 12–40 or more inches tall, with short ascending rhizomes.**

Leaves 2–4 on the lower $1/4$ of the stem, $1/16$–$1/4$ inch wide, slightly M-shaped in cross-section; outer basal sheaths red-brown, the lower sheaths conspicuously bladeless, the inner band whitish- or yellowish-translucent, smooth; ligule as long as wide or somewhat longer.

Spikes erect, narrowly oblong to cylindrical; male spikes 1–2; female spikes 2–3 (occasionally male at the apex), $5/8$–$1 1/4$ inches long, approximate or a little separate, densely flowered; bracts leaf-like, the lowermost normally shorter than the inflorescence. Male scales reddish-brown; female scales brown or dark red-brown, with a broad lighter center, at least the lower ones strongly exceeding the perigynia. *May*

Perigynia olive green with red-brown spots, nearly circular to inverse-egg-shaped in outline, $1/16$–$1/8$ inch long, biconvex at maturity, rounded at the apex, abruptly contracted to a minute beak. Stigmas 2, rather short. *End of May–July*

WI range and habitats Widely distributed and locally common except northeast, in medium to wet prairies and sedge-dominated, often sandy or marshy, wet meadows, less commonly in swampy deciduous woods and thickets (especially along borders, moist openings, sloughs) and moist open depressions and marshy spots generally, e.g., streamside pastures, swales, pond edges, ditch bottoms. ***Curtis Prairie, Greene Prairie***

Similar species See common tussuck sedge (*Carex stricta*).

Field Oval Sedge
Pest sedge, troublesome sedge
Carex molesta

This tall slender sedge has a distinctive stiff terminal cluster of few spherical, bur-like spikes. The broad perigynia with winged margins and a veined inner face also help distinguish it. **Clumped perennial, 14–40 inches tall, from short black tough rhizomes.**

Leaves 3–6 per stem, light green, 5–15 inches long, $1/8$ or less inch wide, very rough on the margins. Sheaths tight about the culm, green- and white-mottled on the back, white-transparent on the inner surface. Ligule wider than long.

Spikes 2–4 (5), in an erect congested head $1/4$–$1^3/8$ inches long, rounded at both ends (the terminal with or without a short narrowed base), with a few male flowers at the base. Female scales light brown with a green or pale center and white margins, egg-shaped, blunt or slightly pointed, much shorter than the perigynia. *Early June*

Perigynia pale brown (green until ripe), elliptic to orbicular, $1/8$–$3/16$ inch long, rather thick (membranous only at the very margins), gradually tapering into a toothed beak less than $1/2$ the length of the body, conspicuously to faintly veined on both faces or veinless on the inner face. Stigmas 2, light reddish-brown, short. *Mid June–mid August*

WI range and habitats South, in medium to wet prairies, moist fields, and swales, also oak openings and wooded ravines, rarely on road cuts and railroad ballast; dry to wet, neutral or calcareous soils. *Dump just northwest of buildings, Gardner Pond, Curtis Prairie, Greene Prairie, Juniper Knoll*

Similar species Field oval sedge most closely resembles fescue sedge (*Carex brevior*) and greater straw sedge (*C. normalis*). It differs most conspicuously from the former in having round-based spikes and perigynium bodies that are elliptic and nerved on the inner face, and from the latter in having fewer, more densely clustered spikes and wider perigynia. Fescue sedge has egg-shaped spikes with a tapering brown base and a perfectly circular, veinless perigynium with an evenly wide white wing.

Pennsylvania Sedge
Common oak sedge
Carex pensylvanica

This familiar, early-flowering sedge is identified by the combination of low stature; reddish-brown, fibrous sheaths; strongly developed, shallow rhizomes; and minutely hairy perigynia with nearly spherical bodies above a stalk-like base. **Loosely clumped perennial, 4–14 inches tall, spreading slowly by creeping rhizomes with fibrous sheaths, forming soft low colonies or (in sand barrens) stiffish circular clones.**

Leaves medium green, the widest ones $1/16$–$1/8$ inch broad, smooth but rough along the margins. Basal sheaths decomposing easily, leaving longitudinal fibers. **Stems** nearly smooth, pale green, clothed in old sheaths, equaling or shorter than the leaves.

Terminal spike male, stalkless or short-stalked, $1/4$–$7/8$ inch long. **Lateral spikes** 1–3, female, stalkless, spherical-egg-shaped to ellipsoid, $3/16$–$3/8$ inch long, contiguous with each other and the male spike. Bracts sheathless, scale-like or the lowest needle-like. *Mid April–late May*

Perigynia 3–13 per female spike, pale green to gray, about $1/8$ inch long, with only 2 ribs (otherwise veinless), tapering into a spongy, stalk-like base and a manifest but very short, straight, 2-toothed beak; often attacked by a black smut fungus. Female scales dark reddish-brown with a green center and narrow white margins, blunt to pointed or sharp-pointed, equaling the perigynium body. Stigmas 3. *Mid May–early July*

WI range and habitats Our most abundant and widespread sedge, in dry to medium woods and forests of all kinds, even in lowland forests, as well as open, sandy or rocky ground, such as dry to dry-medium prairies, barrens, and dunes; full sun to moderate shade, sometimes the dominant ground cover in sandy oak and pine barrens. *East Knoll, Gallistel Woods, Noe Woods, Upper Greene Prairie, Wingra Woods*

Similar species The spikes and perigynia of colonial oak sedge (*Carex communis*) and early oak sedge (*C. umbellata*) are similar to those of Pennsylvania sedge, but the former has wider leaves with longer ligules and a more slender male spike, and the latter has some short stems crowded among the tufted leaf bases (instead of only elongate stems).

Dry-spiked Sedge
Hay sedge, hillside sedge, running savanna sedge
Carex siccata
Synonym *Carex foenea*

Slender stems, arising singly or in small clumps at intervals along prominent rhizomes, combined with strongly flattened perigynia with thin margins and prominent beaks, distinguish this species. The spikes appear similar but are variable in composition; the terminal one is female or has male flowers at the apex. **Perennial, 6–35 inches tall, from slender rhizomes clothed in brownish fibrous sheaths, forming colonies.**

Leaves 4–7, somewhat bunched on the lower $1/4$ of the stem, light green, $1/16$–$1/8$ inch wide, stiff. Sheaths tight, overlapping, the ligule wider than long. **Stems** 1 or 2–3 together, surpassing the leaves. Sterile shoots very leafy.

Spikes 4–12, the lowest female and small, the middle ones narrowly oblong, $1/4$–$3/8$ inch long, largely or entirely male, the upper bisexual, the different types commonly mixed in the same inflorescence. Inflorescences $3/8$–2 inches long, rather dense, the lower 1–3 spikes

often a little separate but all overlapping, the upper ones closely aggregated and often indistinguishable. Bracts similar to the scales. *First to last day of May*

Perigynia pale green or tinged reddish-brown, oblong-lance-shaped, $3/16$–$1/4$ inch long, with narrow green, sharply toothed margins, strongly veined on the back, veined or sometimes veinless on the inner face; beak nearly or fully as long as the body, with 2 firm sharp teeth. Scales lance-egg-shaped, pale reddish- or yellowish-brown with green center and broad translucent margins, about equaling the perigynia. Stigmas 2. *End of May to late July*

WI range and habitats Mostly south, sporadic northward, characteristic of dry, very sandy prairies, oak and pine barrens, open sandy or rocky woods, and abandoned sandy fields, also low sandy prairies and sedge meadows, occasionally on sandbars, thin soil over quartzite or basalt, grassy embankments, roadsides, and railroad ballast. *Grady Tract knolls, low prairie south of Teal Pond, oak woods east of Gardner Marsh*

Common Tussock Sedge
Carex stricta

This moderately robust, wetland sedge forms dense raised tussocks to 8 or more inches tall. In early spring these are topped by distinctive bluish pointed shoots that grow out into a fountain of thin but stiffish, narrow leaves. **Densely clumped, 20–40 or more inches tall, with long, cord-like, horizontal rhizomes.**

Leaves 3–5, somewhat bunched on the lower $^1/_4$ of the stem, $^1/_{16}$ – $^1/_4$ inch wide, slightly M-shaped in cross-section; outer basal sheaths light reddish- to purplish-brown, the lower sheaths conspicuously bladeless, ladder-fibrillose, the inner band often very minutely roughened (detectable with the tip of the tongue); ligule much longer than broad.

Spikes erect, narrowly oblong to cylindrical; male spikes 2–3; female spikes 3–4 (more often than not male at the apex), $^1/_4$ –3 inches long, densely packed with scales and perigynia; bracts leaf-like, the lowermost overtopping the inflorescence. Male scales reddish-brown; female scales reddish- or purplish-brown, blunt to sharp-pointed, usually rather shorter than (rarely exceeding) the perigynia. *May–early June*

Perigynia pale green (brown when falling), elliptic-egg-shaped, $^1/_{16}$ –$^1/_8$ inch long, broadest at or slightly below the middle, flattened, pressed closely against the spike axis, 2-ribbed (the marginal veins) and obscurely veined on the back, short-tapering to the nearly beakless apex. Stigmas 2, rather short. Achenes lens-shaped. *End of May–early August*

WI range and habitats Abundant throughout (the dominant sedge of most wet meadows), in sedge meadows, wet prairies, fens, bogs and conifer swamps, shrub carrs and wet thickets, marshes, shores, edges of creeks, pools, and springs, ditches, and other open wetlands. *East Curtis Prairie, Greene Prairie, Wingra Fen*

Five close relatives are abundant in Wisconsin's open wetlands and wet woods (two or three others are rare), but of these, only common tussock sedge and long-scaled tussock sedge (*Carex haydenii*) occur on Arboretum prairies. They belong to a large subgroup in which the identification of species is particularly difficult.

"blasted" or infected perigynium   normal perigynium   achene

Marsh Straw Sedge
Narrow-leaved oval sedge
Carex tenera

This slightly weedy sedge is best recognized by its characteristically bent or arched cluster of spikes, which are well separated, necklace-like, and tapered at the base. The beaks of the relatively firm, narrowly winged perigynia vary from lying pressed against one another to ascending or spreading in the spikes. **Clumped perennial, (8) 12–35 inches tall, from very short, black rhizomes.**

Leaves 3–5 per stem, the broadest ones $1/16$–$1/8$ inch wide, flat. Sheaths tight about the culm, uniformly green on the back, conspicuously white-transparent on the inner surface. Plants with slender leaning stems, brown bases, and inconspicuous sterile, stem-like shoots.

Spikes 3–8, all alike, egg-shaped, $1/4$–$3/8$ inch long, in a loose head 1–2 inches long, the slender wiry axis curved or nodding between the lowest 2 spikes, rounded at the apex, with a few male flowers at the base. Female scales greenish-white or tawny-tinged, egg-shaped, blunt to pointed, shorter than the perigynia. *Late May–mid June*

Perigynia straw-colored to brown, egg-shaped in outline, $1/8$–$3/16$ inch long, $1/16$ or less inch wide, 2 or more times as long as wide, rather long-beaked, similarly veined on both faces or veinless or faintly veined on the inner face, gradually tapered into a toothed beak $1/2$ the length of the body. Stigmas 2, light reddish-brown, short. *Mid June–early August*

WI range and habitats Common throughout, in low prairies, meadows, fields, swales, wetland edges, and grassy clearings, also thickets and borders of, or trails in, medium to swampy woods; dry to moist, open or wooded ground. *Curtis Prairie, Greene Prairie, Juniper Knoll, low prairie south of Teal Pond, Noe Woods, southeast of Gardner Marsh, Sinaiko Overlook Prairie*

Similar species Marsh straw sedge most closely resembles greater straw sedge (*C. normalis*), which, however, has wider leaves (broadest ones $1/8$–$1/4$ inch wide), loose sheaths, and 3–10 usually crowded spikes (lowermost ones nearly or quite overlapping) on a stiff straight axis.

Common Stiff Sedge
Rigid sedge
Carex tetanica

Hiding in tall grass, this low, very slender sedge is rather inconspicuous. It may be distinguished from most other Wisconsin sedges with beakless perigynia by the combination of elongate rhizomes, long-sheathing bracts, and perigynia with tapering, wedge-shaped (instead of rounded) bases. **Tufted perennial, 6–18 inches tall, with loosely scattered stems from deep-seated, whitish rhizomes**.

Leaves bluish-green, the blades 1–8 inches long, to $3/16$ inch wide. Plant bases pale brown or ± purplish-tinged, the dried-up leaves of the previous year conspicuous, the lowest sheaths with well-developed blades; ligule as long as or longer than wide.

Spikes erect or the lateral ascending, the terminal one wholly male, long-stalked, club-shaped; lateral spikes 1–3, widely separate, female, egg-shaped to narrowly oblong, $1/4$–$1\,1/2$ inches long, containing 8–30 green perigynia. Female scales tinged purplish- to reddish-brown, from blunt to pointed at the apex or with a short sharp point or awn. *May*

Perigynia inverse-egg-shaped or narrowly so, $1/8$ inch long, bluntly 3-angled, broadly tapered or rounded at the apex (hardly beaked), strongly many-veined, densely covered with microscopic rounded projections. Stigmas 3. *June–July*

WI range and habitats Locally frequent southeast (reappearing in Trempealeau County), in fens, low prairies, and sedge meadows, less commonly in other open wetlands, such as edges of conifer bogs and marly lakeshores; wet calcareous peats. *Greene Prairie*

Similar species Common stiff sedge is sometimes difficult to distinguish from Mead's sedge (*Carex meadii*), especially in the Great Lakes region. Mead's sedge is common in dry, usually sandy or rocky prairies and barrens. Prairie gray sedge (*C. conoidea*) is similar to these, but its perigynia are shaped something like an old-fashioned, 7-ounce Coca-Cola bottle: oblong-egg-shaped, round in cross-section, rounded toward the base, sometimes slightly pinched, straight-beaked, and with impressed veins (not raised above the surface).

Fox Sedge
Brown fox sedge
Carex vulpinoidea

This abundant sedge produces fine narrow leaves that spray out in all directions. The inflorescence is composed of numerous short stalkless spikes that are all alike and scarcely distinguishable from one another. The slender firm stems and transversely wrinkled inner band of the leaf sheaths are key vegetative characters. **Densely clumped perennial, 8–35 inches tall, from short black tough roots.**

Leaves 4–5 per stem, yellow-green, to 20 inches long and $1/8$ inch wide, flat to slightly channeled at the base. **Stems** shorter than the leaves, sharply triangular, wiry, rough on the upper $1/4$. Sheaths tight-fitting; ligule wider than long.

Spikes crowded into a narrowly oblong, unbranched or branched, spike-like panicle 1–4 inches long, 1 or 2 lower branches or spikes sometimes separated from the rest; male flowers above the female in the same spike, inconspicuous; bracts variously developed, bristle-like, numerous, those subtending at least the lowest branches conspicuous. *June*

Perigynia up to 20 per spike, pale green to brown, egg-shaped to somewhat circular or elliptic, barely $1/8$ inch long and $1/16$ inch wide, low-convex on the back, with 2 corky wings, gradually tapered into a beak $1/3$–$1/2$ as long as the perigynium. Scales translucent-yellowish-brown with a green midvein and awned tip equaling or exceeding the perigynia. Stigmas 2, elongate. *Late June–mid October*

WI range and habitats Abundant throughout, in marshes, meadows, lakeshores, stream edges, ditches, openings and depressions in swamp forests (deciduous and coniferous), less often in bogs and roadsides; open, damp to very wet soils of all kinds; an important colonizer of disturbed spots in wetlands. *Curtis Prairie, Greene Prairie, low meadow south of Teal Pond, southeast corner of Lake Wingra, Wingra Fen*

Similar species Fox sedge is sometimes confused with common fox sedge (*Carex stipata*), but the latter has stems that are wider, softer, and wing-angled, and perigynia that are larger and taper more gradually into the beak. Also similar is the earlier-blooming yellow-headed fox sedge (*C. annectens*), which differs in having the leaves shorter than the stems and more compact inflorescences with broadly egg-shaped, unwinged perigynia with shorter, more abrupt beaks.

Flat-stemmed Spike-rush
Eleocharis compressa

The strongly flattened, slightly twisted stems and broadly white-margined scales make this one of our most easily field-recognized spike-rushes. **Tufted perennial, 3–18 inches tall, from reddish- to purplish-black, somewhat woody rhizomes.**

Stems $1/16$ or less inch wide; sheath tight-fitting, straight across (as if cut off) at the summit.

Spikes egg-shaped to ellipsoid, $1/4$ inch long; scales medium to dark reddish-brown, those in the lower and middle part of the spike with shallowly to deeply 2-cleft, thin colorless tips that are mostly longer than wide; sepals and petals represented by 0–5 bristles, these variable in length and falling very early. *Mid May–early June*

Achenes inversely pear-shaped, bluntly 3-sided to nearly round in cross section or some biconvex, golden-yellow to orange or brown. Style-base enlarged, persisting as a low,

pyramid-shaped cap (tubercle) perched atop the achene. *Mid June–August* (some achenes often persist after the scales fall)

WI range and habitats Locally frequent, mainly in east and south (rare in a few west-central and far northwestern counties), chiefly in either low prairies or fens, also shallow soil and crevices of dolomite along Lake Michigan, occasionally in other moist openings or boggy spots; wet calcareous clay, sand, or peat. *Greene Prairie*

This genus is composed of rush-like plants without obvious leaves or bracts, but with bladeless sheaths confined to the lower part of the stem. The inflorescence is a single erect terminal spike (hence the common name) with spirally overlapped scales, bisexual flowers, and achenes capped by a tubercle. Superficially, species of *Eleocharis* are very similar to one another, making casual recognition of most species almost impossible.

Similar species Flat-stemmed spike-rush seems to intergrade with elliptic spike-rush (*Eleocharis elliptica*). The stems of the latter are usually round in cross-section but are sometimes compressed. Wisconsin's most common spike-rush, bald spike-rush (*E. erythropoda*), differs from these species in having 2-sided achenes with a smooth shiny surface.

Tall Nut-rush
Whip-grass
Scleria triglomerata

The nearly spherical, hard achenes, smooth and lustrous like white enamel, are exposed and often conspicuous at maturity, making this one of our most readily recognizable sedges. The small inflorescences contain bisexual spikes intermixed with clusters of few-flowered male spikes. **Somewhat coarse perennial, 20–40 inches tall, from hard knotty rhizomes.**

Leaves several, scattered along the lower $1/3$ of the stem, the larger ones $1/8$ – $1/4$ inch wide, V-shaped, abruptly tapering to a short narrow pointed tip; sheaths buff to red-brown, hairy or hairless, the inner band hairless below the sharply demarked ligule. **Stems** clustered, erect to ascending, leafy, relatively stout, sharply triangular.

Spikes few-flowered, brown, oblong, small, $1/8$ – $3/8$ inch long, in a few small bundled clusters, the clusters terminal (often in 3s) and often lateral, subtended by a few leaf-like, ascending to spreading bracts. Scales loosely overlapping, with thin margins, notched tips, and a short sharp point or awn, the lower 2–4 empty, the lowest fertile scale female, the rest male or empty; scales of male spikes narrower. Flowers all unisexual; bristles absent. *June*

Achenes nearly spherical, approaching $1/8$ inch long, elevated on a whitish, minutely pebbled (under magnification), cushion-like collar; style 3-cleft. *June–late September*

WI range and habitats Rare to locally common southwest, in moist to wet prairies (very rarely on dry prairies) and sandy meadows and swales, also borders of marshes and oak openings; characteristically on seasonally moist, sandy loam or silt loam soils. *Greene Prairie*

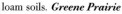

The genus name is derived from the Greek *skleros,* hard, in reference to the bony achenes. The species epithet means "with three clusters."

OTHER CYPERACEAE

YELLOW-HEADED FOX SEDGE
Carex annectens
Native; perennial sedge, 1–3 feet.

WOOD SEDGE
Common wood sedge, eastern woodland sedge
Carex blanda
Native; perennial sedge, 8–24 inches.

FESCUE SEDGE
Plains oval sedge
Carex brevior
Native; perennial sedge, 9–40 inches.

PRAIRIE GRAY SEDGE
Open-field sedge
Carex conoidea
Native; perennial sedge, 7–27 inches.

Cyperaceae • Sedge Family

INLAND SEDGE
Inland star sedge
Carex interior
Native; perennial sedge,
7–31 inches.

UMBRELLA SEDGES
Cyperus species

false nut sedge,
straw-colored cyperus
Cyperus strigosus

GREAT BULRUSH
Soft-stem bulrush
Schoenoplectus tabernaemontani
Synonym *Scirpus validus*
Native; perennial wetland sedge, 3–6 feet.
Flowers May–August.

Common Horsetail
Field horsetail
Equisetum arvense

This and related equisetums have many whorls of branches at the nodes, making them somewhat bushy and thus deserving of the name horsetail. The stems take two dissimilar forms. The fertile ones emerge in early spring; they are whitish or beige to brown, unbranched, and wither quickly once the cones reach maturity, at which time the green branched sterile stems come up from the same rhizome. These last until winter before withering and dropping off. **Perennial, 2–10 (fertile stems) or 12–27 (sterile) inches tall, from tough, dark brown or black rhizomes, forming colonies.**

Sterile stems 4–14-ridged, branched; sheaths appressed to stem, the teeth as many as the grooves, brownish or blackish throughout. Ridges moderately rough to the touch.

Fertile stems soft and somewhat fleshy; sheaths with a fringe of 8–12 separate or basally united, dark brown, persistent teeth.

Cones oval, to 1¼ inch long, long-stalked, blunt-tipped. Spores mature *April–May*

WI range and habitats Abundant throughout, in low fields, damp open woods, stream banks, lakeshores, wet ditches, railroad embankments, roadside fill, and soggy waste sites; often weedy; very tolerant of soil conditions, in full sun or part shade. *Curtis Prairie, Greene Prairie, Sinaiko Overlook Prairie*

The genus name is derived from the Latin *equus,* horse, and *seta,* bristle. This is our most common and most variable species of *Equisetum.* Horsetails are supposed to be poisonous, but there is disagreement in the literature on the symptoms and causes of horsetail poisoning, which may be due to the presence of alkaloids or thiaminase, which catalyzes the breakdown of vitamin B1, or both.

Similar species The much less common meadow horsetail (*Equisetum pratense*) is a more delicate plant with shorter, white-margined teeth on the sheaths. It does not occur in the Arboretum.

Smooth Scouring Rush
Smooth horsetail
Equisetum laevigatum

This equisetum belongs to a subdivision of the genus with evergreen (except in smooth scouring rush), usually unbranched stems. All stems are alike, the spore-bearing cone being produced at the tip of the vegetative shoot. The green fertile stems of smooth horsetail are not very firm and only slightly roughened, and the sheaths have a black band only at the apex. **Reed-like perennial, 16–32 inches tall (but usually shorter), spreading by widely creeping rhizomes and often forming large colonies.**

Stems annual, yellow- or waxy-green with (10) 16–32 ridges, barely roughened owing to an external deposition of silica in granules that are only slightly developed.

Sheaths the same color as the stem, longer than thick, with a narrow black band at the tip or in old stems the lower ones sometimes becoming girdled with brown at the base as well; teeth falling promptly (sometimes a few persistent), dark with transparent borders.

Cones terminal, yellow to brown, ellipsoid, 3/8–3/4 inch long, blunt or often with an inconspicuous short point at the tip. Spores mature *May–June*

WI range and habitats Common in moist or seasonally dry prairies and depressions, along streams, rivers, lakeshores, and seepage slopes, in floodplain thickets, and sandy barrens, invading roadsides, embankments, sand pits, and other disturbed areas; shaded or open, neutral or somewhat acid, sandy soil. *Grady Tract knolls, Juniper Knoll*

These plants were used for cleaning wood and metal utensils, particularly in colonial and frontier times, giving rise to the name "scouring rush." They were also used to smooth items made of wood and bone, such as arrow shafts.

Similar species This species superficially resembles common scouring rush (*Equisetum hyemale*), which is more common in Wisconsin and grows in Curtis and Greene prairies. It differs by its stiff evergreen stems that are rough to the touch, like a medium grade of sandpaper, and its ashy, gray-green sheaths that often have a blackish band in the middle or at the base as well as the top. Ferriss' horsetail (*E.* ×*ferrissii*) is a frequent hybrid of *E. hyemale* and *E. laevigatum*. It recombines features of the two parental species and has abortive white (instead of green) spores.

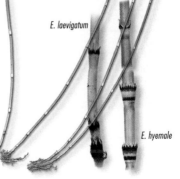

E. laevigatum

E. hyemale

Flowering Spurge
Euphorbia corollata

Scattered white "flowers" in an umbrella-shaped inflorescence of forked branches and small narrow leaves mark this relatively conspicuous member of the Spurge Family. Blooming throughout the summer, like baby's-breath in a florist's bouquet it laces together the matrix of other prairie wildflowers. The actual flowers are minute and borne within a unique cup (involucre) of fused bracts that mimic a flower. **Slender perennial with acrid milky juice, 10–36 inches tall, from a deep root, often growing in colonies.**

Leaves alternate but whorled beneath the inflorescence, stalkless, elliptic, oblong, or linear, to $2\frac{1}{4}$ inches long, entire. Plants with a pale bluish or yellowish-green cast.

Flowers unisexual, without sepals or petals; involucres containing 1 stalked female flower and 10–15 male flowers, the female consisting of a single pistil, each male flower of a single stamen; nectar glands 5, attached to the rim of the cup, yellowish-brown, with a white petal-like appendage that makes the plant somewhat showy when in bloom. *Late May–mid September*

Fruits spherical, 3-lobed capsules, on a long stalk extending from the cup; seeds normally 3, white, egg-shaped. *Early July–late September*

WI range and habitats Common central and south, in prairies, thin jack pine or scrub oak woods, barrens, cedar glades, bluffs, sand flats, blowouts, and lakeshores (including dunes along Lake Michigan), weedy in abandoned fields, roadsides, railroads, fencerows, and occasionally quarries; dry to moist, sandy, gravelly, or loamy soils. *Curtis Prairie, Marion Dunn Prairie, Grady Tract knolls, Greene Prairie, Juniper Knoll, Sinaiko Overlook Prairie, Visitor Center parking lot*

diseased buds

leaf autumn color

Although frequently poisonous to some degree, the characteristic milky juice in leaves and stems of spurges is a source of drugs and dyes. The root of flowering spurge once furnished crude drugs used to induce vomiting (emetic), defecation (cathartic), and perspiration (diaphoretic).

Leafy Spurge
Wolf's-milk
Euphorbia esula
Invasive weed

Leafy spurge belongs to a subgroup of *Euphorbia* in which the involucral cups that surround the group of tiny stamens and pistil regularly have 4 glands that lack petal-like appendages. These euphorbias are commonly annuals with pitted or sculptured seeds, but leafy spurge is a perennial with smooth seeds. **Perennial with milky juice, mostly 16–24 inches tall, from horizontal rhizomes and deep roots, vigorously colonial.**

Leaves alternate, linear to lance-linear, $3/4$–3 inches long, $1/8$–$5/16$ inch wide, those subtending the umbel broader, those of the umbel opposite pairs of broadly triangular-egg- or kidney-shaped bracts about $3/8$ inch long.

Flowers extremely small, in small clusters of 12–25 male flowers and 1 female flower per involucral cup, the cups in turn clustered at the ends of long stalks in umbrella-like inflorescences; each involucre with 4 (5) greenish-yellow or -brown, crescent-shaped glands. *Mid May–early July (mid October)*

Fruits warty, 3-chambered capsules $1/8$ inch thick, each chamber 1-seeded; seeds ellipsoid, silver-gray, often mottled with brown, with an outgrowth near the scar marking the point of attachment. *Mid June–October*

WI range and habitats Now locally abundant throughout, in grassy fields, roadsides, railroad ballast, and waste places, invading open woodlands and prairie remnants; extremely competitive, fully capable of completely displacing desirable plants. *Curtis Prairie, Marion Dunn Prairie, Visitor Center parking lot, West Greene Prairie, West Knoll*

Leafy spurge in the broad sense is a complex of closely related Eurasian species, forms, and hybrids that includes Waldstein's spurge (*Euphorbia waldsteinii*) and twiggy spurge (*E.* ✗*pseudovirgata*), a presumed hybrid between leafy spurge and Waldstein's spurge. Splitting *E. esula* in two by recognizing *E. waldsteinii* results in many areas where the hybrid is the commonest taxon.

A noxious weed under Wisconsin law, leafy spurge is virtually impossible to eradicate. Host-specific beetles that feed only on leafy spurge are now being reared and released in an attempt to implement biological control over this feared and despised invader. The acrid milky sap (latex) can inflame and blister sensitive skin.

Similar species In Wisconsin this species is confused only with cypress spurge (*Euphorbia cyparissias*), a smaller plant only 12 or less inches tall with more densely crowded, narrowly linear leaves $1/16$ or less inch wide.

LEAD-PLANT
Amorpha canescens

Shrubby habit, gray to silvery-green foliage, and pinnately compound leaves make lead-plant stand out against its prairie background. Bright violet spikes, dotted with orange anthers, are equally distinctive. The corolla consists of only one petal, hence the genus name from the Greek *amorphos,* deformed. **Diminutive, semi-woody shrubs, 1–2 feet high, with deep roots and erect branches.**

Leaves nearly stalkless, the numerous leaflets (15–37) elliptic, oblong, or egg-shaped, less than 1 inch long, obscurely gland-dotted. Plants with a permanent, sometimes thin or weather-beaten covering of silver-gray hairs.

Flowers in dense, spike-like racemes clustered in the axils of the uppermost leaves; sepals united; petal (standard) circular or heart-shaped, inward-curving, to $1/4$ inch long; wing and keel petals absent; stamens 10, projecting beyond the petal. *End of May–mid September*

Fruits non-opening, 1-seeded pods (legumes), about $1/8$ inch long, slightly protruding from the persistent calyx, whitish- or grayish-fuzzy. *Early July–early October*

WI range and habitats
Locally frequent southwest, on medium to dry or occasionally low prairies, sand prairies, prairie patches on sandy or rocky ground, oak openings, pine barrens, cedar glades, and along sandy roadsides and railroads; in remnants and unplowed soil. *Curtis, Greene, Marion Dunn, Sinaiko Overlook prairies, Grady Tract knolls, Juniper Knoll*

Lead-plant lacks the butterfly-shaped flower so characteristic of the family. The single petal forms a hood over the lower portion of the stamens, leaving them visible.

One of the most important native legumes of the prairie, lead-plant was named by early miners in southwestern Wisconsin, to whom the leaden-gray hairiness suggested the presence of lead ore deposits. Native American tribes used the dried leaves to make a tea.

Hog-peanut
Amphicarpaea bracteata

Hog-peanut is recognized by its twining habit, 3-foliolate leaves, small nodding racemes of pea-like flowers, and small flat pods. **Annual, 8–80 inches long, from a taproot.**

Leaves on stalks 1–4 inches long, the 3 leaflets broadly lance-, egg- or diamond-egg-shaped, the terminal one on a conspicuous stalk $1/4$–$5/8$ inch long. Paired appendages at base of the stalk persistent and conspicuous.

Flowers of two kinds, **normal** ones in stalked axillary racemes, and **reduced** subterranean, self-pollinating ones on thread-like runners from lower stem nodes. Normal flowers whitish and lavender-tipped to pale or dark blue-purple, $1/2$–$5/8$ inch long; 2 upper sepals completely united, the 4 lobes lance-shaped, shorter than the tube; corolla $3/8$–$9/16$ inch long, the keel enclosing 10 stamens arranged as in *Lathyrus* and *Vicia*. ***August–mid September***

Fruits from normal flowers linear-oblong pods, straight or sickle-shaped, $5/8$–$1 1/8$ inches long, containing 2–4 brown, kidney-shaped seeds. Subterranean fruits ellipsoid to somewhat round, only $1/2$ as big, fleshy, 1-seeded. ***September–early October***

WI range and habitats Infrequent to locally common throughout except for the extreme north-central part, in dry to moist woods, thickets, oak savannas, and sometimes brushy prairies; especially common where disturbed and sunny (trails, borders, banks, clearings, fencerows, overgrown fields, unmowed ditches). ***Curtis Prairie, Grady Tract, Upper Greene Prairie, Noe Woods, Gallistel Woods, Wingra Woods***

Amphicarpaea alludes to the two kinds of fruit. The underground "peanuts" are edible when cooked and were used as a food source by Native Americans.

Most genera of peas and beans have an easily recognized flower structure, described as *papilionaceous,* meaning butterfly-like, and consists of 5 unequal petals: an upright standard, 2 lateral wings, and 2 joined inner petals forming a canoe-like keel, which envelops the 5–10 stamens and the single pistil.

White Wild Indigo
Large-leaved wild indigo, milky wild indigo
Baptisia alba
Synonym *Baptisia leucantha*

This stately plant forms a ball-like mass of blue-green foliage near the middle, out of which emerges a beautiful spire of white, pea-shaped flowers discernible across the prairie landscape. Later in the season, inflated seed pods change from green in summer to black with a bluish bloom in autumn. In early winter the whole plant dries up and becomes a tumbleweed. **Robust perennial, 3–6 feet tall, from a very deep root system.**

Leaves 3-foliolate, short-stalked; leaflets inversely egg- to inversely lance-shaped, mostly $3/4$–$2 1/4$ inches long. **Stem** commonly purplish, covered by a whitish waxy bloom.

Flowers numerous, to 1 inch long, in an erect raceme on a long naked stalk, the bracts falling off early; calyx persistent, slightly 2-lipped, the tube bell-shaped and 4-lobed; corolla commonly dark-blotched, the standard slightly shorter than the wings and keel; stamens 10, separate. *Late May–early August*

Fruits oblong-ellipsoid pods, $1/4$–$1 1/2$ inches long, $3/8$–1 inch thick, hard and smooth, with an abrupt slender beak. *Late July–early October*

WI range and habitats Rare to locally frequent south and central, predominantly on deep-soil, wet-medium prairies and moist sand prairies, rarely in dry prairies, open woods, shores, edges of bottomland fields, swales, and sloughs; generally on deeper, moister soils than cream wild indigo. *Curtis Prairie, Greene Prairie, Juniper Knoll, Marion Dunn Prairie, Sinaiko Overlook Prairie, Visitor Center parking lot, Wingra Oak Savanna*

Opening a ripe seed pod will often reveal weevils, which were deposited as eggs when the pod was young, and as larvae had eaten and thus destroyed many if not all of the seeds.

This species has a symbiotic relationship with bacteria that live in nodules on its roots. These bacteria "fix" or reduce atmospheric nitrogen to ammonium, a form that can be used by cells. Virtually all living organisms are dependent on nitrogen fixation, just as they are ultimately on photosynthesis.

Baptisia comes from the Greek *baptizein*, to dye, because the sap contains a blue pigment once used as a substitute for true indigo dye. True indigo, *Indigofera tinctoria*, is also a legume.

Fabaceae • Pea or Bean Family

Cream Wild Indigo
Long-bracted wild indigo,
plains wild indigo
Baptisia bracteata
Synonym *Baptisia leucophaea*

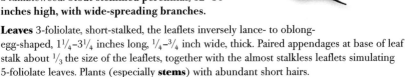

Easily recognized by its low stature, compact habit, and lush racemes of large creamy flowers, this *Baptisia* creates a stunning effect in the spring and early summer prior to growth of warm-season grasses. The plant forms a low "bush" wider than tall that dries up late in the year and becomes a tumbleweed. **Stout-stemmed perennial, 12–30 inches high, with wide-spreading branches.**

Leaves 3-foliolate, short-stalked, the leaflets inversely lance- to oblong-egg-shaped, $1\frac{1}{4}$–$3\frac{1}{4}$ inches long, $\frac{1}{4}$–$\frac{3}{4}$ inch wide, thick. Paired appendages at base of leaf stalk about $\frac{1}{3}$ the size of the leaflets, together with the almost stalkless leaflets simulating 5-foliolate leaves. Plants (especially **stems**) with abundant short hairs.

Flowers numerous, about 1 inch long and wide, in 1-sided racemes 5–15 inches long, each subtended by a persistent large bract; calyx tubular, $\frac{1}{4}$–$\frac{3}{8}$ inch long, the 2 upper lobes almost completely fused; corolla pea-like, $\frac{1}{4}$–$\frac{7}{8}$ inch long, consisting of an upright notched standard, two lateral wings, and two lower petals joined to form the boat-shaped keel. *Late April–early July*

Fruits inflated, dull black pods (legumes), asymmetrically ellipsoid, to 2 inches long, ± hairy, tapering into a slender beak; seeds brown to olive-colored. *Mid July–late October*

WI range and habitats Infrequent southwest, characteristically in sandy, dry to medium prairies, dry limy prairies, open oak woods, oak openings, and pine relicts, on bluffs, sandy or gravelly hillsides, and rarely pastures; full sun or light shade. ***Curtis Prairie, Greene Prairie, Juniper Knoll, Sinaiko Overlook Prairie, West Knoll***

The roots of this slow-growing, long-lived plant can go down to 6 feet deep.

Early spring shoots

White Prairie-clover
Dalea candida
Synonym *Petalostemon candidum*

The tiny white flowers of this prairie-clover are compacted in heads like those of true clovers, but the corolla is not butterfly-shaped. Blooming from the bottom up, they create a bright wreath that ascends the dense green spikes as flowering progresses. **Slender, straight-stemmed perennial, 1–3 feet tall, from a long taproot.**

Leaves pinnate-compound, powdery-bluish, small, with semi-persistent but inconspicuous, awl-shaped stipules; leaflets 5–9, narrowly elliptic to inversely lance-shaped, $1/4$–$7/8$ inch long, to $1/4$ inch wide, flat or partly folded, the lower surface dotted with small glands. Entire plant hairless.

Flowers nearly regular, in long-stalked, non-elongating, egg-shaped to cylindrical spikes $1/2$–$2 3/8$ inches long; calyx $1/8$ inch long, 10-veined, the teeth nearly equal, shorter than the tube; standard and 4 smaller petals all very much alike; anthers pale yellow. *Late June–September*

Fruits 1-seeded, non-opening pods enclosed in the persistent calyx, somewhat flattened, with prominent, teardrop-shaped glands that dry dark. *Mid July–early October*

WI range and habitats Infrequent southwest, characteristic of dry to medium prairie remnants, sand prairies, open oak woods, and cedar glades, occasionally along roadsides and railroads in sandy soils. *Curtis Prairie, Grady knolls, Greene Prairie, Sinaiko Overlook Prairie*

White and purple prairie-clovers overlap geographically and ecologically. They have root systems up to 6 feet deep and can survive long droughts. Great Plains and southwestern tribes chewed or ate the sweet roots, and leaves of both were used to brew tea. A tea-like beverage made from the roots is an old herbal remedy for reducing fever, especially fever associated with measles.

Prairie-clovers are close cousins of leadplant, having pinnately compound leaves, terminal spikes or heads of numerous small flowers, corollas lacking the butterfly-shaped structure typical of the Pea Family, and 1–2-seeded fruits. Although there is a standard, the other petals are not differentiated into wings and keel. The cone-like spikes are somewhat larger than the heads of the true clovers (*Trifolium* species).

Purple Prairie-clover
Dalea purpurea
Synonym *Petalostemon purpureum*

A midsummer favorite, the bright, orange-dotted, red-purple to lavender or rose spikes ("flower heads") of purple prairie-clover color the prairie for many weeks. Growing with black-eyed Susans, white prairie-clover and short prairie grasses, it puts on a stunning floral display in Greene Prairie. **Slender perennial, 8–30 inches high, from a prominent crown.**

Leaves pinnately compound, often with condensed bundles of leaves in the axils; leaflets 3–7 (usually 5), narrowly oblong to linear, $1/4$–$5/8$ inch long, $1/16$ inch wide, tightly folded or rolled inward (toward the upper side); lower surface dotted with small dark glands (visible when dried). **Stems** erect or ascending, simple or branched above, ± hairy.

Flowers tiny, in dense, egg-shaped (in bud) to cylindrical (in fruit) spikes to nearly 3 inches tall and about $1/2$ inch thick, these borne on stalks to 6 inches long; calyx $1/8$ inch long with fine silvery hairs pressed flat against the entire surface; corolla not pea-like, one blade broadly elliptic and concave (the standard, probably the only true petal), other 4 oblong (probably modified stamens); stamens 5, united by their filaments, the anthers orange. *July–late September*

Fruits non-opening legumes, 1-seeded, enclosed in the calyx. *Late July–late October*

WI range and habitats Locally frequent southwest, characteristic of medium to dry prairie remnants, often in scrub oak-jack pine woodlands and dolomite or quartzite glades, rarely spreading along roadsides, railroads, and old fields; once common on flat to rolling prairie, now generally seen in prairie openings on bluffs, outcrops, gravelly hillsides, and sand terraces. *Curtis Prairie, Grady knolls, Greene Prairie*

Once widely abundant, this is a typical plant of the plains and prairies, and farther west, open rangeland, foothills, and buttes. Like most members of the Pea or Bean Family, purple prairie-clover can fix nitrogen from the air and trap it in the soil, thus improving soil fertility. The plant is used as a food source by the larvae of some of the blues butterflies.

CANADIAN TICK-TREFOIL
Showy tick-trefoil
Desmodium canadense

The prettiest flowers of any native *Desmodium* belong to Canadian tick-trefoil, our most common species. Tall and somewhat weedy, it can be identified by its large dense terminal racemes, large, red-violet flowers, and bean-like leaves composed of 3 leaflets. **Perennial, 3–6 feet tall.**

Leaves short-stalked, the leaflets elliptic to lance-shaped, the terminal one $1\frac{1}{2}$–$3\frac{1}{4}$ inches long, slightly to evidently hairy below with fine incurved hairs, most copious on main veins.

Flowers $\frac{3}{8}$ inch long; calyx bell-shaped, 5-lobed; corolla pea-like, blue-violet, quickly turning pale, the standard with 2 yellow spots at the base; stamens in 2 sets, 9 fused by their filaments, the uppermost 1 partly separate. *Late June–early September*

Fruits oblong flattened pods, constricted between the seeds along the bottom and thus divided into a series of 1-seeded joints, each $\frac{1}{8}$–$\frac{1}{4}$ inch long, densely short-hairy, the hairs minutely hooked. *Mid July–late October*

WI range and habitats Frequent and widespread except in the Northern Highlands, on dry to wet prairies, meadows, and fields, oak savannas and woodlands, borders of upland woods, thickets, fencerows, and sandy or gravelly, prairie-like sites (hillsides, causeways, dikes, lake shores), often along roadsides and railroads; more shade-tolerant than Illinois tick-trefoil. *Curtis, Greene, Marion Dunn, Sinaiko Overlook prairies, Grady Tract knolls, Juniper Knoll, Visitor Center parking lot*

Owing to their rapid growth, deep roots, and shade-tolerant leaves, the prairie tick-trefoils compete well with native grasses, establishing readily and growing quickly when introduced into restorations. Like most prairie legumes, tick-trefoils add nitrogen to the soil. Canadian tick-trefoil flowers and fruits prolifically, attracting many insects, birds, and small mammals, which eat the nectar, pollen, and seeds

Desmodium pods are composed of a series of 1–several joints, which at maturity easily separate. Anyone who has walked through the woods or prairies in late summer will be familiar with these flat, fuzzy fruit segments, which stick tenaciously to fabric and fur.

Illinois Tick-trefoil
Prairie tick-trefoil
Desmodium illinoense

Although handsome, this tick-trefoil is not particularly showy, and because of the propensity of its "seeds" (fruit segments) to stick like ticks to clothing or hair, some people consider it a nuisance. Its characteristic pea-like flowers, small and slender-stalked, are borne in a long loose raceme at the top of the plant. **Stout perennial, 2–6 feet tall, from a crown surmounting a taproot.**

Leaves 3-foliolate, on stalks 1–1$^{3}/_{4}$ inches long, the leaflets lance-egg-shaped, the terminal one 2–4 inches long, thick-textured, sticky to the touch above, conspicuously network-veined on the lower surface, covered with tiny hooked hairs. **Stem** and inflorescence axis with spreading glandular hairs and also minute hooked hairs.

Flowers frequently appearing somewhat whorled, on stalks $^{1}/_{4}$–$^{5}/_{8}$ inch long, in simple or slightly branched, somewhat glandular-sticky racemes; calyx bell-shaped; corolla pink to whitish, fading purplish, $^{1}/_{4}$–$^{5}/_{16}$ inch long, the standard erect, flaring; stamens in 2 sets, 9 fused by their filaments, the uppermost 1 mostly separate. *July–late August*

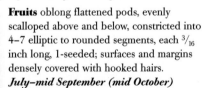

Fruits oblong flattened pods, evenly scalloped above and below, constricted into 4–7 elliptic to rounded segments, each $^{3}/_{16}$ inch long, 1-seeded; surfaces and margins densely covered with hooked hairs. *July–mid September (mid October)*

WI range and habitats Locally frequent southwest, on dry to moist prairies, oak openings and nearby open slopes, ridges, and bluffs, sometimes persisting along roadsides and railroads with mixtures of weeds and prairie species; usually in sandy or gravelly, neutral to acid soil. *Curtis Prairie, Grady knolls, Upper Greene Prairie, Juniper Knoll*

Desmodium is derived from the Greek *desmos,* chain, referring to the connected segments of the pods, which break apart into individual seed cases. Covered with minute hooked bristles, these annoying little freeloaders hitch rides on anything they touch, thus obtaining transportation to potential new habitats.

Marsh Pea
Marsh vetchling, slender-stem pea-vine
Lathyrus palustris

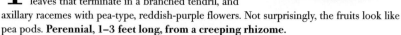

The often 2-winged stems of this sprawling or low-climbing vine bear pinnately compound leaves that terminate in a branched tendril, and axillary racemes with pea-type, reddish-purple flowers. Not surprisingly, the fruits look like pea pods. **Perennial, 1–3 feet long, from a creeping rhizome.**

Leaves with 4–8 leaflets, these elliptic or oblong to narrowly oblong or occasionally lance-linear to linear, $1/2$–$2\,1/2$ inches long. Paired appendages (stipules) at base of leaf stalk conspicuous, each shaped like a half-arrowhead, constricted at the middle of one side into a backward- and an upward-directed lobe. **Stems** angled, narrowly winged or unwinged.

Flowers 2–9 per raceme, $1/2$–$3/4$ inch long, loosely disposed; calyx $3/8$ inch long, the tube oblique, the lowest lobe considerably longer than the others; corolla drying blue to blue-purple, the wings and keel about equaling the standard; stamens 10, 9 united to form a tube open along the top and the 10th lying in this cleft. *(Late May) June–late August*

Fruits linear-oblong, flat, $1\,1/2$–$2\,3/8$ inches long, hairless; seeds 3–8, brown. *July–September*

WI range and habitats Occasional to locally common throughout (rare in the Northern Highlands), in sedge meadows, wet prairies, fens, shrub carrs, bogs, damp grassy edges of woods and swamps, along margins of lakes, streams, and ponds, and in swales and ditches. *Curtis Prairie, Greene Prairie*

The seeds of *Lathyrus* species, including cultivated ones like the sweet pea *(L. odoratus)*, are poisonous. However, the whole pod or peas of some species, including marsh pea and beach pea *(L. maritimus)*, are edible when cooked and were used for food by Native Americans.

Peas, vetches, and vetchlings are superficially very similar to one another. Botanists find it useful to subdivide the species among smaller genera based on technical morphological differences related to the styles and the degree to which the wings cohere with the keel. In any case, the distinctions between our species are clear, and by paying attention to seemingly minor details, the conscientious wildflower lover can soon learn to identify these plants.

Veiny Pea
Forest pea, veiny pea-vine
Lathyrus venosus

The numerous leaflets and flowers and unwinged squarish stems help distinguish this relatively stout species from other sprawling or climbing legumes in our region. The purplish flowers are typical of members of the Pea Family, having a showy upper standard, a pair of lateral wings, and the lower boat-shaped keel. Pods begin forming quickly after the flowers fade. **Perennial, 16–40 inches long, from creeping rhizomes.**

Leaves pinnate-compound, the 8–14 leaflets elliptic to elliptic-egg-shaped, mostly 1–2 inches long, $1/2$–1 inch wide, terminating in an unbranched or branched tendril; veining on the underside prominent. Paired stipules of each leaf half-arrowhead shaped, the 2 lobes triangular to lance- or linear-lance-shaped, untoothed. **Stems** lightly 4-angled. All Wisconsin specimens have a ± uniform covering of minute fine hairs.

Flowers 8–30 per raceme, bluish to red-purple, $1/2$–$7/8$ inch long; calyx obliquely bell-shaped, the upper side much shorter than the lower, $1/4$–$3/8$ inch long; standard differentiated into an inverse-heart-shaped blade and a narrow, stalk-like base; stamens in 2 sets, 9 with united filaments and 1 separate. *Mid May–June*

Fruits linear-oblong legumes $1 1/4$–$2 1/4$ inches long, about $1/4$ inch wide; seeds 3–8, brown. *Mid July–early October*

WI range and habitats Frequent except north-central and northeast, in dry to wet-medium prairies, savannas, and open woods (especially recently cut or burned), as well as a variety of such open habitats as sandy ridges, lightly wooded, rocky slopes and bluffs, talus, and embankments, occasionally in pastures, fencerows, and shores. *Curtis Prairie, Greene Prairie*

This species is variable in hairiness and flower size.

Lathyrus and *Vicia* are closely related genera. All are trailing herbs with weak stems that clamber over the surrounding vegetation with the help of tendrils growing out of the tips of their leaves. Generally, the stems of *Lathyrus* are angled or winged, tendrils are always present (present or absent in *Vicia*), leaflets and flowers are larger than those of most vicias, and additional details of the staminal tube and style.

ROUND-HEADED BUSH-CLOVER
Lespedeza capitata

The subtle, cream-colored flowers of this common prairie and savanna legume are barely noticeable, but the plants are easily recognized by their unbranched rigid stems and congested bur-like clusters of flowers, which turn deep bronzy brown in the fall and stand strongly well into the next growing season. **Perennial, 2–5 feet tall.**

Leaves 3-foliolate, short-stalked to stalkless, the leaflets variable in shape, usually elliptic to oblong, $3/4$–$1 1/2$ inches long, the covering of silky hairs scant to ample. Plants silvery to tawny with soft hairs.

Flowers in rounded, spike- or head-like racemes clustered at the top of the stem; calyx bell-shaped, to $1/2$ inch long, the 5 lance-linear lobes longer than the corolla and fruit; corolla pea-like, the standard longer than the wing and keel petals, with a maroon spot at the base; stamens 10, 9 united into a sheath around the pistil, 1 (the uppermost) free. *(Mid) late July–early October*

Fruits oblong-egg-shaped pods, $1/5$ inch long, papery, hairy, surrounded by the persistent calyx; seeds flat, brown to black, shiny. *September–late October*

WI range and habitats Common and widespread except in the Northern Highlands, characteristic of dry to medium prairies and open woodlands, especially jack pine-scrub oak barrens and juniper glades, less often in oak, pine, and hardwood stands, also on bluffs, cliffs, mounds, and bedrock glades, invading roadsides, railroads, old fields, and shores. *Curtis Prairie, Grady knolls, Greene Prairie, Juniper Knoll, Sinaiko Overlook Prairie*

This wide-ranging, highly variable species is the most common native legume in the north-central states. Like many Pea Family members, round-headed bush-clover enriches the soil through its ability to fix nitrogen.

Bush-clovers can be separated easily from their closest relatives, the tick-trefoils (*Desmodium* species), by their legumes ("pods" reduced to 1 segment, without hooked hairs) and stamens (2 sets, 9 + 1). Their flowers are of two kinds: conspicuous, petal-bearing flowers, and inconspicuous, self-fertilizing flowers with a reduced calyx and no petals. In round-headed bush-clover, the few self-pollinating flowers are the inner ones in the dense clusters and are not easy to see.

Fabaceae • Pea or Bean Family

WILD LUPINE
Sundial lupine
Lupinus perennis

This favorite wildflower features handsome spires of deep lavender, pea-like flowers. The dark green leaves are also attractive, especially when it rains, for the center of the whorl of leaflets captures and holds a bright, jewel-like drop of water. **Perennial, 8–30 inches tall, from a deep taproot, forming patches.**

Leaves palmately compound, with elongate stalks and 6-10 inversely lance-shaped leaflets $5/8$–$1 3/4$ inches long. **Stems** and leaf stalks with sparse long spreading hairs.

Flowers numerous and crowded in an elongate raceme; calyx 2-lipped, the upper lip 2-toothed, the lower lip entire; corolla $1/4$–$1/2$ inch long, at first partly pinkish or whitish, soon becoming shades of blue and blue-violet, the standard bent back at the sides, the 2 wings united toward the apex, the keel strongly incurved; stamens 10, all united by their filaments into a closed sheath. *Mid May–June (sporadically to early August)*

Fruits oblong pods, $1 1/4$–2 inches long, densely stiff-hairy, violently splitting along the 2 edges and throwing the seeds some distance; seeds 4–6 in 2 rows. *July*

WI range and habitats Locally common, in very sandy, open and sunny (sometimes partly shaded) borders and clearings in black oak-Hill's oak-jack pine woods, oak barrens, prairies, old fields, and roadsides, infrequently on sandstone bluffs and ridges. *Curtis Prairie, Grady knolls, Upper Greene Prairie, Sinaiko Overlook Prairie*

Lupine attracts bees, butterflies, and hummingbirds and is the only food plant for the larvae of the endangered Karner blue butterfly. A combination of fire suppression and habitat loss have made both the lupine and the butterfly less common to the extent that this lovely little butterfly is nearly extinct over much of its range. Game birds love the smooth seeds.

The Latin genus name is derived from *lupus*, wolf, because the ancients believed the lupine to be a wolf among the plants, consuming ("wolfing") the nourishing elements of the soil and leaving it barren. This superstition could not have been more wrong, for the roots of lupines bear nodules containing colonies of symbiotic bacteria that replenish nitrogen in the soil by nitrogen fixation.

WHITE SWEET-CLOVER
Melilotus alba

YELLOW SWEET-CLOVER
Melilotus officinalis
Invasive weeds

The vanilla-like fragrance of sweet-clover reminds us of new-mown country hayfields. There are two abundant species, both ubiquitous as part of the roadside and field flora, neither greatly resembling true clovers (*Trifolium* species). They can be easily recognized by the 3-foliolate leaves with finely toothed leaflets, slender racemes of small narrow, pea-like flowers, and short, non-opening pods. Annuals (rarely) or biennials, 1–6 feet tall, from strong taproots.

Leaves compound, the 3 leaflets inversely egg-shaped to elliptic (those of lower leaves) to elliptic or oblong-elliptic (middle leaves), $3/4$–$1 1/4$ inches long, edged with minute sharp teeth. Stipules grown together at the base with the leaf stalk, linear to bristle-like (white sweet-clover) or linear-lance-shaped (yellow sweet-clover).

Flowers numerous, erect in bud, bent downward after flowering, $3/16$ inch long, in spike-like racemes mostly $1 1/2$–$4 1/2$ inches long; calyx bell-shaped; corolla of an upper petal (standard), 2 lateral petals (wings), and 2 fused, boat-shaped lower petals (keel), falling after flowering; stamens in 2 sets, 9 united into a tube and 1 (the uppermost) separate. White sweet-clover has white petals, the standard slightly longer than the wings, whereas yellow sweet-clover has bright yellow petals (paler with age) and the wings and standard of equal length. White sweet-clover: *mid June–July (resprouts flower into October);* yellow sweet-clover: *end of May–early August*

Fruits short, egg-shaped pods, $1/8$ inch long, flattened, leathery, projecting beyond the withered calyx (never curved, coiled, or spiny); seeds 1 or 2, greenish-yellow. Mature pods of white sweet-clover are dark brown to blackish and have a network of raised veins; those of yellow sweet-clover are greenish to tan or light brown and notably cross-wrinkled. *Mid July–October*

WI range and habitats Occasionally cultivated and abundant as weeds, on roadsides, railroads, abandoned fields, old pastures, and degraded prairies, frequently in woods, oak openings, lakeshores, dunes, sand flats, and cliffs; almost anywhere, generally in agricultural and waste areas (e.g., quarries, gravels pits, dredge spoils, around buildings). *Arboretum buildings, Curtis Prairie, Greene Prairie, Juniper Knoll, Marion Dunn Prairie, Visitor Center parking lot, West Knoll, also scattered throughout the Arboretum*

These two species are nearly identical in vegetative characters but are readily recognized by flower color. If only faded flowers are available, they can be separated on the basis of mature pods and relative length of the petals.

Eurasian clovers and sweet-clovers were introduced into temperate regions over the entire world as forage plants, cover crops, and nectar sources for honey. They are also excellent for adding nitrogen to the soil. However, they produce abundant seeds that disperse easily and lie dormant for years; they soon became naturalized weeds over much of the U.S. and southern Canada.

Fabaceae • Pea or Bean Family 191

Prairie-Turnip
Breadroot scurf-pea, pomme-de-prairie, shaggy prairie-turnip
Pediomelum esculentum
Synonym *Psoralea esculenta*
Wisconsin special concern

This delicate-appearing plant has simple or rarely branched stems and long leaf stalks that are conspicuously covered with long spreading white hairs. It is further recognizable by its palmately compound leaves and congested racemes of rather large, pea-like flowers that are whitish in bud, blue to violet in flower, and orangish to straw-color when faded. **Perennial, 4–15 inches tall, from a deep tuberous taproot.**

Leaves palmately compound, basal, the 4–7 leaflets oblong or narrowly inverse-egg-shaped, $3/4$–$2\,3/8$ inches long, the upper surface smooth, the lower surface loosely hairy, not gland-dotted. **Stems** very short to well-developed. By mid or late summer the leaves and stems break off and are blown about by the wind.

Flowers numerous, about $5/8$–$3/4$ inch long, in spike-like racemes arising in lower leaf axils; calyx persistent, rather coarsely hairy; keel blunt, much shorter than the other petals; stamens 10. Bracts leafy, broadly egg-shaped, each subtending 2 or 3 flowers but falling off early. ***June***

Fruits non-opening pods entirely included within calyx remnants, egg-shaped, $5/8$ inch long, papery, terminated by a persistent beak $3/8$–$3/4$ inch long; seed 1, dark brown or purple-spotted, oblong. ***Late July–August***

WI range and habitats Very rare, restricted to relict dry prairies in 7 southwestern and 2 far western counties, on south- to west-facing, gravelly to rocky knolltops, hillsides, and bluffs, very rarely persisting on roadsides; often on limestone. ***Curtis Prairie*** (limestone knoll)

This distinctive legume grows throughout the Great Plains but is rarer in the Southwest and Midwest. Also known as Indian breadroot, its starchy wholesome roots were once a staple food for the plains Native Americans and early travelers. They were eaten raw, cooked, or pounded into flour and made into cakes. The plant also served several medicinal purposes. Roots should never be dug nowadays, for doing so would reduce natural populations of a rare and attractive member of our flora.

Goat's-rue
Rabbit-pea
Tephrosia virginiana

Goat's-rue is unlike any other legume in our flora, combining gray-hairy leaves with bicolored blossoms into a color scheme unique and beautiful. The individual flowers are pea-like and quite cheerful, the standard lemon-yellow outside and whitish within, the wings rose to pink or purplish, the keel yellow- and pink-striped. **Perennial, 8–24 inches high, from a branched woody crown surmounting long tough roots.**

Leaves pinnate-compound with 13–23 elliptic to narrowly oblong leaflets, these $3/8$–$1 1/4$ inches long, with an abrupt small hair-like tip at the apex, silky-hairy on both sides.

Flowers $5/8$–$3/4$ inch long, in compact, pyramid-shaped racemes mostly terminal on the main stems; calyx persistent, broadly bell-shaped, the tube with dense straight soft hairs pressed against the surface; wings and keel converging but not united; stamens 10, joined to form a tube (the uppermost partly free from the others) surrounding the pistil. *June–July (August)*

Fruits linear-oblong pods $1 1/4$–2 inches long, flat, with a covering of dense short matted hairs and long soft shaggy hairs impressed between the dark-spotted seeds. *August–late October*

WI range and habitats Locally frequent southwest, most prevalent in oak barrens, characteristic of scrub oak-jack pine woodlands, sand prairies, old fields, roadsides, railroad cuts, slopes, banks, and sandstone outcrops, occasionally in oak openings or clearings in dry woods, rarely on lake shores; requires very well-drained soil and full sun. *Upper Greene Prairie, West Knoll*

The genus name is derived from *tephros*, Greek for ash-colored or hoary. At night the leaves "sleep" by twisting to lie against the stem. The root contains small quantities of rotenone and related compounds that have cancer-fighting and cancer-causing activity. The Cherokee and other tribes used the roots in shampoo to prevent hair loss, in a tonic given to children to make them strong, and in teas to improve male potency and to treat urinary problems and irregular menstruation. Quail and wild turkey eat the seeds.

AMERICAN VETCH
Vicia americana

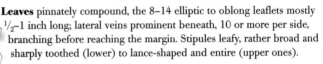

This trailing or low-climbing herb clambers over the surrounding vegetation with the help of simple or forked tendrils growing out of the tip of each pinnately compound leaf. Its pea-like flowers, comparatively few but rather large for the genus, are deep reddish-lavender but become purplish-blue with age. **Slender perennial, 15–40 inches long, from rhizomes, forming colonies.**

Leaves pinnately compound, the 8–14 elliptic to oblong leaflets mostly $1/2$–1 inch long; lateral veins prominent beneath, 10 or more per side, branching before reaching the margin. Stipules leafy, rather broad and sharply toothed (lower) to lance-shaped and entire (upper ones).

Flowers 2–9 or more in relatively short, loose racemes arising in the angle between leaf and stem, each $5/8$–$3/4$ inch long; calyx less than $1/2$ the length of the standard, the tube oblique at the apex; wing petals adherent to the short keel. Stamens 10, 9 united to form a tube open along the upper side, the 10th lying free in this cleft. *Mid May–September*

Fruits oblong flat hairless pods $3/4$–$1 3/8$ inches long, enclosing 1–5 round, olive- to purplish-brown seeds. *July–mid August or later*

WI range and habitats Common throughout in a variety of medium to dry, open to lightly shaded habitats, most often in disturbed deciduous woods, old clearings, degraded prairies, fields, and thickets, also riverbanks, shores, roadsides, railroads, and quarries with mixtures of weeds and woodland or prairie plants. *Greene Prairie*

American vetch is bee-pollinated and self-fertile. As a bee lands, its weight forces open the corolla by depressing the interlocked wings and keel. The stigma, exposed from within the keel, brushes against the bee's body and picks up pollen brought from another flower. The bee then picks up a new pollen load from the flower being visited. The flower springs back to its normal shape when the bee flies off.

Some species of *Vicia* are edible, but others contain compounds that produce hydrocyanic acid and cause cyanide poisoning. Although not poisonous, the small, pea-like seeds of American vetch are generally considered inedible; nevertheless, seeds and pods, both immature and mature, were eaten by Native Americans.

Hairy Vetch
Winter vetch
Vicia villosa

This sprawling vine garnishes roadsides and fields throughout the summer, producing masses of tangled stems. Spike-like racemes of pea-shaped flowers perch jauntily on wayside grasses and weeds. Usually described as purple, the flowers are at first white to pinkish and become light to dark shades of blue-violet as they age. **Annual or biennial, 2–6 feet long.**

Leaves pinnate-compound, each with (10) 14–18, narrowly oblong leaflets to 1 inch long, the terminal one modified into a branched tendril. **Stems,** leaves, and calyces with long white, often tangled hairs.

Flowers mostly 10–30 in dense, 1-sided racemes; calyx tube swollen at base on the upper side, the linear-triangular lobes unequal; corolla $1/2$–$3/4$ inch long, slender, the blade of the standard less than $1/2$ as long as the stalk-like base. *Late May–mid October*

Fruits ellipsoid to oblong pods, $5/8$–$1 5/8$ inches long by $1/4$–$3/8$ inch wide, flattened, obliquely beaked, enclosing small round black seeds. *Mid July–September*

WI range and habitats Frequent to abundant throughout, on roadsides and railroads, dry to damp, grassy fields, hillsides, and embankments, edges of woods, sometimes in prairies, also sand flats, pastures, fencerows, cemeteries, and waste ground. *Curtis Prairie, Sinaiko Overlook Prairie*

The small pod has an explosive mechanism of seed dispersal. As it matures, unequal drying of the cells places a strain on the whole fruit. Further drying increases the tension until the valves suddenly split violently apart, scattering the seeds.

The non-native vetches were introduced as forage and cover crops and attained additional importance in agriculture through their ability to add nitrogen to the soil. Better able to withstand the northern climate than common vetch *(Vicia sativa)* and cow vetch *(V. cracca)*, hairy vetch was once more widely cultivated in Wisconsin than its relatives and has become more thoroughly naturalized.

Vicias should not be confused with the milk-vetches *(Astragalus* species), of which there are hundreds of species in the West and five (four of them rare) in Wisconsin.

OTHER FABACEAE

CROWN-VETCH
Coronilla varia
Introduced, naturalized; loosely ascending perennial forb, 1–2 feet.
Flowers May–September

BIRD'S-FOOT TREFOIL
Bird's-foot deer-vetch
Lotus corniculata
Introduced, naturalized; ecologically invasive, prostrate to nearly erect perennial forb, 6–24 inches.
Flowers June–August

BLACK MEDICK
Medicago lupulina
Introduced, naturalized; prostrate or ascending annual/biennial forb, 1–12 inches.
Flowers May–September

ALSIKE CLOVER
Trifolium hybridum
Introduced, naturalized; erect to sprawling perennial forb, 1–3 feet.
Flowers May–September

RED CLOVER
Trifolium pratense
Introduced, naturalized; ecologically invasive, ascending to nearly erect perennial forb, 1–3 feet.
Flowers May–September

WHITE CLOVER
Trifolium repens
Introduced, naturalized; potentially invasive, creeping perennial forb, 10–30 inches.
Flowers late May–October

Dryopteridaceae • Wood Fern Family

Sensitive Fern
Onoclea sensibilis

Although rank in appearance, this medium-sized fern is easily killed by light frosts. The sterile and fertile leaves are produced singly or several together in a row along a widely creeping, freely rooting rhizome. They are markedly different, readily permitting identification: the sterile is broad, green, and leafy, whereas the less conspicuous fertile leaf does not look like a leaf at all. **Coarse perennial, 1–2½ feet tall, often forming thick colonies.**

Sterile leaves broadly triangular, 5–14 inches long and broad, cleft nearly to the midrib (truly pinnate only at the base) into opposite papery segments; segments 5–11 per side, lance-shaped or lance-oblong, the margins entire, wavy, or coarsely toothed. Veins of the sterile blades branch and rejoin to form a fine network rather than remaining free as in most true ferns. **Fertile leaves** persistent over winter, the blades to 7 inches long, much contracted, twice compound, with modified segments that are tightly inrolled into leathery, berry-like or bead-like bodies arranged in two rows on each branch-like division, the whole forming a very dark green, narrow "panicle" that soon becomes brown and eventually blackish.

Sori completely hidden within modified, berry-like leaf segments. Spores mature *October–November*

WI range and habitats Common almost throughout in damp or wet places: marshes, swamps, thickets along streams and lakes, low woods, muddy banks, seepage slopes, wet meadows, and roadside ditches; neutral to slightly acid soil, in sun or shade but fruiting best in sunny habitats. *Greene Prairie, Gallistel Woods*

The name *onoclea* is Greek and means "to close a vessel," alluding to the margins of the fertile segments. The distinctive fertile leaves appear from mid summer to early fall and last in dried-up condition for 2–3 years. They crack open in early spring (before the sterile leaves expand) to allow the green spores to escape.

Osmundaceae • Royal Fern Family

Royal Fern
Osmunda regalis

This is a large handsome fern with pale green, twice pinnate leaves. The leaves form a tall erect clump, the sterile ones growing in a circle around the fertile ones within, which are divided into sterile and fertile parts. The large, well-spaced divisions give the leaves an airy appearance; at first glance they look something like the leaves of a seed plant. **Perennial, 2–5 feet tall, from the exposed crown of a short stout, semi-erect rhizome.**

Sterile leaves about 15–30 inches long by 10–20 inches wide (occasionally smaller or larger), with 5–8 pairs of primary divisions; each primary division with 14–20 secondary divisions, these oblong-oval to lance-oblong, finely toothed and finely veined. Rhizome covered with masses of old leaf stalk bases and black fibrous roots. **Fertile leaves** similar to the sterile ones; several terminal blade divisions replaced by densely crowded masses of green sporangia, turning yellowish-, then rusty-brown, soon withering. Both kinds of leaves clothed with a woolly covering when young, this falling off as they unfold and leaving them quite smooth.

Sori absent, the relatively large, pear-shaped sporangia borne openly in highly compound clusters, these not associated with any green leafy covering as in many kinds of ferns. Spores mature *May–June*

WI range and habitats Throughout although not particularly common, in bogs, swamps, wet thickets, marshes, shallow pools, and especially along streams; usually in acid soil or on hummocks in non-acid habitats, in both shade and sun. Formerly in wooded wet places in the Arboretum, now all but disappeared owing to the depredations of white-tailed deer. *Greene Prairie*

Similar species Two other species of *Osmunda* occur in Wisconsin. Both cinnamon fern (*O. cinnamomea*) and interrupted fern (*O. claytoniana*) have once-pinnate leaves with the divisions cleft into lobes. The contracted, cinnamon-colored fertile divisions are borne on a separate fertile leaf in the former or produced near the middle between sterile divisions of the fertile blade in interrupted fern.

Dennstaedtiaceae • Hay-scented Fern Family

BRACKEN FERN
Pteridium aquilinum

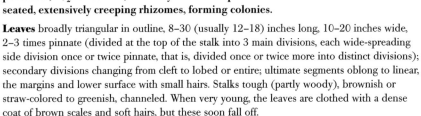

This is one of our commonest and most distinctive ferns. It sends up large solitary compound leaves in a row from elongate forking rhizomes. The pale green, triangular blades are usually divided into 3 parts that are very much alike in size, shape, and cutting. Fertile leaves are similar to the sterile ones. **Coarse perennial, 1½–3 feet tall, from a system of deep-seated, extensively creeping rhizomes, forming colonies.**

Leaves broadly triangular in outline, 8–30 (usually 12–18) inches long, 10–20 inches wide, 2–3 times pinnate (divided at the top of the stalk into 3 main divisions, each wide-spreading side division once or twice pinnate, that is, divided once or twice more into distinct divisions); secondary divisions changing from cleft to lobed or entire; ultimate segments oblong to linear, the margins and lower surface with small hairs. Stalks tough (partly woody), brownish or straw-colored to greenish, channeled. When very young, the leaves are clothed with a dense coat of brown scales and soft hairs, but these soon fall off.

Sori confluent, forming a continuous strip just inside of and covered by the downward-rolled margins of the leaf segments. Spores mature *July–September*

WI range and habitats Abundant throughout in a variety of open or wooded habitats: dry open woodlands, barrens, thickets, burned-over land, abandoned fields, and pastures; prefers dry sterile soil but sometimes growing in damp or loamy soil, ranging from strongly acid to neutral; native but weedy, growing equally well in sun or light shade. *Curtis Prairie, East Knoll, Evjue Pines, Greene Prairie*

The genus name is a diminutive of *pteris,* the classical Greek name for ferns. Somehow "bracken," an Old English name applied to many different kinds of ferns, eventually became attached to this variable and almost worldwide species. It is a common and conspicuous fern in Europe, where it may become such a noxious weed that agronomic conferences have been devoted to discussing how to control it. The very young leaves (crosiers) are cooked and eaten like a vegetable, especially in the Orient, despite evidence that this fern contains mutagenic and carcinogenic substances.

Thelypteridaceae • Marsh Fern Family

Marsh Fern
Thelyptris palustris

Marsh fern is distinguished by its smooth leaf stalk and medium-sized, narrowly lance-shaped blades that are barely narrowed toward the base and taper gradually toward tip. The sterile and fertile leaves are only a little dissimilar. The early leaves are always sterile; the fertile leaves do not appear until midsummer. **Rather delicate, tufted perennial 8–24 inches tall, from long slender blackish rhizomes, growing in colonies.**

Leaves dark (almost bluish-) green, mostly 7–15 inches long and 3–6 inches wide, both the sterile and fertile once pinnately compound into distinct divisions set at right angles to the rachis, each linear-lance-shaped division deeply cut into short blunt, lobe-like segments, ± downy above and below; ultimate segments oblong, blunt, entire; veins few, forked between the midvein and the margins (sterile blades) or partly so (fertile blades). Stalks dark brown to purplish below, straw-colored above, from shorter than to almost twice the length of the blade, without hairs or scales.

Sterile blades with up to 40 pairs of primary divisions, very thin.

Fertile blades slightly contracted, more erect, narrower, and firmer than the sterile ones, with up to 25 pairs of primary divisions; lobes with downward-rolled margins that nearly cover the sori.

Sori small, round, in 2 rows near the midvein on the underside of each lobe. Spores are dark brown or almost black. Spores mature *late summer*

WI range and habitats Locally frequent to abundant in a wide variety of wet habitats: marshes, sedge meadows, wet prairies, fens, and bog-margins, also swamps, thickets, and depressions in deciduous woods, occasionally in ditches; prefers open rather than shady habitats. *Greene Prairie*

CREAM GENTIAN
Gentiana alba
Synonym *Gentiana flavida*
Wisconsin threatened species

The flowers of this rather curious gentian are white or cream-colored with greenish veins. Not only do they lack the exquisite blue color for which the genus is famous, but also they seemingly do not open. As in the case of bottle gentian, however, sturdy bumblebees often struggle halfway in to pollinate them. The leaves are paired and lack stalks. **Relatively robust, hairless perennial, 1–3 feet tall.**

Leaves opposite, each pair arising at right angles to the pair above and the pair below, yellowish-green, egg- to lance-shaped, $1\frac{1}{4}$–4 inches long, 1–2 inches wide, the base somewhat heart-shaped and clasping, the margins smooth.

Flowers stalkless or nearly so, 2–12 in a tight cluster at the top of the stem (sometimes 1 or 2 on stalks from upper axils), subtended by bracts the same size as the leaves; calyx tube $\frac{1}{4}$–$\frac{1}{2}$ inch long, the lobes triangular-egg- to egg-lance-shaped, the clefts between them connected by membranous tissue; corollas tubular, tapering at the apex but ± open, $1\frac{1}{2}$–$1\frac{3}{4}$ inches long,

the 5 lobes broadly egg-shaped, longer than the small rounded folds between them; stamens 5, inserted on the corolla tube, alternate with the lobes. *Flowers early August–late September*

Fruits capsules included or barely projecting from the withered persistent corolla, breaking into 2 parts (valves); seeds numerous, small, flattened, and winged, delicately netted.

WI range and habitats Rare within the limestone region of the southeast, south-central, and southwest, on dry to moist relict prairies along railroads, old unmowed cemeteries, oak openings, edges of oak woods, and on gravelly hills, open wooded ridges, and brushy roadsides. *Greene Prairie, Juniper Knoll*

Beloved throughout the world, especially by botanists and gardeners, gentians are famous for their showy flowers. Ours are essentially all prairie and fen species that are becoming rarer year by year. They should never be picked or transplanted except as a last resort in the face of impending destruction.

Bottle Gentian
Gentiana andrewsii

Modest yet beautiful, this favorite of poets and photographers is a well-known and well-liked plant. Curiously, its deep blue, barrel-shaped blossoms appear to lack an outside opening, yet larger bees and bumblebees at the cost of considerable effort are able to shoulder the front half of their bodies inside (they never enter completely), enticed by the fragrant nectar and translucent white patches at the bottom of the corolla. **Unbranched hairless perennial, 12–31 inches tall.**

Leaves opposite, each pair arising at right angles to the pair above and the pair below, lance- to egg-lance-shaped, mostly 2–3¼ inches long (lower ones as short as ¾ inch), with minutely roughened margins, 3- or 5-veined.

Flowers 3–15 in a dense terminal cluster, sometimes also axillary at upper nodes, blue (occasionally pale blue, pinkish, or rarely white), about 1½ inches long; calyx tube ½ inch long, the lobes lance- to lance-egg-shaped, spreading, the clefts between the lobes connected by membranous tissue; corolla tubular, of 5 united petals, their thin lobes much reduced, about equaling the pale toothed folds between them; stamens 5, inserted on the corolla tube, the anthers united. ***August–mid October***

Fruits capsules, fitting snugly within the persistent calyx and included or barely projecting from the withered persistent corolla, splitting into 2 units (valves); seeds innumerable, small, oblong, broadly winged.

WI range and habitats
Throughout except far north-central, in sedge meadows, wet-medium prairies, streamsides, grassy lakeshores, swales, and damp thickets, sometimes in swampy woods and on roadsides; sun or partial shade. ***Curtis Prairie, Greene Prairie***

Similar species The white-flowered form of bottle gentian can be distinguished from cream gentian (*Gentiana alba*) by its more nearly closed flowers, minutely hairy-fringed calyx lobes, near absence of corolla lobes, and later flowering. Both differ from the gorgeous downy or prairie gentian (*G. puberulenta*) in having hairless stems and the corolla nearly closed during the flowering period.

Downy Gentian
Prairie gentian
Gentiana puberulenta
Synonym *Gentiana puberula*

This gorgeous plant differs from other gentians in having relatively small leaves and funnel- to bell-shaped corollas with open throats and large ascending pointed lobes. The flowers remain closed on dark and rainy days to protect their nectar and delicate sexual parts, opening wide when it is warm and sunny by flattening out the expanded portion of their blue-purple corollas. **Slender perennial, 6–20 inches tall, from a deep root.**

Leaves lance-shaped, $3/4$–2 inches long, $1/8$–$3/8$ inch wide, blunt or rounded at the base, minutely fringed with marginal hairs (under 10✕ magnification). Stems minutely hairy, at least in lines extending downward from the point of leaf insertion.

Flowers 1–5 in an open cluster at the end of the stem (occasionally a few others axillary at the uppermost nodes), stalkless or nearly so; calyx tube $1/4$–$5/8$ inch long, the lobes linear to lance-shaped; corolla $1 1/2$–2 inches long, the 5 lobes broad, bluntly pointed, the folds between them shorter and somewhat toothed. *(Mid August) early September–mid October*

Fruits capsules on stalks to 1 inch long, often projecting beyond the persistent corolla; seeds winged, persisting in some capsules as late as the following April.

WI range and habitats
Characteristic of seasonally damp to dry prairies in the southwest, from very dry, steep rocky prairies to low flat calcareous prairies, rarely in dry upland woods (former oak savannas); rare to locally common.
Greene Prairie (southeast corner)

Blooming activity of the Arboretum's small population has lessened over the last two decades; only one or two plants have been seen every few years. A few plants are probably still present but may be suffering from unusually marshy habitat conditions or may go dormant as litter levels build up during intervals between prescribed burns.

The genus is named for Gentius (d. 167 B.C.), king of an ancient country located on the Adriatic Sea. "Downy gentian," a translation of the species epithet, is something of a misnomer; the plants are essentially hairless.

Gentian roots contain glucosides and small amounts of tannin and were used by European settlers in America as a simple bitter to flavor gin, brandy, and vermouth.

Stiff Gentian
Ague-weed
Gentianella quinquefolia
Synonym *Gentiana quinquefolia*

Slender, much-branched stems and short broad leaves with rounded or clasping bases distinguish stiff gentian. In bloom the clusters of relatively small, short-stalked flowers with the calyx and corolla only slightly cleft are unmistakable. The petals are fused more than two-thirds of their length into a narrowly funnel-shaped tube and lack extra lobes or plaits between the bristle-tipped lobes. **Winter annual or biennial, 4–35 inches tall, from a taproot, usually in large colonies.**

Basal leaves in a rosette $1\frac{1}{4}$–2 inches across, spoon-shaped, withering. **Stem leaves** opposite, each pair arising at right angles to the pair above and the pair below, lance- to egg-shaped, $\frac{3}{8}$–$1\frac{1}{2}$ ($2\frac{3}{8}$) inches long, 5-veined, clasping at the base.

Flowers solitary and in umbel-like clusters of 2–7, borne on short but definite stalks at the ends of the stem and branches and often also in the upper axils, ordinarily purple to pale violet (rarely yellowish-white); sepals united for more than $\frac{1}{2}$ their length, forming a calyx tube to $\frac{3}{16}$ inch long (our plants), without a continuous membrane connecting the spaces between the 5 wholly green, small lobes; corolla $\frac{5}{8}$–1 inch long, lacking folds between the 5 egg- to lance-egg-shaped lobes; stamens 5, inserted deep within the corolla tube alternate with the lobes. *Early September–late October*

Fruits capsules invested by the persistent calyx and withered corolla, breaking into 2 parts (valves) to release numerous light brown seeds, these small, round, and wingless.

WI range and habitats Restricted to the region of limestones (north to Brown, Sauk, and Pierce counties), from dry, medium, and moist prairies to oak openings, edges of oak-hickory woods, and shaded earth banks, also in marshy gravelly places and moist clay bluffs along Lake Michigan; requires calcareous soil and an ecologically open or slightly disturbed microhabitat. *Greene Prairie*

Wisconsin plants belong to the western phase of the species, *Gentianella quinquefolia* subspecies *occidentalis,* an element of the Ozark and Cumberland plateaus, geographically isolated from the Appalachian subspecies *quinquefolia.* The former taxon presumably entered the state directly from the south or southeast following melting of the Wisconsin ice sheets.

FRINGED GENTIAN
Greater fringed gentian
Gentianopsis crinita
Synonym *Gentiana crinita*

Among the last of the parade of autumn wildflowers to bloom are the fringed gentians. Their blue-violet, vase-shaped flowers are distinctive in the quality of the color and the delicate fringe along the margin of each corolla lobe. The combination of relatively broad leaves and longer fringe segments distinguishes the eastern greater fringed gentian from the Great Plains fringed gentian (*Gentianopsis procera*). **Slender annual or biennial, 4–27 inches tall.**

Leaves opposite, stalkless, clasping, egg- to lance-egg-shaped, $3/8$–$2 1/4$ inches long, 2–4 (6) times longer than broad or over $3/8$ inch wide, or both.

Flowers solitary on stalks $3/4$–5 inches long; calyx tube $3/8$–$3/4$ inch long, 4-angled, the lobes of 2 dissimilar pairs, the outer narrow, the inner wider, all with thin dry, non-green margins; corolla of 4 united petals, without folds between the spreading lobes, the hair-like processes comprising the fringe to $1/8$ inch long; stamens 5, inserted on the corolla tube. *Mid August–October*

Fruits stalked ellipsoid capsules enveloped within the persistent calyx, breaking into 2 parts (valves) to release numerous small seeds, these conspicuously covered with pimple-like projections.

WI range and habitats Rare to locally common, mostly east and south, widely distributed in wet prairies, marshes, and sedge meadows, also swales, gravelly flats, and rock pavements, rarely in damp open woodlands and shaded cliffs. *Greene Prairie*

Now rare due to habitat loss and over picking, fringed gentians are mysterious because of the erratic behavior of populations. One year they may occur in a particular area but in the next year may present themselves in reduced numbers or not at all. They should never be picked (picking prevents seed production) or transplanted (transplants almost invariably die) for any purpose except to try to save them from impending destruction.

Similar species The leaves of Great Plains fringed gentian (*Gentianopsis procera*) are narrowly oblong to narrowly linear (5–20 times longer than broad), and the corolla lobes are merely toothed (teeth usually less than $1/16$ inch long) around the broad apex. Although listed in the past as being in the Arboretum, its occurrence has never been confirmed.

West Grady Knoll trail, springtime

WILD GERANIUM
Crane's-bill, spotted geranium
Geranium maculatum

This familiar species blooms about the time spring passes into summer, its rose-purple (rarely white) blossoms shining among the new green growth of the woodlands and woods edges. **Perennial, 8–24 inches tall, from thick branched rhizomes reminiscent of those of wild ginger.**

Basal leaves long-stalked, 4–5 inches long and about as wide. **Stem leaves** only 1 pair per plant, palmately parted into 3–7 (usually 5) divisions, each of which is again sharply and irregularly cut and toothed. Rhizomes show the remains of stems of previous years.

Flowers 2–10 in a loose, flat-topped cluster, 1–1½ inches wide; sepals 5, free, with a slender sharp tip; petals 5, free, ⅝–1 inch long, strongly veined; stamens 10 in 2 sets; ovary 5-chambered, the style 5-cleft. *May–mid June*

Fruits schizocarps. At maturity, the fertile portion of each carpel body opens, exposing the single seed within. Under increasing tension as it dries, the fruit then springs open lengthwise from the base into 5 slender parts (mericarps). Eventually the persistent style splits away from the elongate central column, then curls violently upward and outward, catapulting the seed into the surrounding habitat. *Early June–July*

WI range and habitats Very common throughout except far north, in dry to wet forests, oak, aspen, and pine woods, and thickets, occasionally in medium prairies and borders of sedge meadows; somewhat weedy. *Woodlands and savannas throughout the Arboretum*

The name *Geranium* comes from the Greek *geranos*, crane, and refers to the prominent, beak-like column of styles that characterizes the fruits of this species and its relatives. The family is best known for the cultivated "geraniums," members of the South African genus *Pelargonium,* which are related, but not too closely, to the native wild geraniums of Europe, Asia, North America, and elsewhere.

Similar species Wild geranium is superficially similar to the other Geranium Family members in the state. Its few large leaves and large flowers will easily distinguish it from herb-Robert (*Geranium robertianum*) and Bicknell's crane's-bill (*G. bicknellii*), neither of which occurs in the Arboretum.

Kalm's St. John's-wort
Hypericum kalmianum

This beautiful bushy shrub produces a profusion of bright golden flowers and somewhat crowded, leathery leaves. When in fruit, it displays clusters of conspicuous brown capsules having the form of small bronze candelabras. **Small, semi-evergreen shrub, 1–2 (3) feet high.**

Leaves opposite, lance- to inverse-lance-shaped, mostly $3/4$–$1 1/2$ inches long and $1/8$–$1/4$ inch wide, firm, the margins sometimes rolled under, often with bundles of smaller leaves in their axils, containing glands that make translucent spots. **Bark** whitish, papery; branches 4-angled, twigs 2-edged.

Flowers $3/4$–$1 1/2$ inches across, 3–7 (rarely 10) in open clusters at ends of the stems; sepals 5, (some, at least) leafy, $1/4$–$1/2$ inch long, persistent; petals 5, without black dots, $1/4$–$1/2$ inch long; stamens numerous, not joined, falling off soon after flowering; ovary with 1 chamber and usually 5 styles. *Late June–early October.*

Fruits egg-shaped capsules $1/4$–$3/8$ inch long, beaked by the persistent style base; seeds light brown, oblong. *Early July–October*

WI range and habitats Locally common central and east, its distribution associated with the old beds and outwash plains of major glacial lakes, in fens, calcareous low prairies, sphagnum-sedge meadows, swale borders, and interdunal pools and rocky shores along Lake Michigan and Green Bay; moist rich calcareous soil and full sun. *Greene Prairie*

Kalm's St. John's-wort is increasingly being used as a landscape plant. It was discovered by Pehr Kalm at Niagara Falls in 1749 and subsequently named in his honor by Linnaeus. Its history is of particular interest, because all its stations are in territory covered by continental ice sheets during the Wisconsin glaciation. This distribution suggests to F. H. Utech and H. H. Iltis "...either a pre-glacial origin with subsequent survival...between differentially advancing glacial lobes, or a recent, post-glacial origin from a more widespread southern species. The last hypothesis seems to us the most reasonable."

Twelve *Hypericum* species have been reported for our area. Their identification can be difficult, because some of them are variable and form hybridizing complexes of intergrading plants.

Common St. John's-wort
Klamath-weed
Hypericum perforatum
Invasive weed

St. John's-worts always have pretty, bright yellow flowers and leaves marked with numerous translucent oil glands. Common St. John's-wort is conspicuous during the summer months, producing a profusion of flowers on ± 2-edged branches. Its leaves and occasionally the petals are conspicuously black-dotted, and the plant stays in flower a long time, attracting a number of butterflies. **Tough-stemmed, leafy perennial, 12–30 inches tall, from a somewhat woody crown and branching taproot, producing leafy basal offshoots.**

Leaves opposite, stalkless, elliptic or oblong, mostly $1/2$–$1 1/2$ inches long, averaging less than $3/8$ inch wide, entire, hairless, with oblong translucent oil glands (plainly seen if leaves are held up against the light). **Stems** slender, much-branched above and often at the base.

Flowers numerous, $5/8$–1 inch across, in open leafy clusters at the branch tips, the clusters collectively rounded to somewhat flat-topped; sepals 5, persistent, equal, lance-shaped, $1/4$ inch long; petals 5, twice the length of the sepals; stamens 45–60, weakly united basally into 3 clusters, long, showy, persistent. *Early June–early September*

Fruits oblong-conic capsules about $1/4$ inch long, with 3 persistent separate styles; seeds numerous, dark brown, narrowly oblong, the shiny coat roughened with a net-like pattern. *Late June–early October*

WI range and habitats Abundant throughout, in old fields, degraded prairies, pastures, sand barrens, beaches, roadsides, railroads, and waste ground; mostly in the open on poor or worn soils, growing with the likes of common ragweed, knapweed, Queen Anne's-lace, ox-eye daisy, and mullein. *Curtis Prairie, Grady Tract knolls, Greene Prairie, Juniper Knoll, Marion Dunn Prairie, Sinaiko Overlook Prairie*

Naturalized from Eurasia, this noxious, world-wide weed is now the most common St. John's-wort in North America. Common St. John's-wort blooms in England on or near June 24, St. John's Eve, and is the species that has given the name St. John's-wort to the entire genus. In medieval England, plants were hung in windows and over doors to avert evil spirits and ward off thunder. The flowering tops have been used in folk medicine since early times and were believed to be efficacious in the treatment of many ailments. St. John's-wort is being used at the present time to relieve depression.

OTHER HYPERICACEAE

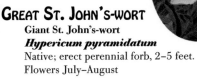

GREAT ST. JOHN'S-WORT
Giant St. John's-wort
Hypericum pyramidatum
Native; erect perennial forb, 2–5 feet.
Flowers July–August

young fruits

Southern Blue Flag
Shreve's iris
Iris virginica
Poisonous

The gorgeous flowers of *Iris* are well known to everyone, as are the slightly fleshy, sword-shaped, flattened leaves. The arrangement of the leaves, appearing basal but actually alternate and sheathing the stem, each folded so as to embrace the next younger one, also characterizes other genera of the family. **Stout perennial, 20–40 inches tall, from a ± tuberous, horizontal branched rhizome, forming colonies.**

Leaves mainly basal and in a fan, those on the stem similar but few and reduced upward, gray-green, 16–21 inches long, $3/8$–$1 1/4$ inch wide.

Flowers 1–several, partly enclosed by sheathing bracts, pale to deep bluish; sepals ("falls") 3, bent downward, with an inverse-egg-shaped blade over 1 inch wide, variegated with yellow, green, or white and marked with purple veins, their midribs ornamented with a bright yellow, hairy blotch at the base; petals ("standards") 3, erect, paler and narrower.
Late May–July

Fruits leathery, obscurely 3-angled capsules 3–4 inches long; seeds in 2 rows in each chamber, tan, round to irregularly D-shaped, about $1/4$ inch wide, with a brittle corky coat.
(Mid July) early August–October

WI range and habitats South, suddenly declining in frequency northward, common in marshes, sedge meadows, low prairies, fens, swamps, bottomland forests, stream margins, lakeshores, ponds, and ditches.
East Greene Prairie

Iris species are poisonous, especially the rhizomes, and should never be eaten. In the past herbalists made a drug from the rhizome and roots to treat such maladies as heartburn, liver and gall bladder problems, syphilis, and other diseases. French naturalist Joseph de Tournefort (1656–1707) named the genus after the mythological Iris, messenger between humans and gods on Mount Olympus. Trailing a rainbow wherever she went, Iris created bridges of color between heaven and earth. King Louis VII of France adopted the iris as an heraldic emblem known as the fleur de Louis, which soon became corrupted to fleur-de-luce and fleur-de-lys.

Similar species This species is readily confused with the more northeastern, barely differentiated northern blue flag (*Iris versicolor*), distinguished from southern blue flag in having the base of the tufted leaves purplish (instead of brown), the spathe shorter and papery (rather than firm), and the relatively thin seed coat more finely and regularly pitted.

Iridaceae • Iris Family

PRAIRIE BLUE-EYED-GRASS
Sisyrinchium campestre

Although superficially grasslike, their 2-edged stems and 2-ranked leaves, each folded upon itself like those of *Iris*, and the small flowers with true sepals and petals and an inferior ovary that matures into a capsule, readily identify members of this genus as belonging to the family of irises and not grasses. The flattened leafless stems are very similar to the leaves, and the delicate, star-shaped flowers soon fade and disappear. **Low tufted perennial, (4) 8–12 (16) inches tall, from fibrous roots.**

Leaves basal, light to olive green, linear, $1/2$–$2/3$ as long as and the same width ($1/16$ inch) as the stems. **Stems** unbranched, bearing 1 (seldom 2) leafy sheathing structures (spathes) at the top.

Flowers in small, umbel-like clusters partly enclosed in a large sheathing spathe of 2 unequal bracts, bluish-purple with a yellow "eye" or white; tepals (3 petal-like sepals and 3 petals) widely spreading, shallowly notched and abruptly pointed; stamens 3, their filaments united; ovary inferior (that is, appearing to be below the point of attachment of the sepals and petals). *Early May–June*

Fruits somewhat spherical capsules $1/8$ inch high, on spreading stalks projecting just beyond the spathe; seeds black. *Late May–mid July*

WI range and habitats Common to abundant southwest, on dry to dry-medium prairies, less frequent on medium and wet-medium prairies, also bluffs, sand plains, open or wooded hillsides, sandy fields, roadsides, railroads, and quarries. *Curtis Prairie, Greene Prairie, Juniper Knoll, West Knoll*

Similar species Six other species of blue-eyed-grass occur in Wisconsin, but besides prairie blue-eyed-grass, only common blue-eyed-grass (*Sisyrinchium albidum*) is frequent in prairies. *Sisyrinchium* is a taxonomically difficult genus, and concepts of the species differ among botanists. Characteristics are highly variable and may overlap greatly among species.

Path Rush
Poverty rush, roadside rush
Juncus tenuis

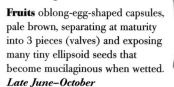

This widespread variable weed is native throughout North America and introduced worldwide. It has decidedly grasslike leaves—long, narrow, and parallel-veined—and greenish to straw-colored flowers with scale-like parts. Though diminutive, the flowers are almost lily-like in that they are radially symmetrical and have true sepals and petals (tepals). **Densely tufted perennial, 4–18 inches tall.**

Leaves all or almost all basal, $1/16$ or less inch wide, flat or channeled; sheaths with broad thin dry margins produced upward into lance-shaped, very thin tongues (auricles) to $3/16$ inch long, lacking a ligule at the junction with the blade.

Flowers scattered or approximate in a compact or generally open inflorescence, each subtended by 2 small opposite bractlets; inflorescence subtended by 2–4 inconspicuous, leaf-like bracts; tepals 6, greenish, becoming straw-colored, lance-awl-shaped, about $1/8$ inch long, pressed against the fruit; stamens 6. *Mid June–August*

Fruits oblong-egg-shaped capsules, pale brown, separating at maturity into 3 pieces (valves) and exposing many tiny ellipsoid seeds that become mucilaginous when wetted. *Late June–October*

WI range and habitats Common throughout, mostly in trampled grassy habitats, especially paths, vehicle tracks, roadsides, parking areas, campgrounds, waysides, etc., also medium to wet woods, edges of marshes, swamps, and bogs, pastures, fields, shores, and waste places (e.g., sand pits); in damp to wet or dry, often compacted soil. *Curtis Prairie, fields, Greene Prairie, Juniper Knoll*

The true rushes (*Juncus* species, *Luzula* species) are grass- or sedge-like in appearance, but their floral and fruit characters are very different. The small flowers have 2 whorls of 3 tepals each (these may be called sepals and petals, distinguishable by position); they are not associated with scales or glumes, lemmas, and paleas. The fruits are 3- to several- or many-seeded capsules rather than a 1-seeded grain or achene.

Similar species Dudley's rush (*Juncus dudleyi*) and inland rush (*J. interior*) are practically identical except that the auricles of the leaf sheath are rounded and short, cartilaginous and yellowish to orange-brown in the former, papery and pale brown to drab in the latter (see Dudley's rush, following page).

Juncaceae • Rush Family

OTHER JUNCACEAE

DUDLEY'S RUSH
Juncus dudleyi
Native; perennial rush, 6–40 inches.

Juncus dudleyi   *Juncus tenuis*

auricles

Common Water-horehound
American water-horehound
Lycopus americanus

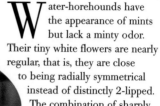

Water-horehounds have the appearance of mints but lack a minty odor. Their tiny white flowers are nearly regular, that is, they are close to being radially symmetrical instead of distinctly 2-lipped. The combination of sharply toothed or even lobed leaves, 4-lobed corollas, and spine-tipped calyx teeth distinguishes American water-horehound from other members of the genus. **Perennials, 4–35 inches tall, with rhizomes and sometimes slender runners.**

Leaves lance- to narrowly egg-shaped or oval, the upper becoming very narrow, ⅝–4 inches long, shallowly to prominently toothed or the lowest often linear-lobed in pinnate fashion, covered with minute translucent pits. Herbage very sparingly hairy to hairless.

Flowers in dense axillary whorls at nearly every node, the floral leaves little reduced; calyx ⅛ inch long, with a bell-shaped tube and 5 narrowly triangular to awl-shaped, rigid teeth distinctly exceeding the nutlets; corolla funnel-shaped, ⅛ inch long, with a pattern of purple dots at the mouth and 4 small spreading lobes; fertile stamens 2, projecting slightly beyond the corolla. *Early July–August*

Fruits 4 brownish nutlets 1/16 inch long, slightly 3-sided, flat on top, with a thickened rim around the margin and a smooth crest (not knobby); top and inner faces dotted with beads of resin. *Mid July–mid October*

WI range and habitats Very common throughout, in any moist to saturated soil: marshes, prairies, fens, bogs, streambanks, lakeshores, springs, and sloughs; prefers full sun. *East Curtis Prairie, Lower Greene Prairie, Sinaiko Overlook Prairie*

Similar species Northern water-horehound or northern bugleweed (*Lycopus uniflorus*) is similar to and often grows with or near common water-horehound. Both northern bugleweed and the less common Virginia water-horehound (*L. virginicus*) have broadly triangular, barely pointed calyx teeth that are shorter than to barely equaling the nutlets. Corollas of northern water-horehound have 5 spreading lobes, and the top of the nutlets is unevenly wavy. Corollas of Virginia water-horehound are 4-lobed, and the crest of the nutlets is covered with tooth-like knobs (use magnification to see these features). The water-horehounds are easily distinguished from wild mint (*Mentha arvensis*), which differs by its aromatic foliage, fuzzy-hairy stems, and whitish to lavender flowers with 4 stamens.

Wild Mint
Field mint
Mentha arvensis

This widespread complex species blooms inconspicuously in damp to wet, open habitats around the globe, the plants perhaps attracting more attention by reason of their aromatic foliage than their small, nearly regular flowers. **Perennial, 7–43 inches high, from spreading rhizomes, forming clones.**

Leaves opposite, short-stalked, narrowly egg-shaped to somewhat angled-elliptic, $3/4$–3 inches long, slightly reduced upward, sharp-toothed, finely hairy to infrequently nearly hairless.

Flowers in dense whorls from middle to upper leaf axils; calyx tubular-bell-shaped, the 5 short teeth nearly equal, hairy throughout; corolla weakly 2-lipped, the upper lip barely notched at the tip, light purple to pink or whitish, funnel-shaped, less than $1/4$ inch long, the short tube not protruding from the calyx, the 4 spreading lobes of about equal length. Stamens 4, almost equal. *End of June–mid September*

Fruits 4 small, yellowish-brownish nutlets in the bottom of the calyx, ellipsoid-egg-shaped, smooth. *Mid August–mid October*

WI range and habitats Common throughout, in sedge meadows, wet prairies, fens, tall shrub thickets, and shallow marshes, on lakeshores and stream banks, also lowland forests, boggy or marshy ponds and pastures, and damp ditches. *East Curtis Prairie, Gardner Marsh, Lower Greene Prairie*

Native peoples used mint tea to treat colds, coughs, fevers, upset stomachs, and many other conditions. Leaves were packed around aching teeth, and cooked plants were added to poultices for aching joints.

Similar species This, Wisconsin's only native mint, is readily confused only with various other species and hybrids of *Mentha*, several of which, such as spearmint, lemon mint, and peppermint, are cultivated for their fragrant volatile oils. These occasionally escape and turn up in wet sites and waste ground. Whereas field mint produces its whorls of flowers in the axils of ordinary leaves separated by internodes of normal length, all other Wisconsin taxa develop whorls that are crowded toward the summit of the stem, forming what appear to be heads or spikes with most of the internodes obscured and the bracteal leaves definitely smaller than the foliage leaves. Most closely resembling field mint is little-leaved mint (*M. ×gracilis*), the hybrid of field mint with spearmint (*M. spicata*). Its leaves are stalkless.

Wild Bergamot
Bee balm
Monarda fistulosa

Although somewhat coarse, wild bergamot is a handsome plant, readily identifiable by its showy round heads of pinkish-lavender (rarely white) flowers. The leaves emit a strong pleasant scent when merely brushed against. **Perennial, 18–46 inches tall, spreading vigorously by slender creeping rhizomes.**

Leaves opposite, short-stalked, slightly ash-colored or gray, triangular-egg- to lance-shaped, 1–3 inches long, rounded or wedge-shaped to straight across at the base, toothed, the lower surface with long soft hairs along the veins, or with only minute curled hairs, or with a mixture of short and longer hairs, or becoming hairless, covered with minute translucent pits.

Flowers in heads, these solitary at the ends of main stems and branches, subtended by leafy, pale green to grayish or pink-tinged outer bracts; sepals united into a tubular, 5-toothed calyx, densely bearded at the throat, glandular; petals 5, fused into a slender tubular corolla $3/4$–$1 1/4$ inches long that opens into a narrow arching upper lip and a down-curved lower lip; stamens 2, protruding from under the upper lip of the corolla. *Late June–early September*

Fruits 1–4 nutlets, yellow-brown to brown, oblong, 3-sided, smooth. *Late September–October*

WI range and habitats Common throughout, on prairies, fields, edges of woods, oak savannas, cedar glades, pine barrens, and thickets, spreading along roadsides, railroads, and waste places; usually in medium to dry, open grassy ground, occasionally in damp muck. *Curtis Prairie, Greene Prairie, Juniper Knoll, Marion Dunn Prairie, Sinaiko Overlook Prairie, Visitor Center parking lot, Wingra Oak Savanna*

This species was used by Native Americans and early colonists to make a mint tea to treat respiratory and digestive ailments. An herbal paste was applied to pimples, cuts and bee stings, and also used as a wash for sore eyes. Its oil is still in use as a flavoring in Earl Grey tea. True bergamot flavor comes from a small citrus tree, the bergamot orange.

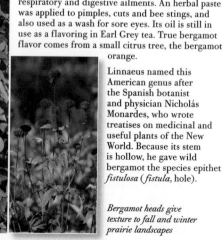

Linnaeus named this American genus after the Spanish botanist and physician Nicholás Monardes, who wrote treatises on medicinal and useful plants of the New World. Because its stem is hollow, he gave wild bergamot the species epithet *fistulosa* (*fistula*, hole).

Bergamot heads give texture to fall and winter prairie landscapes

Dotted Horsemint
Dotted monarda, horsemint, spotted bee balm
Monarda punctata

It is easy to miss seeing the actual flowers of this distinctive mint, partly hidden as they are by conspicuous leafy bracts that are tinged white to lavender or reddish at the base. As with all mints, the flowers are quite pretty when examined closely. The corollas are pale yellow with purple-brown dots and sparsely hairy at the base of the lower lip; the upper lip is helmet-shaped and curved like a sickle. **Perennial, 7–24 inches tall, with a tough forking crown.**

Leaves opposite (alternate pairs at right angles to each other), short-stalked or nearly stalkless, lance-shaped to narrowly elliptic, $3/4$–3 inches long, narrowed at the base, covered with minute translucent pits. Plants with a minty odor when crushed; stems and leaves densely soft-hairy.

Flowers in 1–7 head-like whorls that collectively form an interrupted spike, the bracts spreading or turned slightly downward; calyx tubular, the tiny teeth triangular to lance-shaped, hairy; petals 5, fused to form a long curved, strongly 2-lipped corolla that is slender at the base and abruptly funnel-shaped near the throat. The 2 stamens and 2-cleft style follow the arch of the long curved upper lip. *End of June–mid September*

Fruits 1–4 nutlets (actually schizocarps that separate into halves at maturity), yellow-brown to brown, oblong, 3-sided, smooth. *Late August–mid October*

WI range and habitats Locally common south-central and southwest, very sporadic northward and eastward, on sand prairies, sand barrens, jack pine barrens, oak savannas, sandstone outcrops, and beaches, spreading along roadsides and abandoned fields; open dry sandy soil, mostly in the old bed of Glacial Lake Wisconsin and along the Mississippi tributaries. *West Knoll, Upper Greene Prairie*

The leaves and flowering parts contain the volatile oil of monarda, from which thymol can be derived. Native American groups used the infusion to treat cold symptoms, nausea, and diarrhea, or as a diuretic in urinary disorders. Nowadays herbalists dilute the oil and employ it externally in a liniment to treat chronic rheumatism.

The plant attracts butterflies and hummingbirds and is a favorite nectar source of the endangered Karner blue butterfly.

Heal-all
Self-heal
Prunella vulgaris

Heal-all lacks an aromatic odor when crushed, but its square stem and opposite leaves help identify it as a member of the Mint Family. The short stature and dense spikes of strongly 2-lipped, normally blue-violet flowers make this plant distinctive. **Perennial, 3–24 inches high, from branched crowns or short rhizomes.**

Leaves few, lance-shaped or narrowly elliptic, or the lower sometimes egg-shaped, 1–3 inches long, toothless or nearly so, narrower but not smaller upward, entire. **Stems** weakly 4-sided, usually ascending to erect but sometimes reclining or lying flat on the ground and even rooting at the nodes.

Flowers in clusters that are densely crowded into thick, head-like or oblong spikes $3/4$–2 inches long, usually subtended by a pair of leaves; calyx green or purple, tubular, deeply 2-lipped, with 3 very short upper points and 2 narrow lower teeth, all bristle-tipped; corolla about $1/2$ inch long, the upper petals forming a large, hood-shaped upper lip, the lower ones a short, downwardly bent landing platform (lower lip) for insects. Stamens 4 in 2 sets, shorter than the corolla. Bracts circular-egg-shaped, marginally fringed with hairs. *June–October*

Fruits 4 nutlets, brown or dark brown, ellipsoid, slightly flattened, with a whitish appendage near the point of attachment, enclosed within the calyx. *Mid July–mid October*

WI range and habitats Nearly ubiquitous, in dry to damp woodlands and forests of all kinds (especially along borders, trails, and clearings), fields, prairies, pastures, along streams, on outcrops, roadsides, parking lots, campgrounds, and lawns; a weed, thriving in damp or dry, sunny or shaded, disturbed or undisturbed ground. *Throughout the Arboretum*

Heal-all is one of the commonest and most widely distributed weeds in the world. It has been used for medicinal purposes in many cultures, hence its common names, but little evidence of its effectiveness exists. An introduced variant from Europe with broader leaves with rounded bases has found a comfortable home in lawns across North America, having adapted to being mowed. The native strain with mostly narrower longer leaves with tapering bases presumably grows in both disturbed and natural sites.

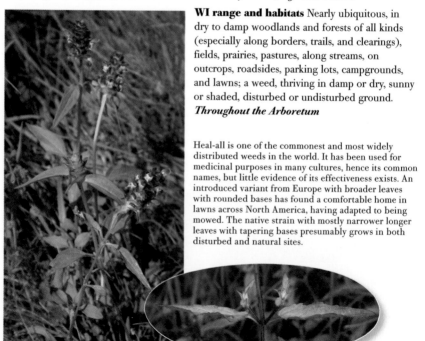

Mountain Mint
Pycnanthemum virginianum

Extremely aromatic mountain mint is familiar to all Midwestern prairie enthusiasts, who know it can be identified simply by crushing and smelling the pungent foliage. The square stem and numerous opposite narrow leaves, combined with the 2-lipped, white or pale lavender flowers, also identify this common prairie plant. **Stiff perennial, 1–3 feet tall, from slender rhizomes.**

Leaves stalkless, lance- to linear-lance-shaped, 1–2¼ inches long, $\frac{1}{16}$–$\frac{3}{8}$ inch wide, with entire but roughened margins and minuscule pits on both surfaces. **Stems** abundantly branched above in flat-topped fashion, hairy on the angles, the sides hairless.

Flowers densely compacted into small head-like clusters ¼–⅜ inch wide, these in turn disposed in a convex-topped arrangement terminating the stems and leafy branchlets; calyx nearly regular, ⅛ inch long, the triangular teeth essentially equal, much shorter than the tube, whitish-hairy; corolla short but much exceeding the calyx, slightly 2-lipped, the tube enlarged slightly upward, the upper lip erect, the lower lip bent downward, 3-lobed, purple-spotted, covered with soft slender hairs and stalkless glands on the outside. Stamens 4 in 2 sets, straight (not curved or inclined toward each other), in different flowers either projecting from or included within the corolla. Flower stalks and flowers subtended by reduced leaves (bracts and bractlets). *Early July–mid September*

Fruits 4 tiny nutlets, brown, oblong, 3-sided, smooth. *Mid August–late September*

WI range and habitats Mostly south, in medium to wet-medium prairies, marsh and bog borders, meadows, and pastures, occasionally in dry prairies or oak-pine woods. *Curtis Prairie, Greene Prairie, Juniper Knoll, Sinaiko Overlook Prairie*

The genus name is derived from the Greek *pycnos*, dense, and *anthemon*, flower, and alludes to the compact inflorescence. The common name is not altogether appropriate; most species occur on the coastal plain or uplands of the eastern U.S. and only a few in the mountains.

The leaves of this mint were used medicinally by Native American groups, either in a tea for the treatment of colds, chills, fevers, indigestion, and colic, or in a poultice in the treatment of headaches.

Leonard's Skullcap
Scutellaria leonardii
Synonyms *Scutellaria parvula* variety *leonardii*, *S. parvula* var. *missouriensis*

The skullcaps are very distinctive, most notably for characters of the calyx, which have a small, helmet-shaped cap or bump on the upper side of the tube and become enlarged and closed after the corolla falls. Size alone helps distinguish Leonard's skullcap. The whole plant is small, purplish, and never glandular. **Bitter-tasting, non-aromatic perennial, 3–10 (14) inches tall, from horizontal underground stems.**

Leaves essentially stalkless, narrowly egg- to oblong-egg-shaped, $1/8$–$3/8$ inch wide, rounded to heart-shaped at the base, slightly thickened, the margins entire or distantly and bluntly toothed, strongly rolled inward. **Stems** 4-sided, very minutely hairy (under magnification) or becoming hairless. Underground stems swollen at regular intervals, producing a bead-like chain of white tuberous thickenings.

Flowers in pairs from upper leaf axils; calyx 2-lipped, $1/16$ inch long in flower, $1/8$ inch long in fruit, the blunt lips entire; corolla bluish-purple, trumpet-shaped, $1/4$ inch long, the tube curving upward, ending in a hooded upper lip and relatively broad, flat lower lip, the lips entire (not lobed). Stamens 4, ascending under the upper lip. *End of May–July*

Fruits 4 tiny, orangish-brown to brown nutlets, somewhat spherical (slightly 3-sided), with a crown of nipple-like pebbling on the back. *Early June–mid September*

WI range and habitats Frequent southwest, in sandy or steep prairies, sandy places in fields, pastures, and river bottoms, also cedar glades, oak openings, jack pine woods, and on bluffs and outcrops. **Curtis Prairie, Juniper Knoll, West Knoll**

The name *Scutellaria* is derived from *scutella,* a small dish, or *scutellum,* a little shield, in either case alluding to the peculiar bump on the upper side of the calyx.

Similar species The Wisconsin-endangered small skullcap (*Scutellaria parvula*) is nearly identical, but its soft hairs are intermixed with either stalked (on the stems) or unstalked (lower leaf surfaces) glands, the leaves are broadly egg-shaped or oval and average wider, and the nutlets are surmounted by a crown of finger-shaped processes. Marsh skullcap (*S. galericulata*) and blue skullcap (*S. lateriflora*) are much larger plants with longer leaves and elongate, blue-violet flowers in axillary or terminal racemes. They grow in wet meadows, swampy thickets, and along shores.

Hedge-Nettle
Woundwort
Stachys palustris

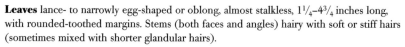

This attractive plant has 4-sided stems, opposite leaves, and distinctly 2-lipped, lavender or rose-purple flowers in prim whorls arranged at regular intervals along the stem. The whorls collectively comprise an interrupted, spike-like inflorescence, to which the genus name, from the Greek *stachys*, spike, refers. **Perennial, 8–40 inches tall, producing rhizomes terminated by whitish tubers.**

Leaves lance- to narrowly egg-shaped or oblong, almost stalkless, $1\frac{1}{4}$–$4\frac{3}{4}$ inches long, with rounded-toothed margins. Stems (both faces and angles) hairy with soft or stiff hairs (sometimes mixed with shorter glandular hairs).

Flowers $\frac{1}{2}$–$\frac{5}{8}$ inch long, 2–8 in each whorl; sepals united into a bell-shaped, downy calyx with 5 narrowly triangular, almost bristle-like teeth; petals fused into a funnel-shaped corolla, mottled with lighter or darker spots, with a straight upper lip barely notched at the tip (representing 2 petals) and a flared, downward-bent, 3-lobed lower lip. Stamens 4, ascending under the upper lip. *Late June–early September*

Fruits 4 nutlets, dark brown or blackish, inverted-egg-shaped, rounded at the summit. *Late July–September*

WI range and habitats Frequent throughout, scattered but mostly south, in wet prairies, fens, marshes, bogs, low pastures, openings in wet woods, edges of thickets, sloughs, streambanks, lakeshores, and ditches. *Greene Prairie*

This is a remarkably complex species of northern Eurasia and most of North America. One standard manual covering the flora of the eastern U.S. divides American specimens into 9 varieties; another recognizes only 3 varieties. Many specimens, however, are intermediate, and geographical differences are obscure, making taxonomic distinctions difficult to apply. The modern tendency is to recognize a single, highly variable North American species or subspecies, perhaps distinct from the European species.

Similar species The less common, narrow-leaved hedge-nettle (*Stachys tenuifolia*) differs in having stems that are hairy only along the angles (glandular hairs absent), leaves on distinct stalks $\frac{3}{16}$–1 inch or more long, and calyces that are quite smooth or bristly on the veins. Both could be mistaken for American germander (*Teucrium canadense*), which is distinguished most readily by its seemingly 1-lipped corolla and felt-like covering of hairs on the leaves and calyces.

Lamiaceae • Mint Family

AMERICAN GERMANDER
Canadian germander, wood sage
Teucrium canadense

Tall stems and opposite leaves, combined with large, reddish-purple, pink, or creamy flowers in a definite terminal raceme, help distinguish American germander. The corolla is particularly distinctive and defines this genus: strongly irregular, appearing 1-lipped because of a deep split along the upper side between two small horns (the lobes of the upper lip), the lower lip conspicuous and spreading. **Rhizomatous perennial, 14–35 inches tall, forming colonies.**

Leaves lance- to lance-egg-, oblong-egg-, or egg-shaped, 1–4 inches long and $5/8$–$1 1/2$ or more inches wide, on definite stalks, the surfaces (our plants) green and sparingly hairy (upper) to white from the covering of dense crooked soft hairs and minute glandular dots (lower).

Flowers in crowded to slightly separated whorls, forming spike-like racemes; calyx bell-shaped, 5-toothed; corolla $3/8$–$3/4$ inch long, the upper lobes appearing lateral and turned forward so that the upper lip seems to be absent, the middle lobe of the lower lip much larger than the 2 tiny lobes on each side; stamens 4, projecting from the cleft between the 2 upper corolla lobes. Calyx and bracts grayish with a felt-like covering of non-glandular hairs or spreading-hairy with non-glandular and gland-tipped hairs mixed. ***Early June–August***

Fruits 1–4 nutlets, golden brown or dark brown, shaped like a quarter of a sphere, almost completely united, network-wrinkled. ***End of July–mid October*** (undoubtedly later)

WI range and habitats Scattered throughout but less common northward, on edges of marshes, thickets, and low woods along rivers, streams, and lakeshores, also wet prairies, fens, and shrub carrs, occasionally in drier habitats, such as dry to medium prairies, oak openings, embankments, dikes, and ditches; moist soils, open and shaded sites. ***Curtis Prairie*** (not reported since 1961), ***McCaffrey Drive opposite Longenecker Gardens***

Similar species
Superficially, American germander resembles the hedge-nettles (*Stachys* species), which, however, have the upper lip of the corolla nearly equaling the lower and the nutlets essentially free (separated by the basal style) instead of almost completely united (style terminal).

Segregate taxa have been recognized in this widespread complex, which is variable in leaf and calyx shape, corolla size, and especially hairiness.

OTHER LAMIACEAE

GIANT HYSSOP
Figwort giant hyssop, purple giant hyssop
Agastache scrophulariaefolia
Native; erect perennial forb, 3–5 feet.
Flowers August–September

CREEPING-CHARLIE
Gill-over-the-ground, ground-ivy
Glechoma hederacea
Introduced, naturalized; ecologically invasive erect perennial forb, 1–8 inches.
Flowers April–June

MOTHERWORT
Lion's-tail
Leonurus cardiaca
Introduced, naturalized; erect perennial forb, 18–60 inches. Fruiting structures are painful burs.
Flowers June–August

CATNIP
Nepeta cataria
Introduced, naturalized; erect perennial forb, 18–40 inches.
Flowers July–August

Wild Onion
Meadow garlic, wild garlic
Allium canadense

Nine true onions and garlics (*Allium* species) occur in our area, 4 of them native, 4 rarely escaped from cultivation, and 1 naturalized. Besides their characteristic odor and taste, onions are distinguished by having leafless stems from a coated bulb and small flowers in terminal umbels. In wild onion some or all of the flowers are typically replaced by bulblets. **Perennial, (6) 8–24 inches tall, from a bulb.**

Leaves 2–6 per bulb, basal, their sheaths on the lower $\frac{1}{4}$ or less of the stem, lustrous green, the narrowly linear blades solid, ± flat, shorter than the stem. **Stems** and leaves die back completely in late June. Outer bulb coats persist as a brownish or grayish, fibrous mesh.

Flowers broadly bell-shaped, $\frac{1}{8}$–$\frac{1}{4}$ inch long, in an erect loose umbel (stalk erect at summit) subtended by a large bract that splits into 3–4 egg- to lance-shaped, membranous segments; tepals (when present) white or pink, just longer than the stamens, withering (not enveloping the capsule). *Beginning of June–early July*

Fruits capsules, broadly inverse-egg-shaped or spherical, bluntly 3-angled; seeds shiny black, triangular, finely honey-combed, with a minute, wart-like protuberance in the middle of each space. Capsules or seeds rarely produced.

WI range and habitats Infrequent to locally common south and central, in low to rich or even dry woods and on streambanks and river shores, weedy on railroad embankments, overgrown quarries, and in damp grassy places, including disturbed prairies and even unkempt yards. *Greene Prairie, Gallistel Woods, Wingra Woods*

The umbel is usually a mixture of bulblets and flowers or completely one or the other. When bulblets are present, the stem may develop several branches from the umbel, each with a smaller umbel; buds and developing bulblets are at first enclosed in a spathe, producing a conical, slender-tipped mass.

The flowers, leaves, and bulbs are edible, raw or cooked.

Nodding Wild Onion
Allium cernuum

This is the most attractive and latest-blooming of Wisconsin's native onions. A true onion, it grows from a bulb and has grasslike foliage with the characteristic odor and taste of onions. Although small, the beautiful, pale pink to deep rose-colored flowers have the same structure as a lily. The tip of the stem is abruptly bent to the side or downward just below the umbel; the flowers are at first nodding but become erect in age. **Perennial, 4–28 inches tall, from clustered oval, membranous-coated bulbs.**

Leaves 3–5 per bulb, shorter than the leafless stem, $1/16$–$1/4$ inch wide, nearly flat to broadly V-shaped or channeled, covered with a whitish waxy bloom.

Flowers bell-shaped, several to many in a nodding umbrella-shaped cluster (umbel), on stalks to 1 inch long, the broad membranous bract splitting into 2 lance-shaped segments that fall at flowering time; tepals pink or white, $1/4$ inch long; stamens 6, projecting beyond the tepals. *Mid July–early September*

Fruits 3-lobed capsules $1/8$ inch long, crested with 6 flattened triangular, entire or toothed processes; seeds dull black, otherwise like those of wild onion. *End of August–mid October*

WI range and habitats
Locally abundant in calcareous, wet to medium prairie and savanna remnants or on grassy wooded banks in the southeastern most corner, rare over dolomite northeast, and in maple-basswood forests or along railroad embankments and roadside ditches southwest. *Greene Prairie*

The nodding umbel distinguishes this species and gives it both its common name and species epithet. The bulbs were eaten sparingly by a few indigenous groups in the West.

Similar species Nodding onion is closely related to prairie onion (*Allium stellatum*). The umbel in both species is nodding in bud but becomes erect by flowering time in prairie onion, whereas it remains permanently recurved in nodding onion. Prairie onion differs further in having egg-shaped bulbs and star-shaped flowers. It is local in northwestern and central Wisconsin, and although rarely adventive farther south, it is not known from the Arboretum.

WILD-HYACINTH
Eastern camas
Camassia scilloides
Wisconsin endangered

One of the handsomest plants of this family in our region is the elusive wild-hyacinth, easily distinguished from other spring wildflowers by its simple leafless stems, arising from the exact center of a rosette of narrow, grasslike leaves, and delicate, sweet-scented, sky-blue to whitish flowers in long loose racemes. It produces an outstanding display when in bloom. **Fairly stout perennial 8–27 inches tall (excluding raceme), from a deep-seated, scaly-coated bulb.**

Leaves 3–8 in a basal tuft, linear, 8–16 inches long and about $1/2$ inch wide, shorter than the stem. Base of plant wrapped in membranous sheathing formed from an extension of the bulb coats.

Flowers somewhat flat and circular in outline, many in cylindrical racemes 4–19 (mostly 7–12) inches long, subtended by narrowly lance-shaped bracts, on slender stalks $3/8$–$3/4$ inch long; tepals 6, in 2 whorls of 3, lance-shaped, free to the base, $3/8$–$5/8$ inch long, 3- or 5-veined near the center, withering after blooming but not falling; stamens 6, the anthers bright yellow, short; ovary superior, with 1 slender style and a 3-cleft stigma. *Mid May–mid June*

Fruits triangular-spherical capsules $1/4$–$3/8$ inch long; seeds several, black and shining, roundish, angled. *Late June–mid July*

WI range and habitats Rare and local, restricted to medium and wet-medium prairie relicts, barely surviving along grassy woodland borders, railroads, and roadsides; prefers half-sun, half-shade areas. This plant will survive in Wisconsin only through the deliberate protection and management of its few surviving stations. *Curtis Prairie*

In past decades this species was locally abundant on Curtis Prairie, after which it seemed to have died out, only to reappear about 10 years ago. Although it has not been seen in recent years, wildflower enthusiasts are hopeful that it will bloom again soon.

Although hardy and easy to grow, our eastern camas is not common in cultivation. Farther east its usual habitats are moist open woods and thickets on calcareous slopes and floodplains, but in our region its two main habitats are savannas and prairies. *Camassia* bulbs were an important staple food for Native Americans, particularly in the Pacific Northwest. They were harvested during or soon after flowering so as not to be confused with death camas (*Zigadenus venenosus*), the bulbs of which are very similar and highly toxic. Bulbs of eastern camas should never be collected in our area, because the plant is so rare and the bulbs so similar to those of white camas (*Z. elegans*).

Yellow Star-grass
Common gold-star, common star-grass
Hypoxis hirsuta

The yellow star-grasses are an interesting group of modest little plants. In Wisconsin we have but a single species of these miniature cousins of the amaryllis. Easily identified by its hairy, grasslike foliage, leafless stems, and 1- to few-flowered inflorescences of bright yellow, star-shaped flowers, it adds interest and variety to prairies and prairie-like habitats. The rosettes of leaves are often overlooked unless the plants are in flower. Usually only 1 flower blooms at a time. **Small perennial, 2–9 inches tall (13 inches in fruit), from a short, bulb-like rhizome (corm).**

Leaves basal with sheathing bases, erect to ascending, 1–23 (mostly 2–18) inches long, about $1/8$–$1/4$ inch wide, soft, flexible, somewhat sparsely hairy. **Underground stem** covered with membranous pale to brown sheaths that do not become fibrous.

Flowers $3/8$–$3/4$ inch across, 1 or 2–7 together in a loose umbel on a stiff, ascending to reclining, thread-like stalk; tepals 6, united below into a tube that is fused to the ovary (which is situated below the attachment of the other floral parts, i.e., inferior), greenish and soft-hairy on the outer surface. When open, the floral envelopes (calyx and corolla) are flat and star-shaped; after flowering, they persist and assume an erect position. *Mid May–late June*

Fruits 3-chambered, ellipsoid capsules that remain closed at maturity; seeds several, black, shiny, the outer coat densely roughened with short sharp projections (use magnification to see this feature). *Late June–July*

WI range and habitats Common across the south in a variety of habitats: dry to wet prairies, on sandy, gravelly, or rocky hillsides, outcrops, and bluff tops, occasionally in old pastures, disturbed sedge meadows, and swamp borders. *Curtis Prairie, Greene Prairie*

Except for the inferior ovary, members of this genus are lily-like in structure. Fire apparently acts as a flowering stimulus for yellow star-grasses; blooming activity seems to be most vigorous 1 to 2 months after their habitat is burned and lessens during subsequent years.

Michigan Lily
Turk's-cap lily
Lilium michiganense

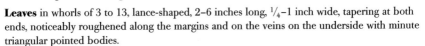

Plants of the familiar genus *Lilium* are mostly tall and leafy with large flowers that clearly display the numerical pattern of the family: 6 tepals in 2 sets of 3 but often all colored alike, 6 stamens, and a 3-chambered ovary that matures into a 3-sided capsule. Michigan lily has several whorls of leaves on the stem and large nodding, red or red-orange flowers spotted with maroon within. **Simple-stemmed perennial, 2–6 feet tall, from yellowish scaly bulbs.**

Leaves in whorls of 3 to 13, lance-shaped, 2–6 inches long, $1/4$–1 inch wide, tapering at both ends, noticeably roughened along the margins and on the veins on the underside with minute triangular pointed bodies.

Flowers 1–5 (or more) at the summit of the stem, bell-shaped; tepals 6, separate, lance-shaped, $2–3\frac{1}{2}$ inches long, strongly curved backwards, not basally tapered into stalks, falling off after flowering; stamens prolonged beyond the tepals, the oblong anthers attached near the middle to the apex of and capable of swinging freely on the long arching filament. *(Mid) late June–early (late) August*

Fruits oblong capsules 1–2 inches long; seeds densely packed in 2 rows in each chamber, roughly circular, flat. As the capsule matures, the stalk becomes erect. *Mid August–September*

WI range and habitats Occasional to locally common throughout, in medium to wet prairies, sedge meadows, fens, borders of and openings in medium hardwood stands and swamp forests, wet thickets, and grassy marshy places such as streamsides, ditches, and fields. *Curtis Prairie, Greene Prairie, Wingra Oak Savanna*

Both of our native lilies are somewhat local and may occur singly or in patches. They are pollinated primarily by swallowtail butterflies, although other butterflies and occasionally hummingbirds visit the flowers. Michigan lily was once common along rural roads and in wet fields and ditches but has decreased greatly as a result of mowing and ditching.

Similar species Michigan lily has been confused with turk's-cap lily (*Lilium superbum*, for which the common name turk's-cap lily is best reserved) and the even more closely related Canada lily (*L. canadense*), neither of which occurs naturally in Wisconsin. The naturalized daylily (*Hemerocallis fulva*) differs from these species in having the leaves basal and tepals partially united into a tube, and by spreading from rhizomes.

Wood Lily
Orange-cup lily, prairie lily
Lilium philadelphicum

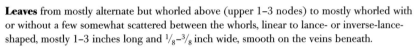

The wood lily of the eastern states is one of our most spectacularly showy plants, its brilliantly colored flowers running the gamut from lemon yellow (rarely) through orange to scarlet. It is further distinguished by its whorled leaves, upright, cup-shaped flowers, and tepals that taper into narrow, stalk-like bases. **Perennial, 1–3 feet tall, from scaly bulbs.**

Leaves from mostly alternate but whorled above (upper 1–3 nodes) to mostly whorled with or without a few somewhat scattered between the whorls, linear to lance- or inverse-lance-shaped, mostly 1–3 inches long and $1/8$–$3/8$ inch wide, smooth on the veins beneath.

Flowers 1–3 (6) per umbel, erect, broadly bell-shaped, 2–3 inches high; tepals 6, orange to scarlet or deep red-orange marked with purple-brown spots toward the base within, tapering at the base into slender stalks (claws), with slightly spreading (not recurved) tips; stamens projecting beyond the tepals, the anthers $3/8$ inch long, on barely spreading filaments 1–2 inches long. *Late June–early August*

Fruits capsules, oblong, oval, or somewhat inverted-egg-shaped, 1–3 inches long, with numerous closely packed, flat seeds. *September–October*

WI range and habitats Absent from the Northern Highlands, infrequent elsewhere in dry, sandy or rocky woods and thickets, pine barrens, dry to moist prairies and swales, wooded bluffs and dunes along Lake Michigan, and grassy rights-of-way and clearings. *Curtis Prairie, Greene Prairie*

Wood lily is continuing to decline in the northeastern part of its range as prairies deteriorate or disappear and white-tailed deer increase in number. Native Americans used the bulbs in many ways, not only in their medicine and ceremonies but also as an emergency food. Nowadays, bulbs should never be collected; doing so only contributes to the decline of the species. See Michigan lily *(Lilium michiganense).*

Death Camas
White camas
Zigadenus elegans
Synonyms *Anticlea elegans, Zigadenus glaucus*
Poisonous

In bloom the creamy-white flowers are unmistakable, having a large green gland near the base of each tepal on the inner side. The somewhat leafless appearance of the pale bluish, hairless stems helps distinguish this species even when it is not in bloom. **Erect perennial, 4–28 inches tall, from a bulb covered with sheathing leaf bases.**

Basal leaves 1–3. **Stem leaves** 1–3, alternate but crowded toward base and rapidly becoming smaller upward, linear to linear-lance-shaped, 4–14 inches long and $3/8$ inch wide. Plants blue-green, evidently whitened with a bloom (like a cabbage leaf), completely hairless.

Flowers 10–50 in a slender loose panicle with ascending branches (seldom merely a raceme), each on an erect long stalk subtended by a bract, bisexual, wheel-shaped; tepals 6, oval to inverse-egg-shaped, $1/4$–$1/2$ inch long, attached to the middle of the ovary, each with a green, inversely heart-shaped gland below the middle; ovary partly inferior, contracted into 3 separate tapering styles. *Late May–July*

Fruits capsules, 3-lobed, 3-chambered, egg- or cone-shaped, at least $3/8$ inch long, separating to the base, then each section opening, subtended by the withered but persistent floral envelopes; seeds inverse-lance-shaped, angular, $3/16$ inch long. *Mid August–early September*

WI range and habitats Occasional southwest, south, and east to the tip of Rock Island, Door Peninsula, in the region of limestones, in oak openings, dry to dry-medium prairies, bluffs, cliffs, outcrops, and dunes. *Greene Prairie*

The name *Zigadenus* is derived from the Greek *zygos,* yoke, and *aden,* gland, and alludes to the glands on the tepals, which are occasionally in pairs. In the inclusive sense (it is sometimes divided into several genera) the genus contains about 15 species, only one of which occurs in Wisconsin. All parts of the plants, fresh or dried, contain toxic alkaloids, including the violently poisonous zygadenine, which causes slowed heartbeat, lowered body temperature, muscular weakness, stomach upset, vomiting, and diarrhea. Ingestion by humans may result in severe illness and occasionally even death. Nonetheless, *Zigadenus* was used in Native American medicine as an emetic as well as a poultice.

Liliaceae • Lily Family 233

OTHER LILIACEAE

ASPARAGUS
Common asparagus, garden asparagus
Asparagus officinalis
Introduced, naturalized; erect perennial forb, 2–7 feet.
Flowers May–June

SOLOMON'S-PLUME
False Solomon's-seal, false spikenard
Maianthemum racemosum
Synonym *Smilacina racemosa*
Native; erect perennial woodland/savanna forb, 16–32 inches.
Flowers May–June

SOLOMON'S-SEAL
Giant Solomon's-seal, smooth Solomon's-seal
Polygonatum biflorum
Native; erect perennial woodland/savanna forb, 1–5 feet.
Flowers May–June

CARRION-FLOWER
Smilax herbacea
Native; climbing annual savanna/wood's-edge vine.
Flowers May–July

Smilax is now often placed in the Smilacaceae, the Cat-brier Family.

Great Blue Lobelia
Lobelia siphilitica

Sometimes called blue cardinal-flower, this showy representative of the lobelias has spike-like racemes of relatively long, blue tubular flowers. The white stripes in the throat are nectar guides and attract bees. An adaptable plant, great blue lobelia can be used for garden naturalizing in sun or partial shade in moist to wet soils. **Perennial by basal offshoots, 1–4 (5) feet tall.**

Leaves alternate, broadly lance-shaped to oblong or inversely lance-shaped, $3/4$–6 inches long, irregularly toothed to almost entire, the upper ones grading into the floral bracts. **Stems** erect, stout, mostly unbranched, somewhat hairy.

Flowers in a leafy-bracted, dense raceme 4–20 inches long; calyx bell-shaped, with leafy basal lobes often $1/8$ or more inch wide; corolla 2-lipped, deep blue (rarely white or pale blue), $5/8$–$1 1/8$ inches long, slit near the base of the tube, the lower lip bent downward. Anthers united into a bluish-gray tube around the brush-like style tip. *(End of June) late July–mid October*

Fruits cup-shaped to spherical capsules $1/4$–$3/8$ inch long and thick, opening from the apex to release minute, chestnut-brown seeds. *September–early October*

WI range and habitats Occasional to common in all but northern counties, in low woods, wet hollows, streamsides, deciduous forests, clearings, marshes, fens, swales, and wet pastures; moist to wet, neutral or somewhat calcareous ground. *Lower Greene Prairie*

Lobelias are noted for the sophisticated irregularity of their flowers, which are strongly bilaterally symmetrical, tubular, and divided into an "upper" lip of 2 lobes and a "lower" one of 3 lobes. They are inverted 180 degrees before expanding by a twisting of the flower stalk so that the 3-lobed true upper lip is on the lower side.

Lobelias are rich in poisonous alkaloids and as a consequence were regarded as important medicinal herbs by Native Americans, who used both great blue lobelia and cardinal-flower as a remedy for syphilis and other diseases. They are still used in homeopathic medicine to treat laryngitis and spasmodic asthma. Most cases of human poisoning result from overdoses of homemade medicinal preparations.

PALE-SPIKE LOBELIA
Spiked lobelia
Lobelia spicata

The more common spiked lobelia also has blue flowers, but they are lighter in color (sometimes nearly white) and less than half as long as in great blue lobelia. In pale spiked lobelia, the corolla tube lacks the lateral slits characteristic of cardinal-flower and great blue lobelia. **Perennial by basal offshoots, 9–36 inches tall.**

Leaves variable, from inversely egg-shaped to elliptic or oblong (lower ones) to mostly inversely lance-shaped to elliptic (middle and upper leaves), and from stalked to nearly or quite stalkless, gradually reduced upward, $1/2$–$2\,3/4$ inches long. **Stems** simple, erect, short-hairy near the base, the hairiness decreasing upward.

Flowers in erect slender, somewhat 1-sided racemes, on stalks shorter than the bracts, $3/8$–$1/2$ inch long; calyx lobes linear-lance- to awl-shaped, often fringed with marginal hairs; corolla tube $1/8$ inch long, lobes of the lower lip slightly bent backward. Stamens united by both their filaments and their anthers, the latter forming a pale bluish-gray (occasionally white) tube $1/16$ inch long, into which the pollen is deposited, only to be pushed out, piston-like, by the emerging style. *Late May–early September*

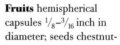

Fruits hemispherical capsules $1/8$–$3/16$ inch in diameter; seeds chestnut-brown. *Late May–early September*

WI range and habitats Occasional to locally frequent south, in dry to wet prairies, thinly wooded bluffs, ridges, and hillsides, clearings, embankments, pastures, and old fields. *Curtis Prairie (limestone knoll), Greene Prairie, Juniper Knoll*

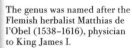

The genus was named after the Flemish herbalist Matthias de l'Obel (1538–1616), physician to King James I.

Similar species Indian-tobacco (*Lobelia inflata*) differs from pale-spike lobelia in having the stem hairy throughout, the floral cup around the ovary nearly as long as the corolla, and the capsule much inflated at maturity.

OTHER LOBELIACEAE

CARDINAL-FLOWER
Lobelia cardinalis
Native; erect perennial forb, 2–5 feet. Prefers wet soil of floodplains, wooded riverbanks, swampy thickets, wet meadows, shores, swales, and ditches. Flowers July–September

WINGED LOOSESTRIFE
Lythrum alatum

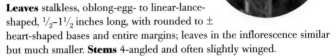

Primarily alternate to opposite or nearly opposite leaves (rarely in 3s) on the same plant help distinguish this innocuous species when not in bloom. The red-violet flowers are utterly distinctive, having a well-developed floral tube and parts in sixes. **Slender perennial, 7–37 inches tall.**

Leaves stalkless, oblong-egg- to linear-lance-shaped, $1/2$–$1 1/2$ inches long, with rounded to ± heart-shaped bases and entire margins; leaves in the inflorescence similar but much smaller. **Stems** 4-angled and often slightly winged.

Flowers solitary or paired in upper leaf axils, on very short stalks, opening a few at a time, somewhat asymmetrical, with 6 tooth-like calyx lobes alternating with accessory, tooth-like appendages and 6 separate petals, all borne on the rim of a cylindrical, calyx-like tube, within which are inserted 6 stamens. Flowers of two forms, having the stamens and style of two lengths (long stamens and short style, or vice versa) on different plants. *Late June–early October*

Fruits capsules, somewhat cylindrical, included within but free from the floral tube; seeds numerous, brown, spindle-shaped. *September–October* (undoubtedly later)

WI range and habitats Chiefly southeast, also Mississippi River bottoms, in alkaline sedge meadows, marshes, wet-medium prairies, fens, and adjoining wet peaty meadows and ditches, also bogs and shores. *Gardner Marsh, Greene Prairie*

Having two types of flowers (three in purple loosestrife), only one type per plant, favors cross- rather than self-pollination. A plant with, say, long stamens can effectively "dust" an insect visitor for transfer of pollen to another plant having a long, intercepting stigma, but also having short stamens that cannot easily consummate self-pollination. This arrangement promotes vigor and induces variability.

Similar species That noxious invader of wet prairies, marshes, shores, and shallow water, Eurasian purple loosestrife (*Lythrum salicaria*), differs from winged loosestrife in its robust habit, larger leaves, and numerous magenta flowers in long dense spikes. Escaped from gardens and thoroughly naturalized, it often grows in large numbers, pushing out native species. A biological control program begun in 1992 is releasing beetles and weevils imported from the plant's native range to feed on its leaves, roots, and seeds. Purple loosestrife occasionally strays into the Arboretum.

Glade Mallow
Napaea dioica
Wisconsin special concern

Glade mallow is easily recognized by its large, maple-leaf-shaped leaves and snow-white, small but numerous flowers in a large branched inflorescence. As in all Malvaceae, the numerous stamens are united by their filaments into a conspicuous cylinder with the anthers in a dense cluster at the upper end. Typically the branched style projects through the mass of stamens (hollyhock, hibiscus), but glade mallow has stamens and pistils in separate flowers on different plants; the male flowers have no pistil and therefore no style. **Robust perennial, 3–8 feet tall, from a thick taproot.**

Leaves palmately 9–11-lobed, 4–12 (basal leaves to 24) inches wide, the lobes again shallowly cut or coarsely toothed; lower surface with straight stiff hairs with or without admixture of star-shaped hairs, in age often becoming hairless.

Flowers in a ± compact panicle at the summit of the stem; sepals 5, united at the base; petals 5, free from each other but fused to the base of the stamen column, $1/4$–$3/8$ inch long. Female flowers with 8–10 ovule-containing organs (carpels) loosely united in a ring around a short column of antherless filaments. Male flowers with 15–20 stamens and no pistil. *Late June–mid August*

Fruits schizocarps with the seed-bearing sections arranged like tangerine wedges, when ripe separating into individual 1-seeded segments; seed reddish-brown, kidney-shaped. *July–early September*

WI range and habitats Rare to locally abundant, in sections of nine southern counties and one far western county (Pierce), in moist borders and openings along streams and rivers, medium to wet prairies, marshy ditches, and moist rank vegetation along rights-of-way; sun or semi-shade; weedy (native). *Curtis Prairie, Visitor Center parking lot, Wingra Oak Savanna*

Glade mallow is one of a kind, the only species contained within its genus. A regional endemic of the north-central U.S., its present-day distribution is restricted very nearly wholly to territory covered by the Wisconsin ice sheet, making its geographic origin and history difficult to explain. Perhaps it survived glaciation in some tiny area beyond the ice margin, from which it has subsequently vanished, then migrated northward or westward after the melting of the Wisconsin ice; or possibly it evolved since the last glacial period.

OTHER MALVACEAE

CHEESES
Common mallow
Malva neglecta
Introduced, naturalized; spreading annual/biennial/perennial forb, 2–18 inches.
Flowers June–October

Virginia Meadow-Beauty
Handsome-Harry,
wing-stem meadow-pitcher
Rhexia virginica
Wisconsin special concern

A striking combination of rose-purple and bright yellow, the flowers of Virginia meadow-beauty are as pretty as they are distinctive. The buds open the night before, and the petals are shed around midday. The floral parts are in 4s and are inserted on the rim of the floral tube. **Low erect perennial, 8–20 inches tall, with tuberous-thickened, fibrous roots.**

Leaves in opposite pairs, lance-egg- to egg-shaped, $3/4$–$2 3/4$ inches long, the margins finely and sharply toothed and fringed with hairs, with 3 main ± parallel veins from the base. Plants with 4-sided, narrowly winged stems and long stiff glandular hairs.

Flowers radially symmetrical except for the stamens, few to several in terminal clusters; bases of sepals, petals, and stamens fused to form a $3/8$-inch-long floral tube that is cylindrical in flower and urn-shaped in fruit, that is, contracted just below the mouth into a short neck; calyx of 4 triangular lobes separate above the tube; corolla of 4 inverted-egg-shaped petals separate above the tube; stamens 8, the filaments reddish, the anthers all on one side of the flower, bright yellow, crescent-shaped, opening by terminal pores. ***July–August***

Fruits partly inferior capsules, retained within the persistent floral tube, opening irregularly; seeds numerous, coiled, tiny, covered with minute, nipple-shaped projections. ***Late August–early September***

WI range and habitats Local central, mainly in the old bed of Glacial Lake Wisconsin, in sedge meadows, edges of bogs, thickets, pine plantations, and ditches; seasonally moist, sandy or peaty soils. ***Upper Greene Prairie*** (planted; not native to the Madison area)

The Virginia meadow-beauty is the only representative in our region of one of the largest plant families. It is what botanists call a coastal plain disjunct, a species disconnected in the Great Lakes region from its primary range on the Atlantic and Gulf coastal plain. It presumably entered Wisconsin late, after the glacial lakes drained, via short- or long-distance dispersal.

The meadow-beauties have ovaries and fruits that are partly inferior in flower but superior at maturity; the lower part of the floral tube, weakly fused to the lower part of the ovary, later detaches from it and becomes free.

Onagraceae • Evening-primrose Family

BIENNIAL GAURA
Biennial bee-blossom
Gaura biennis

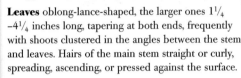

Like the evening-primroses, the gauras have conspicuously inferior ovaries surmounted by a slender, well-developed floral tube and floral parts in 4s, but their flowers are white to pale pink and notably irregular. Biennial gaura is distinguished by its small hard fruits, which taper equally to both ends and are covered with mostly spreading hairs. **Coarse, single-stemmed biennial or winter annual, 2–8 feet tall, with a basal rosette that usually disappears by flowering time.**

Leaves oblong-lance-shaped, the larger ones $1\frac{1}{4}$ –$4\frac{1}{4}$ inches long, tapering at both ends, frequently with shoots clustered in the angles between the stem and leaves. Hairs of the main stem straight or curly, spreading, ascending, or pressed against the surface.

Flowers borne in a branched panicle of several wand-like spikes, the floral tube (below the sepals and petals) about as long as the ovary; petals fading reddish in a day or less; stamens 8, declined below the petals; stigma 4-cleft. *Mid July–September*

Fruits nut-like, non-opening capsules, spindle-shaped, $\frac{1}{4}$ inch long, the basal portion 4-angled, densely short-hairy; seeds (1) 2–4 (5), brown, egg-shaped, small. *August–September*

WI range and habitats Mainly south-central, probably native from Grant to Walworth County, very sporadic and adventive north of Dane County, in medium to moist, often disturbed prairies, roadsides, and railroads, rarely along borders of woods, fields, banks, and shores; native but quite weedy. *Curtis Prairie, Marion Dunn Prairie, Sinaiko Overlook Prairie, Visitor Center parking lot*

The name of the genus comes from the Greek *gauros,* superb, which in the opinion of M. L. Fernald "does not seem always appropriate." The flowers open in the evening and are moth pollinated.

Similar species Scarlet gaura (*Gaura coccinea*) differs from biennial gaura in having pear-shaped fruits that are abruptly constricted to a thick stalk that is round in cross section and squared-off at the base. Furthermore, they are grayish from the covering of incurved or matted hairs. Native to the southwestern U.S., it is rarely adventive in Wisconsin, including Dane County but not the Arboretum.

Common Evening-primrose
Bastard evening-primrose
Oenothera biennis complex

These abundant field weeds have numerous alternate narrow leaves, ± large but ephemeral flowers, and pod-like capsules borne in a prolonged terminal spike. Their bright yellow flowers spring open within minutes near sunset and last until about noon the next day. **Stout biennial or short-lived perennial; 1st-year plants consisting of a rosette of many elongate leaves and strong fleshy roots, 2nd-year plants of an erect stiff stem 2–6 feet tall, from a deep taproot.**

Leaves lance- or lance-egg-shaped to narrowly elliptic or narrowly oblong, $3/4$–8 inches long, averaging more than $3/8$ inch wide, grading into small modified leaves (bracts) of the inflorescence, the wavy margins often red-tinged. Herbage, calyx, and capsule green or grayish-green with sparse to dense, short hairs, these often mixed with long spreading and shorter glandular hairs.

Flowers with 4 sepals (calyx segments), 4 petals, 8 stamens, and a prominent 4-lobed, X-shaped stigma. Floral tube greenish-yellow, stalk-like, much prolonged beyond the ovary, $3/4$–$1 3/8$ inches long below the attachment of the petals; sepals at first united and erect, later becoming ± separated and abruptly bent straight back; petals inverted-egg-shaped, $1/4$–1 inch long. *Early June–mid October*

Fruits linear-oblong capsules, obscurely 4-sided to round in cross-section, $1/2$–$1 1/2$ inches long and a little less than $1/4$ inch thick near the base, the hard valves eventually opening at the tip like a peeling banana, many-seeded. *Late July–mid October*

WI range and habitats Common throughout, in open, especially disturbed habitats: fields (abandoned or cultivated), prairies, pastures, sedge meadows, roadsides, railroads, edges of woods and marshes, riverbanks, lakeshores, open sand, and waste ground in urban areas; native but weedy. *Curtis Prairie, Grady Tract knolls, Greene Prairie, Marion Dunn Prairie, Visitor Center parking lot, Wingra Oak Savanna*

In late summer, fall, and dull weather, the flowers may remain open all day. Their sweet lemony scent increases in the late afternoon and attracts night-flying moths. It reminded European settlers of the Old World primroses (Primulaceae), which have no relation to evening-primroses. The flowers are visited also by butterflies, bees, and beetles. Wildlife, especially goldfinches and other birds, eat the seeds all winter long, when the dead stalks and dried capsules are a familiar sight.

The evening-primrose genus is notoriously confusing taxonomically, primarily because of the peculiar genetics of the subgenus *Oenothera*, consisting primarily of the *O. biennis* complex and represented in Wisconsin by six taxa that only an expert can name with any degree of certainty. Of these, small-flowered evening-primrose (*O. parviflora*), customarily separated as a species on the basis of sepal-tip characters, is almost as common, and in the same habitats, as common evening-primrose.

Onagraceae • Evening-primrose Family 243

Sand Evening-primrose
Cleland's evening-primrose
Oenothera clelandii
Synonym *Oenothera rhombipetala*

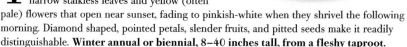

This species, much like common evening-primrose but smaller, is gray-green with narrow stalkless leaves and yellow (often pale) flowers that open near sunset, fading to pinkish-white when they shrivel the following morning. Diamond shaped, pointed petals, slender fruits, and pitted seeds make it readily distinguishable. **Winter annual or biennial, 8–40 inches tall, from a fleshy taproot.**

Rosette leaves (1st year) narrowly inverted-lance-shaped, to 6 inches long. **Stem leaves** narrowly elliptic to linear-oblong or narrowly lance-shaped, $3/4$–$3^{1}/_{2}$ inches long, usually less than $3/8$ inch wide, gradually tapering to a narrow tip, entire to remotely and minutely toothed. Herbage with dense to sparse, short hairs lying flat against the surface.

Flowers numerous in the axils of reduced upper leaves (bracts), forming dense spikes; sepals 4 at the end of a $5/8$–$1^{1}/_{2}$-inches-long floral tube, bent backward and united (mostly) or separate at flowering time, yellowish, with barely developed free tips; petals 4, inversely egg-diamond-shaped, $3/8$–$7/8$ inch long; stamens inserted on the summit of the floral tube, the anthers shedding pollen directly on the stigma, which is at the same level as the anthers at flowering time. *Early July–early October*

Fruits erect cylindrical capsules $3/8$–$5/8$ inch long and $1/8$ inch thick, curved, of uniform thickness from base to apex (not tapered upward), covered with straight stiff hairs pressed against the surface; seeds dark brown, inverted-egg- or spindle-shaped, finely pitted. *Late July–early October*

WI range and habitats Common in the central Wisconsin, Wisconsin River, and Black River sand areas, in abandoned sandy fields, sand barrens, sandy prairies, "goat" prairies, open jack pine-black oak woodlands, river terraces, lakeshores, roadsides, and railroads; generally in disturbed sandy soils; native but somewhat invasive. *West Knoll*

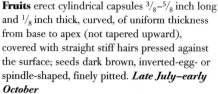

Similar species Long-spike evening-primrose (true *Oenothera rhombipetala*) has virtually identical leaves, stems, and fruits, but the floral tube, sepals, and petals are all longer, and the stigma is elevated beyond the anthers in the fully expanded flower. It is very rare and probably adventive in western Wisconsin. Whereas it is an outcrossing, fully fertile diploid, sand evening-primrose is a self-pollinated, half-sterile "complex heterozygote."

Small Evening-primrose
Small sundrops
Oenothera perennis

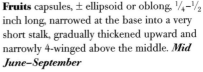

Unlike the oenotheras described previously, small sundrops is a day-bloomer; it opens its flowers in bright sunshine, hence the common name sundrops given to it and some other members of the genus. Small sundrops produces handsome, cup-shaped flowers scattered at the tip of the stem. The flowering part of the plant usually nods in bud but finally straightens. The relatively short floral tube and somewhat club-shaped, narrowly winged capsules are characteristic. **Slender perennial, forming a basal rosette the 1st year and flowering stems the 2nd and successive years, 5–28 inches tall.**

Rosette leaves inverted-lance-shaped, withered by flowering time. **Stem leaves** narrowly elliptic to linear-lance- or inverted-lance-shaped, $3/8$–$2 3/4$ inches long, gradually reduced upward into narrow bracts.

Flowers relatively few, borne in a short raceme; sepals (calyx lobes) 4, backward-bending in 2s or 4s at the tip of the slender, $1/8$–$3/8$-inch-long floral tube; petals 4, bright yellow, inverted-heart-shaped, $1/4$–$3/8$ inch long, broad and notched at the tip; ovaries shorter than the subtending bracts; anthers and stigma lobes very short, borne on the same plane, the anthers shedding pollen directly on the stigma. *Late May–mid September*

Fruits capsules, ± ellipsoid or oblong, $1/4$–$1/2$ inch long, narrowed at the base into a very short stalk, gradually thickened upward and narrowly 4-winged above the middle. *Mid June–September*

WI range and habitats Widespread (missing from far north- and east-central parts of the state), in medium to wet-medium prairies, sedge meadows, pastures, and sandy, muddy, or boggy margins of marshes, shallow bogs, streams, and rivers, also moist cliffs, borrow pits, ditches, roadsides, and rarely dry open woods and railroads. *Curtis Prairie, Greene Prairie*

Flowers of the sundrops attract butterflies and bees, which are active during the day, whereas those of the common and small-flowered evening-primroses are visited by night-flying moths.

Small sundrops is the smallest of the four Midwestern and northeastern species of sundrops, one of which, meadow sundrops (*Oenothera pilosella*), is often cultivated as a garden ornamental. The latter has conspicuous soft spreading hairs and an erect inflorescence of larger flowers, in which the stigma is borne well above the anthers.

OTHER ONAGRACEAE

EASTERN WILLOW-HERB
Cinnamon willow-herb
Epilobium coloratum
Native; erect delicate, often bushy perennial forb, 2–3 feet. Flowers July–October

Oklahoma Grass Pink
Calopogon oklahomensis

This graceful orchid has a single grasslike leaf and relatively large, bilaterally symmetrical flowers distinguished by being apparently inverted, i.e., having the fan-shaped, prominently bearded lip oriented upward instead of in the more usual lowermost position. The male and female organs (stamens, styles, and stigmas) are united into a column in the center of the flower, a distinctive feature of the family. **Slender-stemmed perennial, 4–14 inches tall, from a small hard, bulb-like tuber.**

Leaf 1 (rarely 2), erect, linear-lance-shaped, 3–12 inches long, $1/4$–$5/8$ inch wide, wrapped around (sheathing) the stem near the base, folded lengthwise, strongly veined.

Flowers 2–7 (11) in a loose raceme, $1 1/4$ inches across, with solid pale pink to rose-purple (rarely white) floral envelope composed of 3 outer segments (sepals) and 3 inner segments (petals, longer and narrower than sepals). One petal (lip) different in form and size, strongly bearded on the face with knobbed, yellow-tipped bristles. Column $3/8$–$1/2$ inch long, strongly incurved, winged on each side, at the summit of which is 1 fertile anther, bearing pollen in 4 separate pollen masses called pollinia. *Mid June–early August*

Fruits erect ellipsoid capsules, $1/2$–1 inch long, 6-ribbed, the column persistent on the capsule; seeds very numerous, minute.

WI range and habitats Extremely rare, usually in medium sandy to loamy prairies (Dane and Waukesha counties), also known from a "prairie marsh" and "rich woods" (both Sauk County). ***Arboretum prairies***

ANY ORCHID YOU MIGHT BE PRIVILEGED TO FIND SHOULD BE LEFT ALONE. Native orchids are intolerant of disturbance and difficult to keep in cultivation, and they should never be picked or dug (almost all plants offered for sale have been taken from the wild).

The name *Calopogon* is derived from Greek words meaning "beautiful beard." Three similar, equally showy orchids often grow in the same areas: grass pink (*C. tuberosus*), rose pogonia (*Pogonia ophioglossoides*), and dragon's-mouth (*Arethusa bulbosa*). The first two can form large populations in open sphagnum bogs. Nonetheless, most of our orchid species should be considered rare or endangered.

Similar species *Calopogon oklahomensis* was described as a species in 1995 and may be of hybrid origin. It is morphologically similar to and difficult to distinguish from other species of *Calopogon,* particularly grass pink (*C. tuberosus*), which differs in having flowers that open sequentially (instead of nearly simultaneously), with the dilated middle lobe of the lip usually much wider than long (versus usually much narrower than long in *C. oklahomensis*).

Small White Lady's-slipper
White lady's-slipper
Cypripedium candidum
Wisconsin threatened species

The elusive lady's-slippers (*Cypripedium* species) are unique plants, readily recognized by the curiously shaped lip of the brightly colored flowers. In small white lady's-slipper this trademark pouch is small and pure glossy or waxy white, from which the plant gets both its common names and species epithet. **Small but sturdy, leafy perennial, 6–12 inches tall, from a short crooked rhizome with numerous, cord-like roots.**

Leaves erect to strongly ascending, 3–4 crowded near the middle of the stem plus 2–4 bladeless sheaths below, the lance- to narrowly elliptic blades rather large (except for the smaller uppermost one subtending the flower), 3–8 inches long, often overlapping at base, with numerous prominent parallel veins. **Stems** in clumps of up to 50, sparingly glandular-hairy.

Flower 1 (rarely 2); sepals (apparently 2, the lateral pair united into 1 under the lip) and lateral petals widely spreading, greenish-yellow, often faintly crimson-streaked, $3/4$–$1 3/8$ inches long, the petals often loosely twisted; lip white, veined with violet inside, $5/8$–1 inch long, with a roundish opening at the top; column declined over the opening of the lip, bearing a pair of lateral fertile stamens and a central broad, yellow-spotted-with-purple, non-pollen-producing stamen above. *Early May–mid June*

Fruits ellipsoid capsules $1 1/4$ inches long, opening by longitudinal slits but remaining closed at top and bottom, containing tens of thousands of dust-like seeds.

WI range and habitats Generally rare and local except for one or two impressive colonies, in moist prairies and fens, also sedge meadows, very rarely in dry upland prairies; neutral to distinctly calcareous, usually mucky soils in full sun. *Arboretum prairies*

The genus name means "Aphrodite's sandal" in Greek and refers to the sac-like lip, which together with the flower's mild scent, attracts and temporarily traps small bees. Once inside, a bee finds it difficult to escape without first passing beneath the stigma, then squeezing through one of the two small openings at the back of the pouch, in the process rubbing against one of the anthers.

Similar species White lady's-slipper is readily distinguished from the more familiar large yellow lady's-slipper (*Cypripedium parviflorum* variety *pubescens*) and small yellow lady's-slipper (*C. parviflorum* var. *makasin*), which have leaves more evenly distributed on the stems and the flowers with bright yellow lips striped or blotched with purple.

Small Yellow Lady's-slipper
Northern small yellow lady's-slipper
Cypripedium parviflorum
variety *makasin*

Synonym *Cypripedium calceolus*
variety *parviflorum*

Wisconsin special concern

This dainty orchid's showy flower is usually solitary and very fragrant. Its inflated lip is a clear golden-yellow with magenta-purple spots inside. The leaves are folded like a fan and prominently ribbed. **Perennial, 5–15 inches tall, from a short crooked rhizome, often forming clumps or colonies.**

Leaves 2–5 (often 3), erect to spreading, broadly egg- to elliptic-lance-shaped, 2–6 inches long, sheathing the stem. Plants sparsely glandular-hairy throughout.

Flowers 1–2 (rarely 3), subtended by a leaf-like bract; sepals 3 but apparently 2 (the lateral pair almost wholly united into 1), purple-brown, usually with yellowish-green streaks, wavy-margined or weakly twisted; lateral petals wide-spreading, the color of the sepals, linear-lance-shaped, 1–2 inches long, often loosely twisted; lip $5/8$–$1 1/4$ inches long; column declined over the opening of the lip, the stigma and 2 fertile stamens obscured by a large triangular, yellow-and-red-spotted, sterile stamen in the center. *Mid May–June*

Fruits ellipsoid capsules, $1 1/4$ inches long, opening along longitudinal sutures; seeds very numerous, tiny. Fruits infrequently produced but long-persistent.

WI range and habitats Rare to locally frequent throughout, in moist prairies and fens, conifer swamps, alder thickets, and borders of streams and sloughs, less often in upland forests (coniferous or deciduous); peat or mineral soil, shade or full sun. *Arboretum prairies*

Our cypripediums are all intolerant of environmental disturbances. Burgeoning deer populations, invasive species, thoughtless collection, and especially habitat loss—in the case of this variety, the draining or filling of wetlands—have made them far less abundant now than in earlier times.

Similar species The current trend is to consider North American yellow-flowered lady's-slippers as one species embracing three varieties. Two of these, as well as intermediates between them, occur in our range. Northern small yellow lady's-slipper is essentially a small version of the more familiar large yellow lady's-slipper (*Cypripedium parviflorum* variety *pubescens*), which ordinarily has larger flowers (lip $1 1/4$–2 inches long), sepals less purple and more tightly twisted, flowers less fragrant, a slightly earlier blooming period, and a preference for medium to moist woods. See also comments under small white lady's-slipper (*C. candidum*).

Pale Green Orchid
Tubercled orchid
Platanthera flava
Synonym *Habenaria flava*
Wisconsin threatened species

Although small and not at all showy, the flowers of this grass-green orchid share the complex structure of the larger, more ostentatious orchid species. The upper sepal forms a hood over the column. The bump (tubercle or callus) rising from the floor of the lip near its base is characteristic of this species. **Unbranched hairless perennial, 5–24 (mostly 10–18) inches tall, from fleshy fibrous roots, forming colonies.**

Leaves 2–4, reduced upward into an additional 1–3 bract-like leaves, dark green, spreading to ascending, the blade lance-shaped, 2–14 inches long, the basal part sheathing the stem.

Flowers 15–60 in a rather dense spike 2–8 inches long, green or yellowish-green, small, $3/16$ inch across, very fragrant; lateral petals erect, standing together (but not united) with the upper sepal; lower petal (lip) ± straight across or barely toothed at the tip, but the basal half usually with 1 tooth on each side near the base and a small, nipple-like bump on the upper surface, prolonged backward into a conspicuous slender hollow appendage (spur) about $1/4$ inch long. Lower bracts for the most part much longer than the flowers, divergent. *Late June–late July,* but floral parts persistent on the developing ovaries for much of the summer

Fruits semi-erect, ellipsoid capsules $5/16$ inch long.

WI range and habitats Rare to local, becoming more frequent south, in moist or wet prairies, tall-shrub thickets, floodplain woods, sloughs, riverbanks, and sandy lakeshores; calcareous or circumneutral soils. *Arboretum prairies*

The name of the genus is derived from the Greek meaning "broad anther," referring to the unusually wide anther. Apparently the tubercle at the base of the lip functions to deflect would-be pollinators, in this case mosquitoes and moths, toward one of the two coherent clusters of pollen (pollinia), which adhere to the proboscis for transport to another flower.

Similar species The prominent tubercle on the lip distinguishes pale green orchid from long-bracted green orchid (*Coeloglossum viride*), tall northern bog orchid (*Platanthera huronensis*), and other platantheras with lips neither fringed nor 3-parted.

Prairie White Fringed Orchid
Eastern prairie fringed orchid
Platanthera leucophaea
Synonym *Habenaria leucophaea*
Federally threatened species
Wisconsin endangered species

The strikingly handsome eastern prairie fringed orchid produces conspicuous spires of creamy- to greenish-white flowers rising above the surrounding grasses. At dusk they exude a delicious fragrance, attracting many moths. The lip is deeply parted into three wedge-shaped segments that are dissected one-third to one-half their length into a ragged fringe, and as in the other fringed orchids, it is extended at the base into a hollow appendage (spur). **Single-stemmed perennial, 16–45 inches tall, from thick fleshy roots.**

Lower leaves 1–4, lance-shaped to narrowly oblong, 4–8 inches long, $1/2$–$1\frac{1}{2}$ inches wide, the bases sheathing the stem. **Upper leaves** bract-like.

Flowers 5–28 in a lax to dense raceme 2–8 inches long, 1–$1\frac{1}{2}$ inches long (including ovaries but not spurs); sepals and petals loosely converging and overarching the lip and column; lateral petals fan-shaped, shallowly toothed, slightly longer than the sepals; central division of the lip narrowed into a short stalk-like base; column short; spur slender, 1–$1\frac{3}{4}$ inches long, slightly thickened toward the tip. *Early July–early August*

Fruits ellipsoid capsules $3/4$ inch long.

WI range and habitats Rare and local south, but sometimes developing large colonies, typically in wet prairies, less often sphagnum-sedge bog mats and open tamarack swamps; usually in neutral, rich moist black soil, always in full sun. *Arboretum prairies*

Plants bloom at sporadic intervals, making accurate assessment of their status difficult. Because these orchids may not appear above ground each growing season, populations seem to fluctuate from year to year. In Greene Prairie 12 flowering plants were counted in 1997, 14 in 1998. Other years only 5 plants will appear, or even fewer or none, as in 2005, when a single undersized specimen was found. Local botanists fear that this population will soon be lost, not because conditions are unfavorable, but because white-tailed deer browse so heavily on this species. Only by caging each plant each season will this species survive to bloom again.

Similar species Eastern prairie fringed orchid most closely resembles western prairie fringed orchid (*Platanthera praeclara*), which has shorter, fewer-flowered inflorescences and larger flowers that differ in details of the column. It is more western and does not extend into Wisconsin.

Great Plains Lady's-tresses
Spiranthes magnicamporum

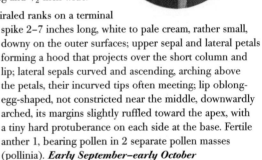

The delicate flowers of this beautiful genus mark the boundary between summer and fall. The inflorescence is unmistakable, consisting of uniformly white, ± tubular flowers closely packed into a spirally twisted spike. Sepal posture, lip color, and fragrance are important characters to note in fresh specimens. Great Plains lady's-tresses is leafless at flowering time, has flowers with a yellowish center on the underside of the lip, and a distinctive vanilla-like fragrance. **Stiffly erect, slender-stemmed perennial, 3–15 inches tall, from a bundle of tuberous roots.**

Leaves 2–3, essentially basal except for 4–6 sheathing bracts on the upper stem, withering and disappearing several weeks before flowering time, inverse-lance- to linear-lance-shaped, to 6 inches long and 1/2 inch wide.

Flowers 20–40, in 3 or 4 gently spiraled ranks on a terminal

spike 2–7 inches long, white to pale cream, rather small, downy on the outer surfaces; upper sepal and lateral petals forming a hood that projects over the short column and lip; lateral sepals curved and ascending, arching above the petals, their incurved tips often meeting; lip oblong-egg-shaped, not constricted near the middle, downwardly arched, its margins slightly ruffled toward the apex, with a tiny hard protuberance on each side at the base. Fertile anther 1, bearing pollen in 2 separate pollen masses (pollinia). *Early September–early October*

Fruits erect capsules with very numerous, very small seeds.

WI range and habitats Local south (north to Pierce County in the west, Milwaukee County in the east), on dry to moist, calcareous prairies, typically on open gravelly ridges and bluffs along the Mississippi River and in rich, medium to moist prairies in the southeast, as well as swales and clay bluffs along Lake Michigan. *Arboretum prairies*

The genus name for lady's-tresses comes from Greek words for spiral and flower. The flowers seem to be woven into intertwining strands, giving the spike axis a braided appearance. Actually, they are borne in vertical rows; only the stem spirals.

Similar species This species was not recognized as distinct until 1973. It resembles nodding lady's-tresses (*Spiranthes cernua*), which has leaves present at flowering time and weakly fragrant flowers, and northern slender lady's-tresses (*S. lacera*), which bears smaller flowers in a single vertical rank. The latter once grew in the dry soil under the Leopold Pines.

Violet Wood-sorrel
Oxalis violacea

This small "stemless" plant is particularly attractive when in bloom. The rose-violet flowers, from which it gets both its common name and species epithet, overtop distinctive, clover-like leaves that are frequently tinged dark reddish above and purplish beneath. **Low perennial, 4–10 (12) inches tall, from brownish scaly bulbs, forming colonies.**

Leaves $1/2$–$4/5$ the height of the flowering stems, palmately compound, the 3 leaflets inverted-heart-shaped, $3/8$–1 inch wide, folding together at night.

Flowers 2–8 in an umbel on a separate stalk from the leaves; sepals 5, each with a firm orange tip; petals 5, rarely white, $3/8$–$3/4$ inch long; stamens 10, of two lengths, their filaments all grown together at the base; pistil with 5 separate styles. Plants flowering sporadically in late summer usually lack foliage. *Late April–June (August)*

Fruits spherical capsules $1/8$–$1/4$ inch long; seeds reddish, enclosed in a large transparent appendage that functions in dispersal, both in expelling the seed from the capsule and attracting ants. *Late May–June (August)*

WI range and habitats Rare to locally common southwest, in dry to medium prairies, dry upland woods (especially grassy openings and edges), bedrock glades, grassy brushy slopes, hills, cliffs, and bluffs; usually in sandy, gravelly, or rocky soil. *Grady Tract knolls, Sinaiko Overlook Prairie*

Violet wood-sorrel produces two types of flowers (distylous). In one, the sigma occupies the lower position in the flower while the two sets of anthers occur in the middle and upper positions; in the other form, the stigma occupies the upper position, and the anthers fill the middle and lower positions.

All wood-sorrels have the same distinctly sour taste owing to the presence of oxalic acid in the foliage, hence the genus name *Oxalis*, the Greek name for sorrel (from *oxys*, sour). This flavor explains the common name sour-clover, given to the weedy, yellow-flowered *Oxalis* species so familiar in gardens, roadsides, fields, disturbed woods, etc.

Similar species A noteworthy relative, mountain wood-sorrel (*Oxalis montana*), has slender rhizomes (not bulbs) and solitary flowers with pink-veined, white or pinkish petals. It is a species of cool damp northern forests and does not occur in the Arboretum.

Other Oxalidaceae

Dillenius's Oxalis
Southern yellow wood-sorrel
Oxalis dillenii
Native; perennial creeping forb, 1–8 inches.
Flowers May–July

Common Yellow Oxalis
Tall wood-sorrel
Oxalis stricta
Native; erect/trailing perennial forb, 4–24 inches.
Flowers May–October

Woolly Plantain
Plantago patagonica
Synonym *Plantago purshii*

Woolly plantain is so-called because plants are silvery- or greenish-white throughout (drying yellowish) owing to the covering of long soft hairs. They are further distinguished by their dwarf habit, leafless stems, narrow leaf blades, and silvery-woolly spikes with short bracts. **Slender annual, winter annual, or biennial, $1^1/_2$–8 inches tall, with taproots terminated by an inconspicuous crown, often forming patches or colonies.**

Leaves linear-oblong to narrowly inverted-lance-shaped, $3/_4$–6 inches long, the largest very rarely exceeding $1/_4$ inch wide, usually with a short abrupt hard tip. Hairs of the stem ascending or closely pressed against the surface.

Flowers very numerous, small, thin and dry in texture; spikes $3/_4$–4 inches long; sepals 4, narrowly inverted-egg-shaped, less than $1/_8$ inch long; corolla long persistent, irregular, its tube covering the summit of the capsule, the 4 lobes spreading; stamens 4, included within or projecting slightly beyond the corolla. Floral bracts hidden or barely projecting, pressed against the calyx, narrowly triangular to almost linear, especially the lowermost with an elongated green vein. *Early May–June* (occasionally into September)

Fruits ellipsoid capsules $1/_8$ inch long, opening at or just below the middle; seeds 2 per capsule, reddish-tan.

WI range and habitats Locally common south and west, in sand prairies, abandoned fields, clearings on upland wooded slopes and knolls, dunes, and sand blows, also road shoulders, railroad beds, road cuts, lanes and paths, old gravel and sand pits, and waste places; dry, exposed or disturbed sands and gravels. *West Knoll lanes*

Plantagos have many of the characters associated with wind pollination, namely small greenish scentless flowers, large hairy stigmas that are receptive before the pollen is liberated, large anthers on long filaments, and spherical pollen that is dry and relatively smooth. Some species, however, may be partly insect-pollinated.

Great Plains plants are apparently identical with those from Argentina. This temperate distribution at comparable latitudes across the equator, but not in the tropics, constitutes an interesting problem in plant geography. The species is naturalized eastward to the western Great Lakes, including Wisconsin, and adventive at several localities farther east. It is also native to the mountain system of western North America, where it is represented by *Plantago purshii* in the narrow sense.

Rugel's Plantain
American plantain, black-seeded plantain, red-stalked plantain
Plantago rugelii

The plantains are well-known weeds with long leafless stems arising from a cluster of low-lying, basal leaves and flowers (and fruits) in a dense elongate spike. The broad leaves of American plantain have stalks that are usually reddish-purple at the base. The tiny greenish flowers are conspicuous in bloom owing to the numerous large lavender anthers. **Perennial, 2–12 or more inches tall, from a short tough crown with stringy fibrous roots.**

Leaves oval or egg-shaped, 2½–8 inches long, with entire or remotely toothed margins and 5–9 prominent parallel veins.

Flowers numerous, in dense spikes that open from the bottom upward; sepals 4, in 2 unequal pairs, egg-shaped to oblong; petals 4, united into a dry chaffy corolla, the tube covering the summit of the ovary, the lobes bent backward after flowering; stamens 4. Calyx and corolla both persisting. Bracts lance-shaped to lance-triangular, terminating in a sharp point. *Early June–November*

Fruits small oblong membranous capsules opening well below the middle by an encircling transverse line, the lid-like top falling to release 4–10 black, very irregular, tiny seeds. Seed coat mucilaginous when moistened. *July–November*

WI range and habitats Ubiquitous, in all kinds of rich-soil habitats, including prairies, fields, pastures, open woods, paths, damp shorelines, roadsides, lawns, gravel pits, parking lots, and cracks in pavement; supposedly native in eastern North America, but rarely, if ever, a component of any undisturbed native community. *Throughout the Arboretum*

The names plantain and *Plantago* come from the Latin *planta,* sole of the foot, and refer to the broad flat, low-lying leaves of some of the common weedy species. American plantain and pestilent related species from England and Eurasia are edible and were popular among immigrants as greens; they are rich in vitamins A, C, and K and are actually more nutritious than chard or spinach. The leaves were also used for making compresses and poultices.

Similar species American plantain is morphologically and ecologically nearly identical to common plantain (*Plantago major*), a naturalized species differing in its thicker, duller-colored leaves, broader, blunt-tipped bracts, and angular-egg-shaped capsules that open near or a little below the middle.

Other Plantaginaceae

English Plantain
Narrow-leaved plantain, ribgrass
Plantago lanceolata
Introduced, naturalized; erect
perennial forb, 6–30 inches.
Flowers June–August

Common Plantain
Broad-leaved plantain
Plantago major
Introduced, naturalized; erect perennial
forb, 3–18 inches. Note very dense
inflorescence and green leaf-stalk bases
compared to *P. rugelii*.

Poaceae • Grass Family

About Grasses

Grasses (Poaceae or Gramineae) compose one of the largest families of flowering plants and probably are of greater economic importance than any other plant group. About 220 species occur in Wisconsin, where they are integral components of prairie vegetation. Although most grasses are indigenous, introduced species are an inevitable part of the flora, especially in the summer-dry, southern and western parts of the state, where the prairies are located. Depending on the amount of disturbance, interval since disturbance, and soil type, dominance in prairie stands may have passed to such alien weeds as Canada bluegrass, quackgrass, Kentucky bluegrass, redtop, or reed canary grass. The presence of native prairie grasses in many areas may be due to plantings made in the name of prairie or wetland restoration. Only the most common prairie species are presented in this guide.

Grass **stems** are usually *round* in cross-section, jointed, and hollow in the internodes. The **leaves** are in 2 distinct ranks on the stem, each consisting of *sheath*, *ligule* (membranous or hairy appendage at the summit of the sheath), *blade*, and if present, *auricle* (ear-like lobe or flange at the base of the blade). The sheath encircles the stem and is usually split (margins free and overlapping) but is sometimes closed. If your specimen has a solid stem or is triangular in cross-section, refer to the section on the Sedge Family (pages 159–173).

The tiny, wind-pollinated **flowers** are usually bisexual. Each is enclosed between 2 small, scale-like bracts, the inner *palea* and larger outer *lemma*. The palea, lemma, and flower collectively form a *floret*. The floret or florets, each typically with a pair of empty bracts, the *glumes*, at the base, and slender axis on which they're borne, are collectively termed the *spikelet*. The spikelets may be arranged in panicles or sometimes spike-like or raceme-like inflorescences. The **fruit** is typically a dry, 1-seeded *grain*.

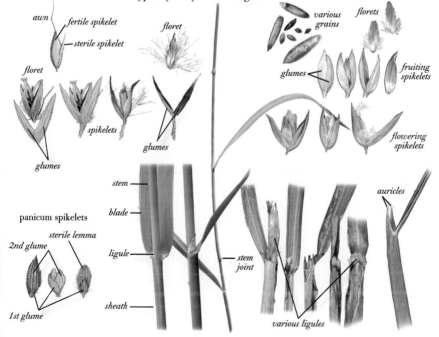

Big Bluestem
Turkey-foot
Andropogon gerardii

This was the quintessential grass of the original tall-grass prairie, distinctive owing to the color of the stems and the often 3-parted shape of the inflorescence (hence its common names), which consists of 2–6 spike-like racemes with two types of spikelets that resemble each other in size and shape. **Perennial, 3–8 feet tall, sometimes from short rhizomes, usually forming large clumps or sods.**

Blades 6–16 (basal ones to 24) inches long, $1/4$–$3/8$ inch wide, flat or rolled inward or outward, usually hairless or with a ring of hairs back of the ligule; ligules short, membranous, with a fringe of hairs on the margin. **Stems** with a waxy bloom and dull pinkish internodes that turn yellow, orange, or red in the late summer in contrast with the often pinkish to bluish or purplish sheaths. Post-frost plants turn a reddish-bronze that slowly fades through the winter.

Inflorescence of racemes arising from a common point at the tip of the stem, usually also smaller inflorescences from a few lateral branches, each slender raceme 2–4 inches long, consisting of a jointed axis bearing numerous pairs of spikelets. Joints of axis and stalks of spikelets sparsely to densely fringed with long white hairs. Flowering *July–late August*

Spikelets paired at each node of the axis and pressed against it, one of each pair stalkless, bisexual, and awned, the other at the tip of a flattened bearded stalk, male, about the same size and shape as the stalkless spikelet but awnless. Stalkless spikelet lance-shaped, containing 2 lemmas, the lower empty and awnless, the upper one fertile and with a bent twisted awn. Grain *mid September–November* (fruits persist on the plant into winter)

WI range and habitats Formerly abundant and still locally common south and west, in wet to dry prairies, fens, and dry open woods, co-dominant with little bluestem on calcareous hill prairies, sandy plains, and old fields; northward in sandy areas and on roadsides and railroads; prefers deep fertile soils and full sun. ***Curtis Prairie, Marion Dunn Prairie, Greene Prairie, Juniper Knoll, Sinaiko Overlook Prairie, Visitor Center parking lot***

Side-oats Grama
Bouteloua curtipendula

Boutelouas are characterized by especially graceful inflorescences that give rise to the name "side-oats." They consist of short, comb-like spikes that in side-oats grama are numerous and uniformly droop to one side of the slightly zig-zag, main axis. During full bloom its protruding, bright purple stigmas and reddish-orange anthers contrast sensuously with the bluish-green spikelets. **Perennial, 10–30 inches tall, from tillers and short scaly rhizomes, tufted or somewhat mat-forming.**

Blades numerous, 2–12 inches long, $1/16$–$1/4$ inch wide, flat, minutely roughened above and on the margins, smooth below, with widely spaced, pale hairs from tiny, blister-like bumps on the margins, especially toward the base; sheaths hairless; ligule an extremely short, membranous band fringed with stiff hairs. Foliage in autumn bronze-purple to orange-red, in winter straw-colored.

Inflorescence a loose, 1-sided raceme 3–12 inches long, consisting of 14–44 spikes, each about $1/2$ inch long, spreading or abruptly bent downward. When mature, spikes fall whole from the intact axis. Flowering *(mid June) mid July–mid September*

Spikelets 3–8 per spike, borne in 2 rows along the lower side of the narrow flat axis, closely overlapping, each of 1 bisexual and 1–2 rudimentary florets above it; glumes unequal, usually somewhat shorter than the shortly 3-awned, fertile lemma. Grain *August–October*

WI range and habitats Rather common southwest, in dry-medium to dry prairies, oak openings, and cedar glades, on sandy plains and knolls, gravelly or rocky hills, and bluffs; prefers somewhat limy, dry to loamy soils and full sun. *Curtis Prairie (limestone knoll), Juniper Knoll*

Side-oats grama is often dominant or co-dominant on uplands throughout the tall-grass prairie region, and it is an indicator species for the dry segment of the prairie gradient in Wisconsin.

The genus is named after the brothers Claudio (1774–1842) and Esteban (1776–1813) Boutelou, Spanish botanists who wrote about floriculture and agriculture. The name grama is a literal translation of the Spanish word for grass.

Hairy Grama
Hairy grama grass
Bouteloua hirsuta

This particularly distinctive grass possesses the same kind of comb-like, 1-sided spikes as its more famous cousin, side-oats grama (*Bouteloua curtipendula*), except that the spikes normally number only 2 on each common axis, and the axis of each spike extends beyond the uppermost spikelet as a short stiff bristle. **Somewhat delicate, low perennial, 4–20 (mostly less than 10) inches tall, tufted or somewhat sod-forming.**

Blades mostly basal, numerous, 1–12 (mostly 5 or less) inches long, very narrow, flat or somewhat inrolled, somewhat curled, bristly on the margins toward the base; sheaths slightly hairy with straight spreading hairs at the throat; ligule an extremely short fringe of hairs. Foliage in autumn pale green to brownish-straw-colored.

Inflorescence a raceme of 1–3 straight or backward-curved spikes, the axis of the raceme extending beyond the last spikelet as a straight stiff point that resembles a bee's stinger; spikes $3/8$–$1 1/2$ inches long. Spikelets breaking up above the glumes, which persist on the axis. Flowering *late July–late September*

Spikelets about 20–50 per spike, closely overlapping, borne in 2 rows along the lower side of each narrow spike axis, green to dark purple, each of 1 bisexual floret below and 1–2 densely hairy, rudimentary florets above; glumes unequal, narrow, the keel of 2nd (upper) with brown or blackish, nipple-based hairs; fertile lemma 3-cleft and 1–3-awned. Grain *late August–November*

WI range and habitats Infrequent southwest (north to Waupaca and St. Croix counties), in upland sand prairies, dry hill prairies, cedar glades, oak openings, and on dry sandstone or limestone bluffs, usually growing with the much more common side-oats grama; shallow, dry, sandy, or gravelly soils, in full sun or slight shade. *Grady Tract knolls*

Similar species Hairy grama should never be confused with side-oats grama (*Bouteloua curtipendula*) or blue grama (*B. gracilis*), both of which have spikes that lack a well-developed, naked bristle at the tip and glumes that are without the small black projections of hairy grama. All three are prominent in the short-grass region of the Great Plains, but hairy grama is infrequent and blue grama rarely adventive in Wisconsin.

BLUE-JOINT
Blue-joint grass
Calamagrostis canadensis

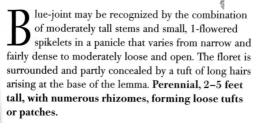

Blue-joint may be recognized by the combination of moderately tall stems and small, 1-flowered spikelets in a panicle that varies from narrow and fairly dense to moderately loose and open. The floret is surrounded and partly concealed by a tuft of long hairs arising at the base of the lemma. **Perennial, 2–5 feet tall, with numerous rhizomes, forming loose tufts or patches.**

Blades long and narrow, usually 4–12 inches long, $1/16$–$1/4$ inch wide, flat or sometimes inrolled toward the tips, minutely roughened on both sides; sheaths hairless; ligules $1/8$–$1/4$ inch long, membranous. **Stems** with a ring of felt-like hairs just below the nodes.

Inflorescence a broadly lance-shaped to oblong panicle 3–10 inches long, becoming sinuous or somewhat nodding, the slender branches spreading during flowering but ascending in bud and fruit. Flowering *late June–July*

Spikelets stalked, about $1/8$ inch long; glumes about equal; lemma slightly shorter and more thinly membranous than the glumes, smooth, with a delicate straight awn attached just below the middle of the back (use a hand lens to see this); hard point at the base of the lemma bearing abundant fine hairs about as long as the lemma. Spikelets breaking up above the glumes. Grain *early July–early October*

WI range and habitats Common throughout, a major dominant in sedge meadows, wet to wet-medium prairies, and fens; a prevalent species of shrub carrs and alder thickets, also swales, cat-tail marshes, bogs, openings in wet woods and thickets, and along sloughs, streams, and shores. *Curtis Prairie, Greene Prairie*

Wisconsin specimens of this extremely variable, circumboreal species are referred to the typical variety *canadensis* and the small-flowered extreme, variety *macouniana*.

Similar species Reed grass (*Calamagrostis stricta*) differs from blue-joint in having a cylindrical panicle bearing densely clustered spikelets with firmer, minutely roughened lemmas. Sand reed grass (*Calamovilfa longifolia*), dune grass (*Ammophila breviligulata*), and common reed (*Phragmites communis*) all have florets that bear a tuft of silky hairs at the base, but they are much more robust and have wider leaf blades and larger spikelets. Of these, only common reed occurs in the Arboretum.

Panic Grasses
Dichanthelium species

In the broad sense *Panicum* is the largest genus in the Grass Family. Twenty-three species, representing the two subgenera or genera *Dichanthelium* and *Panicum*, occur in Wisconsin. Those of *Dichanthelium*, to quote A. Cronquist, "form an intergrading group in which the taxonomic limits are difficult to discern and are still debatable."

Dichantheliums are small perennials, 4–40 inches tall (but usually smaller), forming small clumps. They flower twice, first from May to early July, producing simple stems with terminal or "spring" panicles. These primary panicles are open-pollinated for a brief period and then become self-pollinating. The second bloom, usually after the primary panicle has shed its spikelets, occurs from late June through September. The stems branch from the middle, lower, or upper nodes and produce small secondary, self-pollinating panicles that are partly concealed within the sheaths of upper leaves. In fall, most of the species develop basal rosettes of short stubby leaves that remain green all winter and usually die in the spring. The old terminal panicles may wither and fall, especially if the plants become bushy or mat-like.

The small spikelets of *Dichanthelium* and *Panicum* are all similar. They are compressed front-to-back (versus side-to-side) and at maturity break off below the glumes and fall entire. The glumes are very unequal, the 1st (lower) small, the 2nd one covering the whole back of the spikelet. Above them are 2 florets, the "sterile" floret (represented by a lemma, sometimes also stamens) and the leathery to rigid, shining fertile terminal floret. The sterile lemma is the 3rd bract, on the other side of the spikelet from the 2nd glume, which it resembles (the 2nd glume and sterile lemma might be mistaken for a pair of glumes if the 1st glume is overlooked). The fertile lemma and palea are glistening-white and smooth and tightly enclose the grain.

Leiberg's Panic Grass
Dichanthelium leibergii
Synonym *Panicum leibergii*

Blades $1/4$–$1/2$ inch wide, hairy on both sides or almost hairless above; sheaths shorter than the internodes (nodes exposed), with coarse spreading hairs; ligule a short fringed membrane or nearly obsolete. Terminal (spring) **panicle** 2–4 inches long, the ascending branches few-flowered. **Spikelets** relatively large, rather oblong, $1/8$ inch long, swollen, rounded at the apex, with long soft spreading hairs; 1st glume triangular-egg-shaped, longer than wide, sharp-pointed, reaching to about the middle of the spikelet.
WI range and habitats Southwest, in dry to wet-medium prairies, dry, sandy or gravelly, bluff top or hillside prairies, oak savannas, and prairie-like habitats. ***Curtis and Greene prairies***

SCRIBNER'S PANIC GRASS
Dichanthelium oligosanthes
Synonym *Panicum oligosanthes*

Blades relatively broad, $1/4$–$5/8$ inch wide, the rounded base fringed with nipple-based hairs, the upper surface hairless; ligule a very short band of hairs of varying length, $1/16$ or less inch long. **Stems** often bent at the base and tending to sprawl. Primary **panicle** 2–$3 1/2$ inches long, little-branched. **Spikelets** relatively large, inverted-egg-shaped, $1/8$ inch long, sparsely short-hairy; 1st glume as long as wide, much less than $1/2$ the length of the spikelet, the 2nd (upper) glume usually with an orange or purplish spot at the base. **WI range and habitats** Southwest (north to Burnett and Marathon counties, rarely beyond), a major dominant in dry- to wet-medium prairies, also fields, pastures, and black oak savannas; dry open, usually sandy or disturbed ground, growing in open spaces among bunches of taller grasses. *Curtis, Greene, and Sinaiko Overlook prairies, West Knoll*

STIFF-LEAVED PANIC GRASS
Dichanthelium ovale* subspecies *praecocius
Synonyms *Dichanthelium praecocius, Panicum praecocius*

Blades of main stems $1/8$–$1/2$ inch wide, less than 15 times as long as wide; internodes, sheaths, and blades (both surfaces) with soft spreading straight hairs (not intermingled with minute curled hairs), those on upper sheaths $1/8$ or more inch long; ligule a band of hairs $1/8$–$3/16$ inches long, protruding from the sheath. Primary **panicle** $3/4$–$2 1/4$ inches long. **Spikelets** ellipsoid or inverse-oblong-egg-shaped, $1/16$ inch long, the lower on long stalks (not borne close to the axis); 1st glume triangular-egg-shaped, sharp-pointed, $1/3$–$1/2$ as long as the spikelet. **WI range and habitats** South and central, in sand prairies, sand barrens, prairie hillsides and bluff tops, open woods, borders, fields, and roadsides; dry open, usually sandy soil. *Curtis and Sinaiko Overlook prairies, Grady Tract and Juniper knolls*

Canada Wild-rye
Elymus canadensis

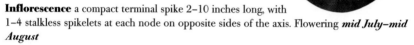

This moderately robust bunch grass has wide leaf blades and a large, arching or drooping spike that resembles cultivated rye. The long stiff, bristle-like awns are spreading to strongly curved outward when mature. **Perennial, 2–5 feet tall, forming a loose sod.**

Blades 4–15 inches long, $3/8$–$3/4$ inch wide, flat or with inrolled margins or the whole blade ± inrolled, the upper surface minutely roughened to sparsely hairy; sheaths with claw-like flanges (when intact) at the summit; ligule membranous, short, finely notched and fringed. Herbage from green to strongly waxy bluish-green.

Inflorescence a compact terminal spike 2–10 inches long, with 1–4 stalkless spikelets at each node on opposite sides of the axis. Flowering *mid July–mid August*

Spikelets (2) 3–6 (7)-flowered, $1/2$–$5/8$ inch long (excluding awns); glumes equal, turned slightly so that both are on the side away from the axis, narrowly lance-shaped and tapering into an awn, firm; lemmas lance-shaped, hairless or minutely roughened to soft-hairy, the awn mostly $3/4$–$1 1/4$ inches long. Spikelets break up at maturity above the glumes and between the florets. Grain *early August–October*

WI range and habitats Frequent throughout, in dry to wet-medium prairies, sandy woods, oak openings, and thickets, particularly common on lakeshores, sand dunes, and riverbanks, weedy along roadsides, railroads, and abandoned fields; in a wide range of soil types, often in places with a history of disturbance. *Curtis Prairie, Marion Dunn Prairie, Greene Prairie, Wingra Oak Savanna*

Canada wild-rye is an excellent, early successional species for stabilizing slopes and roadsides and for restoring prairies, serving as a nurse crop for prairie seedlings before slowly declining as the warm season grasses mature.

Elymus is the Greek name for a kind of grain. The name is derived from *elyo*, rolled up, from the grain's being tightly enclosed by the persistent lemma and palea.

Similar species Canada wild-rye sometimes grows with Virginia wild-rye (*Elymus virginicus*), which differs in having erect straight spikes and smaller spikelets with straight awns.

NEEDLE GRASS
Porcupine grass
Hesperostipa spartea
Synonym *Stipa spartea*

This cool-season bunch grass has conspicuously long glumes (longer than the lemma) enclosing a single hardened, narrowly spindle-shaped floret, which tapers below into a needle-sharp, oblique point and above into a persistent long awn. **Wispy, tufted perennial, 1–3 feet tall, lacking rhizomes, forming rather open populations.**

Blades up to 30 inches long (those of the stem 8–12 inches), $1/16$–$3/16$ inch wide, the margins usually rolled inward, typically somewhat hairy above, hairless beneath; sheaths essentially hairless at the throat; ligules of upper leaves $1/4$ inch long, firmly membranous, notched (longer on the sides).

Inflorescence a gracefully arching, somewhat open, narrow panicle 4–8 inches long, the few slender branches each bearing 1 or 2 large spikelets. Spikelets break up above the glumes. Flowering *late May*

Spikelets 1-flowered and stalked, the glumes nearly equal, 1–2 inches long, pale, papery; lemma light to dark brown, stiff, $5/8$–1 inch long, the lower half hairy, its edges overlapping the palea; awn $4\,3/4$–8 inches long, the basal part very tightly coiled where jointed to the lemma apex, usually twice bent near the middle. Grain *(early) mid June–mid July* (dispersed as early as the first week of July).

WI range and habitats Locally frequent southwest, on dry to dry-medium prairie relicts, sand prairies, open oak woods, and dunes, sometimes spreading to roadsides and railroads; well-drained sandy or gravelly soils. *Curtis and Greene prairies, Grady Tract knolls*

Needle grass florets are self-planting devices. The needle-sharp tip, aided by the upwardly directed hairs, orients the seed to the soil (and sometimes works itself into clothing and the hair and faces of livestock). The spirally twisted awn uncoils when damp and coils when dry, ratcheting the floret into the soil.

The genus name is derived from the Greek *hesperos,* an ancient name for the evening star, and *stype,* tow, as in unraveled fibers, in allusion to the appearance of the awns. The species epithet *spartea* is Latin and means broom-like.

Current taxonomic classification segregates this small genus from the Eurasian *Stipa* (in the narrow sense) based on lemma, palea, and embryo anatomy, which relate hesperostipas to South American genera.

Poaceae • Grass Family

June Grass
Prairie June grass
Koeleria macrantha
Synonym *Koeleria pyramidata*

Stems of this low-growing, cool-season grass are in small, short-lived clumps, and the short flat, dark green leaves are mostly basal. The pale chartreuse, spike-like panicles shine when turned in the sun. **Stiff, slender-stemmed perennial, 8–30 inches tall, with numerous leafy shoots at the base.**

Blades 1–5 inches long, $1/8$ or less inch wide, erect or stiffly ascending, flat (margins rolled backward when dry), coarsely ribbed on the upper surface, hairless or hairy beneath; sheaths (at least the lower) hairy, distinctly veined; ligule very short, membranous, with an irregular margin (as if gnawed).

Inflorescence an erect dense cylindrical panicle 2–$5\frac{1}{2}$ inches long, often interrupted below, the short branches spreading to ascending during flowering but later the inflorescence contracting to less than $3/8$ inch thick. **Stem** (below the panicle) and panicle axis and branches minutely hairy. Flowering ***May–mid July***

Spikelets 2- (rarely 3-) flowered, $3/16$ inch long, compressed, the glumes nearly equal in length but unlike in shape, the first narrow, barely shorter than the first floret; florets uniform, their lemmas about as long as the glumes, narrow, sharply keeled, somewhat thin and shining, sharp-pointed but scarcely awned. At maturity the spikelets break up above the glumes and between the florets. Grain ***August–mid September***

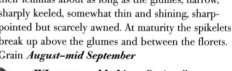

WI range and habitats Regionally common (lacking in north-central and eastern parts of the state), a prevalent grass in dry to dry-medium prairies, sand barrens, oak barrens, and cedar glades, also in abandoned sandy fields and sand dunes, occasionally in wet-medium prairies; prefers poor dry, sandy or gravelly soil; very drought tolerant. ***West Knoll***

June grass is one of the most common grasses of the prairies and plains. The genus name honors Georg Ludwig Koeler (1756–1807), German professor and grass student.

Similar species June grass is confused at times with the wedge grasses (*Sphenopholis* species), which differ in having the foliage and inflorescence hairless and the glumes and stalk falling with the rest of the spikelet. In June grass the glumes tend to remain on the plant. Slender wedge grass (*S. intermedia*) occurs on Greene Prairie.

Switch Grass
Panicum virgatum

This tough vigorous grass is easily identified by its large open inflorescence and purplish, teardrop-shaped spikelets, each with 3 visible overlapping pointed scaly parts enclosing 2 florets. **Solitary or clumped perennial, 2–4 (6) feet tall, spreading by short thick rhizomes, forming a sod.**

Blades 4–18 or more inches long, $1/8$–$5/8$ inch wide, flat, sometimes with a patch of dense hairs on the upper surface near the base; sheath hairless (usually hairy at the throat and on the margins); ligule membranous, $1/16$–$1/4$ inch long, fringed.

Inflorescence an open to diffuse, egg- to pyramid-shaped panicle 4–20 inches long, the branches ascending to spreading, with scattered spikelets near the tips. Flowering *July–early September*

Spikelets mostly $1/8$–$3/16$ inch long, the lower floret male ("sterile"), the upper one bisexual ("fertile"); first glume $1/2$–$4/5$ as long as the spikelet, tapering into a very sharp point; second glume and sterile lemma keeled, extending beyond the upper floret; fertile lemma blunt. Awns lacking. Grain *late August–early October*

WI range and habitats Common throughout except for large north-central and wooded northeastern areas, in medium to low prairies, swales, sand barrens, dunes, and blowouts, spreading into fields, roadsides, railroads, and waste ground; prefers moist prairies but fares well in drier calcareous soils. *Curtis Prairie, Greene Prairie, Juniper Knoll, Marion Dunn Prairie, Sinaiko Overlook Prairie, West Knoll*

The genus name is derived from the Latin *panis*, bread, or *panus*, an ear of millet. The species epithet *virgatum* means wand-like.

Switch grass is the earliest maturing of the native warm season grasses. Although characteristic of and sometimes dominant on the tall-grass prairies farther west, it was not as important on Wisconsin prairies. It is widely used for wildlife cover and erosion control, its extensive root system reaching deep into the soil and providing outstanding soil stabilization.

Unlike the closely related dichantheliums, switch grass is a true *Panicum*. It does not form a winter rosette, and it flowers but once, the panicles appearing after midsummer. The florets are all functional and the upper ones mostly open-pollinated, whereas in *Dichanthelium* they are often sterile or self-pollinating.

Reed Canary Grass
Phalaris arundinacea
Invasive weed

This leafy, deep-rooted grass can be recognized by its growth habit, wide flat hairless leaves, and dense, spike-like panicles. Each flattened spikelet contains 1 hard shiny fertile floret and 2 much smaller, sterile lemmas very unlike and below the fertile lemma. **Clump-forming perennial, 2–5 feet tall, spreading extensively by strong scaly rhizomes, often forming dense stands.**

Blades generally 3–8 inches long, $1/4$–$5/8$ inch wide; sheaths open, the margins overlapping; ligule $1/8$–$5/16$ inch long, membranous, entire to torn or cut, minutely hairy on the back. Herbage light green or yellow-green.

Inflorescence a slender, often lobed panicle 2–7 (10) inches long, greenish-white, often somewhat purplish in flower and turning straw-colored in fruit, the many short branches at first ascending, spreading in flower but later contracting. Flowering *June–July*

Spikelets $3/16$ inch long, apparently 1-flowered but with 2 sterile florets reduced to minute linear hairy scales (easily mistaken for clusters of hairs) either side of and closely pressed against the single fertile floret and falling with it; glumes nearly equal, longer than the florets, boat-shaped; fertile lemma tan, smooth and glossy below, with a few hairs toward the tip, awnless. Grain *late June–October*

WI range and habitat Abundant throughout, in disturbed wet or moist soil: borders of rivers, streams, and ponds, wet meadows, wet thickets, swamps, and ditches; partly adapted to drier ground and to some degree established along roadsides, railroads, fields, and occasionally upland woods. *Throughout the Arboretum*

Phalaris is the ancient Greek name for canary grass. The species epithet *arundinacea* means reed-like.

Because it is extremely aggressive on wet or mucky soils, reed canary grass has been recommended for hay and forage on sites disturbed by agricultural use and for stabilizing gullies and ditches. Seed from invasive European strains has been widely planted in the U.S., and now, persistent monotypic stands of probably non-native plants are crowding out native vegetation everywhere. The species is native to the northern hemisphere of both continents, but there are no known characters to distinguish American plants from introduced ones.

Poaceae • Grass Family

LITTLE BLUESTEM
Schizachyrium scoparium
Synonym *Andropogon scoparius*

Little bluestem grows in dense tufts that are greenish-blue to purplish in early summer but turn reddish to tan or brown when mature and eventually (after frost) bronzy orange. Spikelets are in dissimilar pairs at each node along the straight or sinuous upper part of each stem or branch. **Slender to robust, tufted perennial, 2–3 ft tall, occasionally short-rhizomatous, ± mat-forming.**

Blades 2–12 (16) inches long, $1/8$–$1/4$ inch wide, flat, sparsely hairy at the junction with the sheath; sheaths usually hairless (our variety), keeled; ligules brown, $1/16$–$1/8$ inch long, membranous with a ragged margin. **Stems** somewhat flattened, the upper half branching. **Leaves** mostly basal, often with a whitish or bluish cast.

Inflorescences single racemes mostly $3/4$–2 inches long, consisting of 5–20 pairs of spikelets, these eventually turning silvery-white and fluffy. Flowering *mid July –September (mid October)*

Spikelets in scattered pairs; 1st (lower) spikelet of each pair stalkless and fertile, with 2 florets, the lower reduced, the upper bisexual, lance-shaped, about $1/4$ inch long, its lemma bearing a bent twisted awn to $1/2$ inch long; 2nd spikelet stalked, sterile or occasionally male, with 1 floret, unawned or short-awned. Spikelet set falling together with a joint of the axis, the latter fringed with long grayish hairs on the margins. Grain *late August–October*

WI range and habitats
Widespread and locally common, in dry prairies, open oak woods, pine barrens, cedar glades, dunes, sand plains, and shores, spreading along roadsides, railroads, and into old fields; in a variety of soils, acidic to alkaline; on higher drier sites than big bluestem. *Curtis Prairie, Grady Tract knolls, Greene Prairie, Juniper Knoll, Sinaiko Overlook Prairie*

Abundant and highly variable, little bluestem occurs to some extent over the entire U.S. It once covered the prairies and plains and remains the most common and widespread prairie grass in Wisconsin, where it is co-dominant with big bluestem, side-oats grama, and Indian grass on prairies ranging from extremely dry to medium. It is now being used in ornamental plantings.

Indian Grass
Sorghastrum nutans

Plume-like, silky-soft panicles of Indian grass are at first bronze-yellow but in fruit become chestnut brown, bestowing beauty and character to the autumnal prairie. As in the bluestems, the spikelets are paired although they appear to be single, each stalkless bisexual spikelet being accompanied only by a hairy sterile stalk arising at its base, the stalked spikelet missing. The absence of the stalked spikelet makes Indian grass readily recognizable. **Tall, 1-stemmed or loosely tufted perennial, 3–7 feet tall, from short rhizomes.**

Blades 2–24 (40) inches long, $3/16$–$3/8$ inch wide, flat, hairless; sheaths extending into a pair of distinctive erect, "rabbit-ear" appendages at the summit; ligules firm, $1/16$–$1/4$ inch long. **Stems** hairless except for the silky nodes.

Inflorescences narrowly ellipsoid to narrowly oblong, rather dense, 4–13 inches long, often nodding at the apex, at first loose but contracting after flowering. Inflorescence joints and stalks copiously white-hairy. Flowering *end of July–mid October*

Spikelets stalkless, fertile, with 2 florets, lance-shaped, $1/4$ inch long, each subtending a feather-like, fringed stalk devoid of a spikelet or rudiment; glumes leathery, the 1st hairy and clasping the 2nd (upper) one; 1st (lower) floret reduced, the upper bisexual, terminated by a bent twisted awn longer than the spikelet. Grain *early August–October* (persists on the plant into winter)

WI range and habitats Chiefly south and west (missing from a number of north-central counties), in dry to wet-medium prairies and open pine, oak, or red cedar woods, spreading somewhat into sandy fields, roadsides, and railroads, especially in northeastern and northwestern parts of the state; prefers deep, well-drained soils and full sun. *Curtis Prairie, Greene Prairie, Marion Dunn Prairie, Sinaiko Overlook Prairie, West Knoll*

Prior to agricultural development, Indian grass was second in importance only to big bluestem in the tall-grass prairie region. It is still locally common on Wisconsin's prairie remnants and open savannas. Like big bluestem, it is drought-tolerant, owing to its exceptionally deep root system. It is an attractive plant, useful in dried flower arrangements and roadside beautification projects.

Prairie Cord Grass
Slough grass
Spartina pectinata

Cord grass cuts a refined figure with its tall rigid stems, gracefully arching leaf blades ending in long, extremely slender tips, and a raceme of stiff, 1-sided spikes along a common axis. **Coarse perennial, 3–7 feet tall, with extensively creeping, hard scaly rhizomes, often forming dense clonal patches.**

Blades 1–3 feet long, 1/4–3/8 inch wide, flat at the base but soon inrolled (especially when dry), tough, drawn out into a thread-like, flexuous portion with sharply roughened margins that can cut the skin; ligule a cartilaginous band fringed with a row of short hairs, 1/16–1/8 inch long. **Stems** solitary or in small clumps. Plants turn light straw-color in early fall.

Inflorescence of 6–25 (mostly 10–20) short-stalked, comb-like spikes, ascending or often lying flat against the main axis, each 1–4 (5) inches long and consisting of 20–30 (or more) straw-colored spikelets. Spikelets breaking up below the glumes, falling entire.

Flowering *July–beginning of September*

Spikelets closely overlapping in 2 rows along 2 sides of the 3-sided spike axis, lance-shaped, stalkless, 1-flowered, 1/2–1 inch long, strongly flattened and slightly twisted; glumes narrow, very unequal, short-awned, the lower shorter than the enclosed floret, the upper one surpassing the floret; lemma apex with 2 rounded teeth, awnless, often equaled or exceeded by the palea. Grain very rarely produced.

WI range and habitats Locally frequent almost throughout (not known from some far northern counties), confined mostly to wet prairie remnants, moist depressions in medium prairies, fens, sandy borders of lakes and sloughs, and damp, grassy to wet, marly spots along roadsides and railroads; prefers heavy lowland soils but sometimes grows in dry sand. *Curtis Prairie, Greene Prairie*

Cord grass was formerly abundant and widespread in the tall-grass prairie region and common eastward to New England. It provides good cover for wildlife, and because of its extensive root system, it can be used to stabilize muddy soils.

The genus name is derived from the Greek *spartine,* a cord made from *Spartium junceum,* Spanish broom (a legume), and was probably adopted for *Spartina* because of its tough leaves.

One Spartina *leaf outlined in white*

Prairie Dropseed
Sporobolus heterolepis

This distinctive grass produces an extremely dense, basal fountain of fine leaves and delicately scented, open panicles at the tops of the stems. The small, 1-flowered spikelets become much distended by the yellowish-brown, ball-like seed (achene), which falls readily from the spikelet when ripe. **Perennial, 12–28 inches tall, often in large circular clumps.**

Blades to 12 inches long, less than $1/8$ (mostly less than $1/16$) inch wide, flat, becoming somewhat folded lengthwise, tapering very gradually to a very slender tip; sheaths with a tuft of hairs at the throat and sometimes with long hairs along the margins; ligule a fringe of very short hairs. Plants turn golden-yellow in the fall.

Inflorescence an open but narrow panicle 2–9 inches long, the branches spreading at maturity, naked on the lower $1/3$, few-flowered above. Flowering *August–early September*

Spikelets with 1 (rarely 2) florets, pinkish-brown to blackish, $1/8$–$1/4$ inch long; glumes markedly unequal, the lower $1/2$ or less the length of spikelet and the upper one nearly equaling or exceeding the lemma; glumes and lemma very thin, awnless. Spikelets break up above the glumes. Achenes *late August–early October (November)*

WI range and habitats Infrequent south and west, dominant or prevalent in dry to wet-medium prairies, often toward the bottoms of hills and on river terraces where conditions are more medium than dry; usually an indicator of relict prairies. *Curtis Prairie, Greene Prairie, Sinaiko Overlook Prairie, West Knoll*

The unit of dispersal is usually the fruit but sometimes a floret. Unlike most grasses, the seed coat is free from instead of attached to the ovary wall of the fruit, which botanically makes it an achene instead of a grain. The genus name is derived from the Greek *sporos,* seed, and *bolos,* a throw, and refers to the seeds being sometimes forcibly ejected when the mucilaginous wall of the fruit dries.

This is another characteristic species of the tall-grass prairie. Perhaps the most handsome of the prairie grasses, it adds a touch of elegance to any planting, whether butterfly garden, naturalized meadow, or border. Plains Native Americans made nutritious flour from the achenes of prairie dropseed.

Other Poaceae

Redtop
Agrostis gigantea
Introduced, naturalized; perennial grass, recognized by its characteristically maroon inflorescences (greenish in shade).

Bent Grass
Hair grass, tickle grass
Agrostis hyemalis
Synonym *Agrostis scabra*
Native; filamentous perennial grass.

Bottlebrush Grass
Elymus hystrix
Synonym *Hystrix patula*
Native; perennial grass of savannas and sunny woodlands; unmistakable open spike looks like its namesake.

Quackgrass
Elytrigia repens
Synonym *Agropyron repens*
Introduced, naturalized; ecologically invasive perennial grass.

Poaceae • Grass Family

Purple Love Grass
Tumble grass
Eragrostis spectabilis
Native; perennial grass of roadsides and sandy areas, forming colonies.

Left inflorescence in full bloom; younger stamens on lower half

Timothy
Phleum pratense
Introduced, naturalized; perennial forage grass, 2–3 feet. Easily recognized by its dense, pencil-like heads.

Canada Bluegrass
Wiregrass
Poa compressa
Introduced, naturalized; ecologically invasive perennial grass.

Kentucky Bluegrass
Poa pratensis
Introduced, naturalized; ecologically invasive perennial lawn and pasture grass.

Brome Grasses
Bromus species

This is a genus of chiefly coarse, large-grained grasses with inflorescences that are loose or open panicles, usually drooping or nodding. Although a nearly worldwide grassland genus, *Bromus* occurs at extremely low frequency in the Arboretum. *Bromus ciliatus*, *B. inermis*, and *B. kalmii* have appeared rarely in quadrat surveys over the years.

Bromus is from *bromos,* Greek for oat. Bromes somewhat resemble wheat, and a few species are in fact useful as forage. However, many brome grasses develop sharp-barbed grain at maturity that are injurious to livestock (hence the name ripgut grass).

Fringed Brome
Bromus ciliatus
Native perennial grass, distinguishable by the long hairs along the margins of the lemmas.

Smooth Brome
Bromus inermis
Introduced, naturalized; ecologically invasive.

Kalm's Brome
Arctic brome, prairie brome
Bromus kalmii
Native perennial grass, easily recognized by the ± uniformly hairy lemmas.

Phloxes

Prior to blooming, prairie phloxes are inconspicuous, blending into the grass community; however, during flowering, they produce an incredible display of bright color and sweet scent. They superficially resemble some members of the Pink Family (Caryophyllaceae), having opposite simple entire leaves that are many times longer than broad, and handsome flowers arranged in flattish clusters at the tips of a few inflorescence branches. **Erect perennials. Smooth phlox: 18–32 inches tall, from a short rhizome; downy phlox: 8–25 inches tall, from a somewhat woody base.**

Leaves stalkless, linear to lance- or lance-egg-shaped, mostly 1–4 inches long, averaging $3\frac{1}{2}$ inches long in smooth phlox and 2 inches long in downy phlox. Sterile shoots few, neither evergreen nor prostrate, reclining, or ascending (as in some *Phlox* species). Smooth phlox has 11–18 pairs, whereas downy phlox has 5–14 pairs of leaves on the stem.

Flowers radially symmetrical, with 5 half-united, awl-shaped sepals and 5 bright pink to reddish-purple (or white) petals that are fused into a salver-shaped corolla, that is, one having

Smooth Phlox
Phlox glaberrima
Wisconsin Endangered

a slender tube that is abruptly expanded into a flat spreading part ½–¾ inch wide. The green portions of the calyx tube are separated by thin translucent margins that are distended or ruptured by the developing capsule. The 5 unequal stamens are borne on the corolla tube alternate with its lobes. Smooth phlox: *late June–July;* downy phlox: *mid May–late July*

Fruits ellipsoid capsules, splitting into three pieces (valves) that break away from the central column; seeds with a large basal appendage that functions in dispersal, both in expelling the seed from the capsule and attracting ants. Smooth phlox: no data for Wisconsin; downy phlox: *late June–late July*

WI range and habitats

Smooth phlox: Only in Racine and Kenosha counties, where locally common in rich dense, fen-like prairies, sedgy depressions, and lowland oak openings along or near Lake Michigan. Planted in *Curtis Prairie, Greene Prairie* (not native to the Madison area)

Downy phlox: Widespread except in the Northern Highlands, in dry to wet-medium prairies, prairie relicts along railroads, oak openings, and dry woodlands (especially oak barrens and burned-over jack pine stands). *Curtis Prairie, Greene Prairie, Juniper Knoll*

These two species are likely to be confused only with each other. In downy phlox the stems, flower stalks, sepals, and even the corolla tubes bear abundant fine multicellular hairs that are sometimes glandular. Smooth phlox has very similar foliage but is completely hairless. These species may be further distinguished by the fact that the style in downy phlox is short and that of smooth phlox elongate. Like other phloxes, they lack poisonous or medicinal properties but make excellent garden plants.

DOWNY PHLOX
Prairie phlox
Phlox pilosa

Spreading Jacob's-ladder
Greek-valerian
Polemonium reptans

Bumblebees and gardeners are among the first to celebrate the appearance of this well-known spring wildflower. Its deep blue, bell-shaped flowers, which open to welcome the morning sun, and attractive, ladder-like leaves make Jacob's-ladder distinctive. **Perennial, 8–16 inches tall, the stems 1–several from a tough crown.**

Leaves pinnately compound, the egg- or lance-shaped to oval leaflets thin, mostly in pairs, the basal and lower with 11–17 leaflets, the upper leaves with 7–11 leaflets almost as large as those of middle leaves.

Flowers hanging in long-stalked, few-flowered clusters in which the central flowers bloom first, the clusters making up a rather open inflorescence; sepals 5, united into a green, bell-shaped, minutely hairy calyx, the broadly triangular lobes nearly as long as the tube; petals 5, joined to form a deep blue bell about $\frac{1}{2}$ inch long, the tube short, the lobes ascending to spreading, rounded; stamens 5, included within the corolla. Calyx enlarging in age, becoming papery-textured in fruit. *Early May–mid June*

Fruits capsules that split open along the median line of each seed chamber. *Late May–August*

WI range and habitats Common southwest, in dry-medium forests, rich, often moist hardwood or mixed pine-maple forests, medium oak savannas, and moist meadows, uncommon in wet-medium to wet prairies and fens, sometimes along roadsides and railroads. *Gallistel Woods, Greene Prairie, Juniper Knoll, Sinaiko Overlook Prairie*

The common name refers to the Biblical story of Jacob's vision of angels ascending a ladder into heaven, the paired leaflets climbing the leaf axis like the rungs of a ladder. The species epithet *reptans* means creeping, but this is a misnomer. The erect to loosely spreading stems do not run along the surface of the ground, and they never root.

The flowers are nodding in bud but straighten out as they open. A portion of the corolla reflects ultraviolet light, whereas other parts absorb ultraviolet. As a result, bees perceive an impressive pattern (invisible to humans) of three discreet color zones, which attracts them to the flowers.

Purple Milkwort
Bitter milkwort, racemed milkwort
Polygala polygama

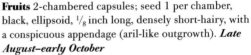

Small but interesting, purple milkwort produces two types of flowers: normal, rose-pink to rose-purple (rarely nearly white) flowers in a terminal raceme, and small, white or pale green flowers in prostrate racemes at the base of the stem, these usually buried but occasionally just above the soil surface. **Biennial or weak perennial, 2–11 inches tall, from a taproot.**

Leaves alternate (lowest sometimes opposite), the lowest blunt to rounded, the rest lance- to linear-oblong and ± sharp-pointed, $3/8$–$1\,1/4$ inches long. **Stems** radiating from a crown.

Flowers $1/8$–$3/16$ inch long, on stalks $1/8$ or less inch long, in loose racemes to 4 inches long; sepals 5, the outer 3 green, small, the lateral 2 colored like the petals, much larger, $1/8$–$1/4$ inch long; petals 3, the lower keel-like, 3-lobed, with a fringe-like crest, united at the base with each other and the 7 stamens with united filaments. *Early June–early August (petal-bearing, aerial flowers)*

Fruits 2-chambered capsules; seed 1 per chamber, black, ellipsoid, $1/8$ inch long, densely short-hairy, with a conspicuous appendage (aril-like outgrowth). *Late August–early October*

WI range and habitats Frequent southeast, central, and north, in dry, sandy, gravelly, or rocky prairies, sand barrens, sand blows, open woods (oak, pine, red cedar), pastured slopes, fields, and roadsides; sometimes in moist sand near lakes and ponds, rarely on sandstone ledges and granite outcrops. *West Knoll*

The intricate flowers of milkworts superficially resemble those of the Pea Family; the inner 2 sepals are petal-like and stalked and are called wings, and the lower petal forms a keel. Besides these showy blossoms, purple milkwort produces an abundance of non-opening, self-pollinating flowers that form small plump capsules along horizontal stems from the base of the plant. The former promote genetic variation in offspring through cross-fertilization; the small colorless flowers ensure seed set by self-fertilization.

Field Milkwort
Blood milkwort
Polygala sanguinea

Our milkworts are all small plants with very irregular flowers. Field milkwort has tightly overlapped flowers in a compact thimble at the top of a low slender stem. **Erect annual, (2) 4–16 inches tall, from a slender taproot.**

Leaves numerous, all alternate, erect or ascending, simple, linear, linear-oblong, or narrowly elliptic, $1/16$ inch wide, with a very small, abrupt tip. **Stem** solitary, simple or branched above.

Flowers greenish and rosy or reddish-purple (rarely whitish), in very dense, head-like racemes, the floriferous portion $3/8 - 3/4$ inch long and $1/4 - 1/2$ inch thick, rounded at the summit; sepals 5, the outer 3 small and greenish, the lateral 2 much larger and colored like the petals; petals 3, the lower keel-like and 3-lobed, united with each other and the 8 stamens, which themselves are also ± united into a sheath. *Early July–early October*

Fruits capsules, 2-chambered and 2-seeded, almost circular (except for the wedge-shaped base) and flattened, falling promptly; seeds blackish, pear-shaped, short-hairy, with an appendage (aril) of 2 flattened linear lobes arising in the area of the scar indicating the point of attachment. *Late July–October*

WI range and habitats Frequent southwest and far west, reappearing near the western shore of Green Bay, in sandy or rocky prairies, cedar glades, open woods, and outcrops, less often on shores and edges of sedge meadows and bogs, abandoned fields, ditches, and trails; dry to moist, open habitats, generally in acidic soils. *Curtis Prairie, Greene Prairie*

The genus name is derived from the Greek words for much (*polys*) and milk (*gala*). Rather than referring to the plant's having milky juice, it commemorates the reputed virtue of the common European species *Polygala vulgaris* to increase lactation in humans.

Polygala flowers are strongly asymmetrical, in structure somewhat resembling those of the Pea or Bean Family, only in their case the 2 wings are large, petal-like sepals, and the keel is comprised of only 1 petal.

Seneca Snakeroot
Polygala senega

Clumps of low, curved-ascending stems (as many as 15 or 20 or more) and solitary petite racemes of small white flowers make Seneca snakeroot distinctive. **Perennial, 4–20 inches tall, with a woody root and tough crown.**

Leaves alternate, crowded, lance- to lance-egg-shaped or elliptic, anywhere from $1/16$–$1 3/8$ inches wide, blunt to sharp-pointed at both ends, hard-margined. Lower leaves reduced or scale-like.

Flowers in a dense, conic or conic-cylindrical, spike-like raceme $1/4$–$3/8$ inch thick, this terminal on a stalk about $1/2$–1 inch long; sepals 5, petal-like, the 2 lateral ones almost circular, larger than the others, tapered into long, narrow bases; petals 3, united into a tube and irregularly cut as if torn at the apex; stamens 7. Lower flowers develop first and have already fruited by the time the upper flowers open. *Early May–June*

Fruits plump rounded capsules, $1/8$ inch broad, each containing 2 sparsely hairy seeds with a white appendage nearly or quite as long as the body of the seed. *Early June–late July*

WI range and habitats
Locally frequent south, in a diversity of dry to moist woods (especially clearings and borders), thickets, and prairies or prairie-like habitats, in fens, damp brushy banks, dunes, and shores. ***Curtis Prairie, Greene Prairie***

Senega is an old genus name, adopted because the plant was the source of a medicinal extract used by the Seneca, Ho-chunk, Dakota, and probably other Native American tribes to cure snake bites. Introduced into European medicine in the early 1700s, Seneca snakeroot became sought after as a remedy for treating respiratory problems, such as pleurisy. The official drug is obtained from the dried root, collected in autumn, and is used as an expectorant for bronchitis and asthma.

WHORLED MILKWORT
Polygala verticillata

This is a low-growing, delicate plant, distinctive but not often seen, having at least the lower leaves whorled. Neither the small, white or greenish-white flowers (rarely purple-tinged) nor the early-falling fruits are prominent. **Slender annual, 2–16 inches tall.**

Leaves mostly if not all in whorls of 2–5, or the upper alternate or scattered, linear-oblong or narrowly lance-shaped, no more than $1/8$ inch wide, tipped with a short rigid point. **Stems** single, usually few-branched above.

Flowers in spike-like racemes borne on stalks $1/4$–$1 1/2$ inches long, the conic to cylindrical floriferous portion to 2 inches long and less than $1/4$ inch thick, tapering to the tip; sepals 5, unequal, the 2 lateral ones larger and colored like the petals; petals 3, united into a tube, the middle lobe of which has a fringe-like crest on each side of the flat part; stamens 8. *Early July–late September*

Fruits tiny oval capsules, $1/16$ inch long; seeds nearly black, minutely hairy, with a lobed appendage $1/3$–$3/4$ as long as the seed. The lower fruits often drop promptly. *Late August–late September*

WI range and habitats Occasional southwest, in dry to moist prairies (especially "goat prairies," hill prairies, and sand prairies), sometimes in sandy black oak savannas or cedar glades, and open grassy sterile ground. *Curtis Prairie, Greene Prairie, West Knoll*

The small fatty seed appendage, an *aril*, implies that milkworts are dispersed by ants, which seize seeds like these with their mandibles, almost always by the arils, carry them to their nest, and feed the arils to their larvae. Afterwards, the otherwise intact seeds are placed into the nest's garbage dump, where, perhaps better protected or in more nutrient-rich soils than seeds of other species, they are free to germinate.

This variable species is represented in Wisconsin by two of the three currently accepted taxa, variety *isocycla* (almost all of our plants) and var. *ambigua* (or *Polygala ambigua*, rarely encountered), which is sometimes recognized as a distinct species. They are distinguished by the shape of the inflorescence, the length of the wings relative to the mature capsule, and the length of the aril relative to the seed.

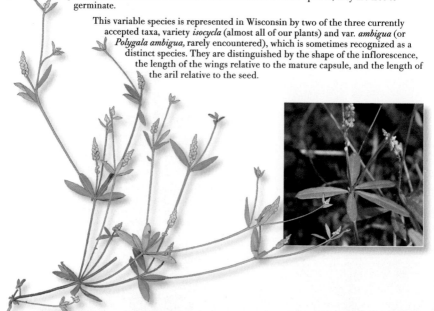

Water Smartweed
Water heart's-ease
Persicaria amphibia
Synonyms *Polygonum amphibium, P. coccineum, P. natans*

This is one of the more attractive smartweeds, having handsome, rose-pink spikes of densely crowded flowers. As connoted by the species epithet, is at home both in the water and on land. **Prostrate or ascending to erect perennial, 1–5 feet (terrestrial plants) or to 10–40 feet (aquatic plants) long, spreading by long tough rhizomes and runners to form colonies in shallow water and on mudflats and banks.**

Aquatic forms: stems and branches floating or submersed, with floating oval leaves $1\frac{1}{2}$–8 inches long and $\frac{3}{8}$–$1\frac{1}{2}$ inches wide, on stalks $\frac{3}{8}$–3 inches long, rounded to blunt or somewhat pointed at the apex, leathery, hairless; stipular sheaths membranous. **Semi-aquatic and terrestrial forms:** leaves spreading to ascending, stalkless or short-stalked, oblong-egg- or lance-egg-shaped, to 19 inches long, the apex forming an acute angle or short-tapering to a point, hairy, the papery sheath with or without a flared green flange.

Flowers in 1 (sometimes 2) terminal spikes, these egg-shaped and $\frac{5}{8}$–$1\frac{1}{2}$ inch long, on hairless stalks (aquatic extreme) or cylindrical and 1–6 inches long (in the water or not, the stalk and axis hairy and glandular); tepals 5, lobed to below the middle, $\frac{1}{4}$ inch long or a little less; stamens or style (not both) protruding. *(Early) late June–September*

Fruits biconvex achenes, broadly egg-shaped to circular in outline, dark brown to black, $\frac{1}{8}$ inch long, shiny or dull. **Mid July–early October**

WI range and habitats Throughout, in or along lakes, ponds, sloughs, bog pools, quiet streams, and ditches, or terrestrial forms in adjacent drier ground (shores, marshes, low prairies, fields, embankments); persisting when its habitat dries up or is drained, and spreading up banks to higher ground. **Curtis Prairie, Lower Greene Prairie**

Our members of this complex species consist of two intergrading ecotypic varieties that were once variously separated into two species and many varieties and forms based mostly on vegetative characters (habit, leaf shape, hairiness). Using these characters, specimens from the same rhizome may be referred to different taxa depending on whether they came from deep water, shallow water, or drier ground. Like smartweeds in general, this species is an excellent food source for ducks and other waterfowl.

WATER-PEPPER
Marsh-pepper knotweed
Persicaria hydropiper
Synonym *Polygonum hydropiper*

Although not considered attractive, the smartweeds *en masse* are capable of producing a display of color. Water-pepper is distinguished from other smartweeds by the combination of sheaths fringed with a row of bristles, stems and inflorescence stalks not glandular, and tepals covered with shiny yellow or dark, glandular dots. **Erect to sprawling annual, 1–3 feet high.**

Leaves alternate, nearly stalkless or short-stalked, narrowly lance-shaped, 1–4 inches long, hairless, peppery-tasting. Sheaths around the stem (derived from stipules) $1/4$–$1/2$ inch long, membranous, those from the middle nodes upward swollen by enclosed, self-pollinating flowers. **Stems** sometimes rooting at lower nodes, branched or unbranched, reddish, smooth.

Flowers numerous, $1/8$ or less inch long, in terminal and axillary, erect to arching, spike-like racemes (actually in small close clusters that collectively simulate racemes); tepals mostly 4 (5), nearly equal, greenish with narrow, rose-colored or white margins; stamens 6. *Mid July–August*

Fruits biconvex or usually bluntly 3-angled achenes, brownish black, small, dull, enclosed by the persistent tepals. *(Mid) late July–late November*

WI range and habitats Infrequent to locally abundant throughout, along muddy or sandy shores of streams, lakes, ponds, and sloughs, in swamps and rich woods (especially along roads), weedy in slightly disturbed areas (low wet fields, pastures, dried-up ponds, ditches). ***Curtis Prairie, Lower Greene Prairie***

This species is probably naturalized from Europe although some authors think our plants are in part native.

As their common names indicate, the knotweeds and smartweeds (often segregated as distinct genera, *Polygonum* and *Persicaria*, respectively) are small, mostly weedy herbs with relatively small flowers lacking showy sepals and petals. The family is easily recognized by the swollen nodes and thin dry sheaths, called *ocrea*, just above the bases of the leaves.

Similar species Water-pepper resembles other common species in our flora, especially dotted smartweed (*Persicaria punctata*) and false water-pepper (*P. hydropiperoides*). The former has interrupted spikes of white flowers and shiny achenes. False water-pepper differs from these in that the calyx lacks glandular dots and the achenes are sharply triangular.

Arrow-leaved Tear-thumb
Persicaria sagittata
Synonym *Polygonum sagittatum*

Tear-thumbs have slender branching stems with long internodes and are armed with downward-pointing prickles on stem angles, leaf stalks, inflorescence stalks, and midribs beneath. The sprawling habit and arrowhead-shaped leaves make arrow-leaved tear-thumb easily identifiable. **Slender, weak-stemmed annual, 1–6 feet long.**

Leaves egg-lance- or oblong-lance-shaped to oval, long- (lower leaves) to short-stalked (upper ones), $3/4$–4 inches long and $1/4$–1 inch wide, arrowhead- to heart-shaped at the base (the short lobes directed downward); stipular sheaths $3/16$–$3/8$ inch long, transparent, with a few hairs on the margins. **Stems** entangled with or reclining on other vegetation, 4-angled.

Flowers in short, head-like racemes to $3/8$ inch long, these on long slender stalks at ends of stems or from leaf axils; tepals white or greenish-white, often with pinkish tinges, 5-parted to below the middle, $1/8$ inch long; stamens 8, the anthers pink. *Early July–early September*

Fruits sharply 3-angled achenes, brown to black, $1/8$ inch long, dull to shiny. *Late July–early October*

WI range and habitats Infrequent to frequent throughout, generally in marshy or boggy places: marshes, wet meadows, shrub carrs, bog margins (including commercial cranberry bogs), burned wetlands, swamps (deciduous and coniferous), and about streams, ponds, sloughs, and swales, occasionally in damp fields and wet roadsides. *Lower Greene Prairie*

Although the Buckwheat Family is easily recognized, within-family classification is much in dispute, particularly the circumscription of *Polygonum* in the broad sense. Revised taxonomies treat one or most of its several species groups as segregate genera, most often separating the smartweeds as *Persicaria* and leaving the knotweeds in *Polygonum*.

Similar species Halberd-leaved tear-thumb (*Persicaria arifolia*) also has reclining stems with hooked prickles, but its leaves are hastate, that is, the basal lobes are triangular and divergent. Moreover, the styles are only $1/3$ as long, and the achenes are 2-sided. It is not found in the Arboretum.

Sheep Sorrel
Field sorrel, red sorrel
Rumex acetosella
Invasive weed

Like the smartweeds and knotweeds, many of the docks (*Rumex* species) are definitely weeds, including the little sheep sorrel. This common species is easily recognizable by its low stature and distinctive foliage. In bloom the numerous tiny, 6-tepaled flowers, variously greenish or suffused with yellowish or reddish and not at all showy, differentiate from everything else in our flora. **Slender perennial, 4–20 inches tall, from spreading rhizomes $1/8$ inch thick, forming colonies.**

Leaves variable, at least some narrowly arrowhead-shaped with spreading lobes, mostly $3/4$–2 inches long and $3/8$–$3/4$ inches wide (basal leaves often larger), smaller and simpler upward. Membranous sheaths prominent but usually disintegrating by fruiting time, brownish at the base, silvery and irregularly cut above. **Stems** single or clumped, rising above a basal rosette. Foliage hairless, sour to the taste.

Flowers unisexual (male and female on different plants), less than $1/16$ inch long, on slender stalks with an evident joint just below the flower, in small whorls that combined form loose narrow branched panicles $1/2$–$2/3$ the height of the stem; female flowers with 3 small scaly tepals below 3 egg-shaped ones (valves); male flowers with 6 erect, scale-like tepals and 6 stamens. *Mid May–mid July* (sporadically to late September)

Fruits yellowish- or reddish-brown, 3-sided achenes $1/16$ inch long, enveloped by the inner tepals (valves), which are scarcely changed in fruit and barely exceed the achene. *Late June–September*

WI range and habitats Frequent to common throughout the state, locally abundant in pastures, fields (abandoned or cultivated), sand flats, rock outcrops, roadsides, and railroads; flourishing in disturbed, cultivated, or waste ground; prefers poor, usually acidic soils, but invades woods and better soils. *Grady Tract knolls, Upper Greene Prairie*

Sheep sorrel is highly variable, comprising a complex of polyploid races centered in Eurasia. Some of these races are now naturalized almost worldwide.

Other Polygonaceae

Heart's-ease
Spotted lady's-thumb
Persicaria maculosa
Synonym *Polygonum persicaria*
Introduced, naturalized;
erect annual forb, 2–6 feet.
Flowers July–September

Curly Dock
Sour dock
Rumex crispus
Introduced, naturalized;
perennial forb, 1–4 feet.
The brown dried "coffee
grounds" inflorescences are
familiar in disturbed areas.
Flowers June–August

Eastern Shooting-star
Dodecatheon meadia

This captivating wild flower is well-named; its swept-back petals give the showy flowers the appearance of a shooting-star. Its colors vary; blooms of deep magenta are seen within the same population as pinks to pure whites. Flowers are erect in bud, nodding when open, and erect again in fruit. **Erect perennial, 8–24 inches tall, with very short rhizomes and white, fleshy-fibrous roots.**

Leaves clustered at the plant base, egg-shaped to oblong-elliptic to inverse-lance-shaped, 4–12 inches long (including the reddish stalk), entire or with hard points along the margin. Foliage disappears after the plants have bloomed, but the dried-up stalks persist.

Flowers in an umbel atop a solitary leafless stem, with a very short, tubular calyx and a regular corolla, the petals partially united into a very short tube, the lobes sharply swept backward from a yellowish collar, several times as long as the tube. Stamens inserted on the corolla tube, coming together (but not fused) into a protruding colorful cone that surrounds the single, slightly protruding style. *Early May–June*

Fruits 1-chambered, firm-walled capsules, dark reddish-brown, egg-shaped to oblong, $5/16$ inch long, opening at the tip into 5 teeth-like segments, filled with many seeds. *Late June–August*

WI range and habitats
Originally abundant southeast and subsequently surviving in prairie relicts but now disappearing; dry to wet-medium prairies, fens, open deciduous woods, oak savannas, bluffs and cliffs, and along rights-of-way. ***Curtis Prairie, Greene Prairie, Gallistel Woods, Noe Woods, Sinaiko Overlook Prairie***

Shooting-star flowers offer neither scent nor nectar. They are pollinated only by bumblebees and certain solitary bees able to perform buzz pollination. A bee grasps the base of the anther cone with its mandibles, curls its body under the anther cone, and by very rapid vibration of thoracic muscles (producing an audible buzz) causes the release of pollen from the anthers. Other flowers built for buzz pollination are tomatoes, potatoes, eggplants, and in our flora, partridge pea, Virginia meadow-beauty, and cranberries.

The genus name is derived from the Greek *dodeca*, twelve, and *theos*, god, and was given by Pliny to the primrose. Modern molecular studies of relationships within the Primrose Family show that *Dodecatheon* should be included in the genus *Primula*.

Narrow-Leaved Loosestrife
Smooth loosestrife
Lysimachia quadriflora

Yellow flowers of this wetland herb resemble those of other widespread loosestrife species, but the combination of solitary nodding flowers and rather rigid leaves that lack marginal hairs at the base distinguish narrow-leaved loosestrife. **Perennial, 1–2 feet tall, from a thickish crown, forming rhizome-like offshoots.**

Basal leaves sometimes persistent as a rosette, elliptic to inverse-egg-shaped. **Stem leaves** opposite (appearing whorled owing to leaf-bundles in the axils), linear to narrowly lance-shaped, averaging less than $1/4$ inch wide, tapering to the stalkless base, firm, with prominent midveins, obscure lateral veins, and margins rolled under (toward the lower side). **Stems** simple or short-branched above, slender, 4-angled.

Flowers solitary in the axils of foliage leaves, on slender stalks that soon become longer than the flowers; calyx 5-parted nearly to the base, the lance-shaped sepals with gradually tapering, slender tips, persistent in fruit; petals likewise basally united into a very short tube, forming a deeply 5-parted, wheel-shaped corolla, the egg-shaped divisions minutely toothed as if gnawed, densely glandular and sometimes with red-purple coloration toward the base on the inner surface. Stamens fused to the corolla opposite the lobes and alternate with small membranous sterile stamens. *Early July–August*

Fruits 1-chambered, egg-shaped (almost spherical) capsules that break into 5 parts to release the many 3-angled seeds. *Late July–October*

WI range and habitats Locally frequent southeast (reappearing in Door County, lacking in the Driftless Area), in marshes, low prairies, fens, and swales, also around calcareous springs; wet grassy limy habitats. *Curtis Prairie, Greene Prairie*

Similar species Narrow-leaved loosestrife is one of 10 *Lysimachia* species in Wisconsin. Lance-leaved loosestrife (*L. lanceolata*) also has solitary flowers, but the slender stems arise from distinctive long, runner-like rhizomes, and the somewhat membranous leaf blades are evidently veined, fringed with bristly hairs at the base, and not rolled under at the margins. Swamp-candles (*L. terrestris*) and swamp loosestrife (*L. thyrsiflora*) have flowers in racemes, terminal in the former, axillary and short in the latter; both have leaves and corollas dark-dotted or streaked.

Other Primulaceae

Fringed Loosestrife
Lysimachia ciliata
Native; erect perennial forb, 10–40 inches. Common in moist or wet habitats such as lowland forests, damp meadows, marshes, and low prairies.
Flowers June–July

Canada Anemone
Meadow anemone
Anemone canadensis

The masses of jagged-edged, deeply cleft or cut leaves of Canada anemone are distinctive even when flowering time is past. Leaf blades are broad and coarse for this genus and bracts of the involucre stalkless or nearly so, distinguishing it from other anemones. The pure white flowers are solitary on each long leafless stalk arising from the involucre.
Perennial, 8–36 inches tall, rhizomatous, forming patches.

Basal leaves long-stalked, 3–5-parted into oblong or wedge-shaped divisions, these in turn divided or merely irregularly cut and toothed. **Stem leaves** stalkless or nearly so, collectively constituting a whorl of bracts (involucre) below and remote from the flower, the segments cleft, coarsely toothed, and prominently net-veined. Plants slightly hairy.

Flowers $3/4$–$1 3/4$ inches broad, the 4–6 (usually 5) petal-like sepals (petals none) hairy on the back; stamens and pistils numerous. Flowers generally not long lasting. ***Late May–July*** (sporadically to early September)

Fruits flattened hairy achenes, nearly circular in outline, $1/8$ inch long, with long tips, many aggregated into a compact spherical head $3/8$–$3/4$ inch in diameter. ***Late June–September***

WI range and habitats Locally common throughout, in sedge meadows, prairies, marshes, swales, shrub carrs, borders and clearings in woods, stony shores, grassy or brushy banks, ditches, and rights-of-way; medium to moist ground, open or partly shaded. ***Curtis Prairie, Greene Prairie***

The flowers of anemones have much the same structure and arrangement as do buttercups, but there are no petals. The principal leaves are basal; the stem leaves are opposite or whorled and form a whorl called an involucre. Canada anemone often survives after its native associates have been destroyed; it therefore can be found growing among non-native weeds in floodplain pastures, drained swamps, ditches, and prairie edges. It is also one of the easiest of our northeastern wildflowers to grow in the home garden.

Many native tribes used the leaves and roots of Canada anemone to treat wounds. The Chippewa chewed the root to stop internal bleeding and placed pieces of it in the nostrils to stop nosebleeds. The Ojibwe used the root to prepare a remedy for sore throats and to clear the throats of singers.

Candle Anemone
Long-headed anemone
Anemone cylindrica

Nearly compound, jaggedly lobed and toothed leaves, whorled at the base of the flower stalks but remote from the flowers, combined with the oblong fruiting heads, distinguish this elegant prairie wildflower. **Single-stemmed perennial, 12–28 inches tall, from a short stout crown.**

Basal leaves deeply 3-parted into broadly diamond- to wedge-shaped segments, these usually cleft and irregularly toothed only above the middle, with straight margins toward the base, almost leathery in texture. **Stem leaves** 3–9 (often 5-7), much like the basal but smaller, the largest with blades 1–2½ inches long. Plants slightly gray-silky.

Flowers ⅜–¾ inch across, single on each of the 2–8 long, mostly leafless stalks, the parts on an oblong, thimble-shaped receptacle 1–2 inches tall, about 3 times as long as thick; sepals 4–6, petal-like, whitish (drab, creamy, or greenish), ⅜–½ inch long; stamens and pistils numerous. *Late May–July*

Fruits small achenes, numerous in dense oblong "heads" ¾–1⅜ inches long and ¼–⅜ inch thick (2–4 times as long as thick) that turn densely white-cottony. *Early July–November* (persisting until early the following spring)

WI range and habitats Throughout all but the Northern Highlands, in dry to medium prairie remnants, scrub oak and/or jack pine barrens, oak openings, dry open woodlands, sandy abandoned fields, also sand plains, limestone flats, bluffs, outcrops, banks, dunes, roadsides, and railroads. *Curtis Prairie, Grady Tract knolls, Greene Prairie*

Like some other genera of the Buttercup Family, anemone flowers lack petals; instead, they feature large, petal-like sepals. The involucral leaves are manifestly stalked, and the flowering stalks are invariably 1-flowered. The numerous pistils in the center of the flower together with the floral axis (receptacle) form a green, thimble-shaped structure that persists when the plant is in fruit.

Similar species Ordinarily this plant can be distinguished from thimbleweed or tall anemone, *Anemone virginiana,* by the number of involucral leaves, typically more than 3, and the elongate flower stalks. When mature, candle anemone's heads of achenes are oblong; heads of thimbleweed are more egg-shaped and thicker.

American Pasqueflower
Prairie-smoke
Anemone patens
Synonym *Pulsatilla patens*

Prior generations living on the northern prairies eagerly sought this plant in flower, because its lovely, pale bluish-purple to white flowers were regarded as the earliest harbinger of spring. After flowering, the short stalks grow several more inches to give the large fluffy seed heads better exposure to the wind. **Tufted perennial, 4–16 inches tall, from a branched crown.**

Leaves divided into 3 main divisions, each division again 2- or 3-parted, the terminal segments linear or narrowly lance-shaped, averaging about $1/8$ inch wide. Basal leaves few to many, developing as or after the flowers bloom; leaves of the upper stem stalkless, constituting a whorl of bracts at the base of the flower stalk. Entire plant clothed with long gray silky hairs but becoming hairless in age.

Flowers erect, cup-shaped, with 5–8 petal-like sepals usually $1-1\frac{1}{2}$ inches long; stamens numerous, some of them reduced to glands and appearing like petals; pistils numerous, separate. *Late March–May*

Fruits silky achenes clustered into a dense head, each with a long feathery tail $3/4-1\frac{5}{8}$ inches long that can be carried for some distance in the wind. *Mid May–early July*

WI range and habitats Locally abundant southwest, in dry prairie remnants and open scrub oak-jack pine barrens, rarely in medium prairies, on sandy, gravelly, or rocky hillsides, glacial outwash, bluffs, cliffs, and undulating land with outcrops. *West Knoll* (planted)

Although known by a number of common names, this species is most commonly called pasqueflower, because it blooms during the Pasque season of Passover and Easter. The genus name, a derivative of the Greek *anemos,* wind, may imply that the plants bloom in early spring when the wind still howls. It could also be a combination of *anemos* plus *mone,* habitat, suggesting that anemones are found most frequently in windy places. Ancient legends tell their own stories about the naming of the genus.

The pasqueflowers are sometimes segregated into the genus *Pulsatilla*. They differ from anemones in the strict sense in that the hairy styles remain attached to the fruits and elongate, forming long feathery appendages similar to those of *Geum triflorum,* also (and more commonly) called prairie-smoke, and *Clematis virginiana,* virgin's-bower.

THIMBLEWEED
Tall anemone
Anemone virginiana

Thimbleweed is like candle anemone in having medium-sized, whitish flowers with a thimble-shaped receptacle in the center, and is similar in other characteristics. They are distinguished by the structure of the involucre and shape of the thimble-like heads, which can be dried and dyed for use in cut-flower arrangements. **Single-stemmed perennial, 1–3 feet tall, from a short thickened crown.**

Basal leaves divided into 3–5 somewhat diamond-shaped segments, broadly wedge-shaped, convex- or concave-curved or straight-sided toward the base, usually shallowly once or twice cleft, coarsely toothed to or below the middle, relatively thin-textured. **Stem leaves** 3 (rarely 4 or 5), similar to the basal leaves, manifestly stalked, whorled at the base of the flower stalks to form an involucre, the blades $2-5\frac{1}{2}$ inches long. Plants lightly hairy.

Flowers 3–9 per plant (solitary or rarely 2 or 3 on each of the 1–3 long slender stalks), all but the leading stalk with a pair of secondary bracts near the middle; sepals usually 5, petal-like, white to greenish- or yellowish-white, $\frac{3}{8}-\frac{3}{4}$ inch long, surrounding numerous stamens and numerous pistils on an egg-shaped to ellipsoid head that is about twice as tall ($\frac{1}{2}-1$ inch) as thick ($\frac{1}{4}-\frac{1}{2}$ inch). *Late May–late July*

Fruits small achenes in an egg-shaped, ellipsoid, or oblong head $\frac{1}{2}-1$ inch long, $\frac{1}{4}-\frac{5}{8}$ inch thick (less than twice as long as thick), concealed by dense cottony hairs. *Mid August–October* (undoubtedly later)

WI range and habitats Throughout, in dry to moist, disturbed or open woods, savannas, and thickets, wooded bluffs, ravines, and roadsides, sometimes in more open habitats like medium prairies, old lowland pastures, and fields. *Curtis Prairie, Gallistel Woods, Greene Prairie, Juniper Knoll, Marion Dunn Prairie, Noe Woods, Sinaiko Overlook Prairie, Wingra Woods*

Similar species See candle anemone (*Anemone cylindrica*). The name *Anemone riparia* (= *A. virginiana* variety *alba*) has been applied to northern thimbleweeds that approach candle anemone in having somewhat slenderer heads and leaf segments more wedge-shaped toward the base. Both types are somewhat common in Wisconsin but are difficult to distinguish.

WILD COLUMBINE
Canadian columbine, red columbine
Aquilegia canadensis
Poisonous

The nodding, curiously shaped flowers that adorn this exquisite eastern columbine are unlike anything else in our flora, each of the 5 petals being prolonged backward into a long spur. To the casual observer, they give no hint of their kinship to more familiar family members like buttercups, anemones, and marsh marigold. **Perennial, mostly 1–3 feet high, from a stout rhizome.**

Leaves compound, each leaf 2 or 3 times divided into 3s, bright olive green above, whitened with a bloom beneath; leaflets wedge-shaped and shallowly to deeply scalloped.

Flowers $1\frac{1}{4}$–$1\frac{1}{2}$ inches long, scarlet outside (the spurs) but yellow within (petal blades), in an open panicle; sepals 5; petals (4) 5, all extended into a conspicuous slender tube with a bulb at the end containing nectar. A tuft of many yellow projecting stamens (plus an inner series of sterile stamens) and usually 5 pistils is in the center. *May–July*

Fruits several pods (follicles) $\frac{5}{8}$–1 inch long, slightly spreading at the tips, enclosing numerous shiny black seeds. *Mid May–early September*

WI range and habitats Throughout, in dry to medium or even low open woods and forests, especially borders and clearings, savannas, cedar glades, on wooded to open hillsides, bluffs, cliffs, outcrops, and ledges, as well as beach ridges, gravelly shores, quarries, and roadsides; almost always in shade, rarely in prairies. *Curtis Prairie, Juniper Knoll, Wingra Woods*

The genus name may be derived from *aquila,* Latin for eagle, likening the spurs to an eagle's claws, or from *aqua,* water, and *legere,* to collect, referring to the nectar that collects at the base of the hollow spurs. The nectar can be reached only by large, long-tongued bees, moths, and hummingbirds, which pollinate the flowers while probing for nectar. Some insects, especially bumblebees, cheat the system, biting holes in the ends of the spurs to gain access to the nectar. Leaf miners frequently eat their way through the inner layers of the leaflets, making tunnels as they go until becoming large enough to drop out of the leaf.

The roots of columbines contain a number of alkaloids, conferring some importance to the plants in folk medicine but also making them poisonous.

Virgin's-Bower
Devil's-darning-needle
Clematis virginiana

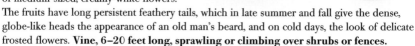

Individual vines, whether male or female, are particularly handsome during the second half of summer, when dressed with small to large inflorescences of medium-sized, creamy-white flowers. The fruits have long persistent feathery tails, which in late summer and fall give the dense, globe-like heads the appearance of an old man's beard, and on cold days, the look of delicate frosted flowers. **Vine, 6–20 feet long, sprawling or climbing over shrubs or fences.**

Leaves opposite, pinnately compound, each with 3 egg- to lance-shaped, stalked leaflets $1\frac{1}{2}$–$3\frac{1}{2}$ inches long, somewhat lobed or coarsely few-toothed. Plants climbing by the bending or clasping of the leaf stalks; stems ± herbaceous, the lower part becoming woody.

Flowers numerous, $\frac{3}{8}$–1 inch across, borne in delicate flattish clusters aggregated into an axillary compound cluster of limited size; sepals 4 or 5, spreading, petal-like, $\frac{3}{8}$–$\frac{5}{8}$ inch long, surrounding a center of many slender, greenish-yellow stamens or white styles. Individual plants produce flowers that are all either male (without pistils) or female (often with sterile stamens). *End of May–late August*

Fruits brown flattened hairy achenes, each attenuated into a single curvy tail $\frac{3}{4}$–2 inches long. *(Mid July) August–October*

WI range and habitats Somewhat uncommon throughout, in medium to wet woods and swamps (deciduous and coniferous), especially in clearings and on borders, low thickets, stream banks, marshy or weedy roadsides, and railroad embankments, also on dry outcrops and Great Lakes beaches. *East Curtis Prairie*

This native vine belongs to the same genus as the popular garden clematises, but its individual flowers are much smaller. Virgin's-bower, indeed all clematises, scarcely twine; instead of having tendrils like grapes, they climb by twisting and wrapping the stalks of their compound leaves around other plants. The genus name *Clematis* is derived from the Greek *klema*, a plant shoot.

Similar species Virgin's-bower is easily distinguished from Wisconsin's other native clematis species, purple clematis (*Clematis occidentalis*), which has solitary, reddish-purple flowers that are twice as large. Purple clematis occurs primarily in northwest Wisconsin (south to Wisconsin Dells, Parfrey's Glen, and Blue Mounds) and is not found in the Arboretum.

Carolina Larkspur
Prairie larkspur
Delphinium carolinianum
Poisonous

Larkspur's oddly shaped flowers are one of Mother Nature's unique designs. Borne well above the finely cut, long-stalked leaves in a terminal raceme, each flower has a slender, backward-projecting tube or "spur" that resembles the spur of the horned lark. The stamens, pistils, and pods resemble those of columbine (*Aquilegia*). **Perennial, 18–32 inches tall, from a bundle of spindle-shaped roots.**

Basal leaves palmately divided into 3–5 primary divisions, these again 2–3-cleft into many linear or narrowly lance-shaped, sharply toothed divisions. **Upper stem leaves** less dissected. **Stems** wand-like, closely and minutely hairy.

Flowers 5–27 or more in a slender, often dense, spike-like raceme, white to greenish- or bluish-white, strongly irregular; sepals 5, the uppermost prolonged into a spur $1/2$–$3/4$ inch long, the other 4 petal-like; petals 4, in 2 unlike pairs, small, the upper pair continued backward into long spurs that are enclosed within the sepal spur; stamens many. *June–mid July*

Fruits 3 oblong pods (follicles) containing several small, light gray seeds that are covered with projecting wavy, scale-like ridges. *Late June–July*

WI range and habitats Far west-central, rare to locally common in dry-medium prairies and prairie-like habitats on limestone, dry sand prairies, cedar glades, and black oak-jack pine barrens, spreading to sandy or gravelly lakeshores, roadsides, and (rarely) abandoned sandy fields; scattered in sunny exposed spots. *West Knoll,* where introduced or adventive

Gardeners make the distinction between the annual and perennial species, calling the former larkspurs and the latter delphiniums. Flowers may be double in some cultivated forms.

The larkspurs contain a number of toxic alkaloids, and except for locoweeds (*Astragalus* species), they cause more cases of cattle poisoning than any other poisonous plant. Beside stomach upset, nausea, and constipation, these alkaloids cause nervous conditions and may be fatal if large amounts are ingested.

The Latin name comes from *delphinus,* dolphin, because of a fancied resemblance of the shape of the flower to classical sculptures of dolphins. This species is quite variable. Recent authors follow M. J. Warnock in recognizing four subspecies, one of which, subspecies *virescens,* of the prairies and plains, enters Wisconsin.

Tall Meadow-rue
Purple meadow-rue
Thalictrum dasycarpum

Large, lacy leaves and showy masses of delicate unisexual flowers identify the tallest of our meadow rues. Standing sentinel-like over midsummer grasses and herbs, its dark reddish to purple stems and bluish-green leaves lend color and texture to the marshes and wet meadows in which it flourishes. **Sturdy perennial, 2–7 feet tall, from a short thick rhizome.**

Basal and lower stem leaves 3–5 times compound into 3s. **Stem leaves** twice compound, stalkless or nearly so, dividing into 3 parts directly above the sheathing base; leaflets numerous, brownish- to dark or bright green, usually shallowly 2–3 (5)-lobed, the margins often rolled backward, prominently veiny beneath; lower surface pale, hairless or with fine jointed hairs.

Flowers many in a terminal, plume-like panicle, either male or female, the two types occurring on separate plants (dioecious), with 4 or 5 small, purplish to white sepals that often fall early; petals none; **male flowers** with many showy, yellowish stamens that soon droop and entangle and no pistils; **female flowers** with 5–14 pistils with long stigmas and no stamens. *Early June–mid July*

Fruits egg- to spindle-shaped achenes, dark-colored, $1/8$–$3/16$ inch long, coarsely ribbed, crowded on a small receptacle. *July–mid September*

WI range and habitats Common throughout, in moist, open to shaded habitats: marshes, sedge meadows, wet to medium prairies, fens, swampy woods (deciduous or coniferous), bogs, edges of forests and thickets, and along shores, swales, sloughs, and ditches. *Curtis, Greene, Marion Dunn, Sinaiko Overlook prairies, Teal Pond,* marshy and swampy ground elsewhere in the Arboretum

The flowers are wind-pollinated but attract bees, butterflies, and other insects.

Some Native American tribes used the dried fruits and leaves as a perfume, and made an infusion from the roots to reduce fever. The Potawatomi smoked dried fruits while hunting to bring good luck.

Similar species One of five species of *Thalictrum* in Wisconsin, tall meadow-rue is most similar to the generally overlooked wax-leaf meadow-rue (*T. revolutum*), which typically has the lower surface of the leaflets covered with minute stalkless or short-stalked glands that are difficult to see (use a 15✕ hand lens).

OTHER RANUNCULACEAE

MARSH-MARIGOLD
Cowslip, yellow marsh-marigold
Caltha palustris
Native; erect perennial forb, 8–24 inches.
Marshes, streambanks, swamps, and other sunny or partly shaded wetlands.
Flowers April–June

BRISTLY BUTTERCUP
Bristly crowfoot, Pennsylvania buttercup
Ranunculus pensylvanicus
Native; annual/biennial forb, 8–20 inches. Sedge meadows, streambanks, ditches, shores, and other wet places.
Flowers May–August

New Jersey Tea
Red-root
Ceanothus americanus

This sprawling shrub has ball-shaped masses of small white flowers that reward a close look, not only at their structure but also for the mixture of interesting insects they invariably entice. Hummingbirds are attracted in turn to eat the tiny insects. **Branching shrub, 1–3 feet tall.**

Leaves alternate, egg-shaped, $3/4$–2 inches wide, strongly 3-veined from near the base, the margins irregularly toothed, the undersides soft-hairy to velvety and with a pale cast.

Flowers fragrant, in small clusters aggregated into elaborate dense panicles, these terminal on relatively long ($1\frac{1}{2}$–4 inches), nearly leafless stalks produced on new branches emerging from the leaf axils; calyx lobes 5; petals 5, stalked, hooded, bent downward between the sepals; stamens 5, opposite the petals. In bud, each stamen is clasped by the infolded sides of the petals; in flower, they become free from the expanding petals. *Late June–August*

Fruits 3-lobed, capsule-like, black, $3/16$ inch thick, eventually separating into 3 single-seeded carpels. Floral cup and disk persisting like a pedestal just beneath the fruit. *Late July–mid October*

WI range and habitats Frequent throughout except lacking from the Northern Highlands, in dry to medium prairies, dry woodlands, oak openings, oak barrens, and sandy, gravelly, or rocky slopes, bluffs, and outcrops, sometimes along railroads, rarely on sandy lakeshores; often in transition zones between prairies and woodlands. *Curtis Prairie, Greene Prairie, East Knoll, Juniper Knoll, Sinaiko Overlook Prairie*

The leaves can be dried and brewed into a tea substitute. This tea contains caffeine and was popular before and during the Revolutionary War, when it was difficult to import real tea from China.

Like many native prairie plants, this shrub tends to disappear when prairies are overgrazed. The seeds survive in the soil for many years before fire stimulates them to germinate. The caterpillars of mottled dusky-wing butterflies feed exclusively on species of New Jersey tea.

Similar species Inland New Jersey tea or prairie red-root (*Ceanothus herbaceus*) is less common and earlier-flowering. Its oblong or narrowly oval leaves are smaller and shiny above, and its short-stalked inflorescences terminate regular leafy branches of the season. A single shrub grows on the West Knoll.

C. americanus   *C. herbaceus*

Wild Strawberry
Fragaria virginiana

Creamy white flowers, peeking out from amid long hairy leaf stalks, are a familiar sight in spring. The fruits, green at first and speckled with minute "seeds," soon turn cream-colored and eventually red, fragrant, and pulpy. **Perennial, 3–14 inches tall, from a crown at the tip of a thick rhizome, spreading by long runners.**

Leaves basal, compound, each with 3 sharply toothed leaflets on short stalks; terminal tooth on each leaflet commonly shorter and narrower that the 2 adjacent teeth on either side. Old leaves firm or leathery, new ones thinner.

Flowers 2–12 in loose broad, umbel-like clusters about equaling or slightly shorter than the leaves, unisexual or bisexual (within a patch, unisexual flowers may occur on the same or different individuals as bisexual ones); sepals 5; petals 5, round, $1/4$–$3/8$ inch long, surrounding a small yellow cluster of stamens and pistils (these sometimes reduced and sterile). *April–June (September)*

Fruits tiny individual achenes partly imbedded in the surface of the greatly enlarged, edible receptacle. *May–July*

WI range and habitats Nearly ubiquitous, in a wide variety of habitats: northern dry forests, woodland edges and clearings, dry to moist prairies, savannas, and old fields, also in boggy or swampy habitats and rocky openings, shores, roadsides, and railroads; sun or shade, preferring open grassy places. *Throughout the Arboretum*

Indigenous peoples from Canada to Chile used the American species for food and medicine for various stomach complaints. The domesticated strawberry originated as a natural hybrid in France. It proved to be a cross between the Chilean *Fragaria chiloense* (female parent) and the eastern North American *F. virginiana* (male parent), both of which had been introduced into that country.

Similar species Wisconsin's only other native strawberry, woodland strawberry (*Fragaria vesca*), differs in having leaflets that are unstalked, calyx lobes that spread or bend backward rather than converge around the young fruit, "berries" that are conical rather than nearly spherical, and achenes that are raised above rather than sunken into its surface. No single character will reliably separate the two; it is necessary to evaluate several characters before naming specimens.

Prairie-smoke
Old-man's-whiskers
Geum triflorum

This is one of the most common wildflowers of natural prairies, and one of the first to bloom. Its nodding, urn-shaped flowers look as though they have swallowed a marble; they have a very narrow tip and seem never to really open, this impression enhanced by the relatively short petals. **Perennial, 6–18 inches tall, forming patches from branching rhizomes covered with old leaf bases.**

Basal leaves pinnately compound, forming tufts (rosettes) 4–8 inches tall, with 7–17 irregularly cut or lobed leaflets. **Stem leaves** 2 (4), opposite, small, cut into segments that are very irregular in size and shape. Plant soft-hairy.

Flowers in open, umbel-like groups of 3 or 4 (9) at the top of a forking stem; floral tube hemispheric; calyx purplish-red, the 5 sepals (calyx lobes) appearing double, alternating with a row of linear bractlets that range from shorter to longer than the sepals; petals yellowish-white, sometimes suffused with pinkish or purplish, about equaling the bractlets; stamens numerous. *April–June*

Fruits fluffy tufts of numerous achenes, each tipped by a very slender, flexuous plume (the persistent style) $1\frac{1}{2}$-$3\frac{1}{2}$ inches long that functions like a sail, dispersing the achene on the wind. *End of May–June*

WI range and habitats Common southwest, on dry prairies, sandy open jack pine woods, poor dry soil of open ridges, bluffs, and hillsides, also in pastures, fields, and occasionally in moist meadows. *Curtis Prairie, Greene Prairie, Juniper Knoll, West Knoll*

Once pollinated, the flower stalk straightens, the "seed head" (fruiting receptacle) turns upright, and the feathery plumes elongate, conferring a smoky or hazy appearance not only to the plant but also to the prairie as a whole. The beautiful fruiting heads resemble those of the pasqueflower, with which prairie-smoke commonly grows, and with which it shares a similar geographical distribution and history.

Similar species Yellow avens (*Geum aleppicum*) and rough avens (*G. laciniatum*) are not uncommon in moist prairies, sedge meadows, fens, low woods, and ditches in much of Wisconsin. They can be distinguished from prairie-smoke by their erect flowers with green calyces and jointed styles with hooked basal segments that attach to fur or clothing.

Shrubby Cinquefoil
Shrubby five-fingers
Pentaphylloides floribunda
Synonym *Potentilla fruticosa*

Flowers of this small twiggy shrub look like bright yellow stars nestling among the soft green leaves. The leaves are not necessarily true to the name "five-fingers," having anywhere from 3–7 leaflets, each with the margins rolled under so as to appear even narrower than they really are. **Bushy-branched shrub, 16–36 inches tall, with shredding brown bark.**

Leaves pinnately compound; leaflets mostly 5, the terminal 3 often confluent at the base, narrowly oblong to elliptic-oblong or narrowly inverted-egg-shaped, $3/8$–$3/4$ inch long, less than $1/4$ inch wide, entire, silky-hairy and grayish beneath. Young branches silky but soon becoming hairless.

Flowers solitary in leaf axils or few in small terminal clusters, $3/8$–$7/8$ inch wide; calyx deeply 5-cleft, the egg-shaped lobes alternating with and equaled or exceeded by 5 lance-shaped bractlets; petals 5, circular to oblong-egg-shaped; stamens 20–30, inserted on the margin of the saucer-shaped floral cup. *June–September*

Fruits ovoid achenes $1/16$ inch long, densely white-hairy, collected on the raised receptacle. *July–October*

WI range and habitats Local southeast, rare elsewhere, in low prairies, fens, sedge meadows, marshes, bogs, and open conifer swamps, along creeks and lakes, calcareous beach flats and rocky shores close to Lake Michigan, rarely on edges of cliffs; ± alkaline, wet open ground and thickets. ***Greene Prairie***

A tea can be made from the dried leaves.

Until recently, shrubby cinquefoil was considered to be our only truly woody member of the large and taxonomically difficult genus *Potentilla*. It is now often segregated with other shrubby species into the separate genus *Pentaphylloides*. North American plants are consistently diploid, whereas Eurasian populations are either diploid or tetraploid. Diploid plants represent *Pentaphylloides floribunda* or *P. fruticosa* in the broad sense. Tetraploid plants are *P. fruticosa* in the narrow sense. Numerous horticultural variants of these species (or subspecies) have been developed, generally from Eurasian strains, as ornamental shrubs.

Silvery Cinquefoil
Silvery five-fingers
Potentilla argentea

This is a low tufted species, its several to many stems flattened to the ground or rising obliquely upward but never rooting at the nodes. The leaflets have only a few irregular coarse teeth or lobes per side and are silvery-white beneath, hence the name silvery cinquefoil. **Branching perennial, 4–12 or more inches tall (reaching nearly 2 feet), woody at the base.**

Leaves palmately compound, the larger ones with 5 (rarely 6) leaflets. Basal and stem leaves dissimilar, the former with wedge-shaped leaflets that themselves are almost compound; individual leaflets $3/8$–$1 1/4$ inches long, divided $1/2$–$3/4$ of the way to the midvein into 1–5 sharp teeth or lobes arranged on each side in a pinnate pattern; margins rolled under. Upper surface dark green, the lower with a white, felt-like covering of short woolly hairs.

Flowers $1/4$–$3/8$ inch across, few to many in a leafy, much-branched, ± flat-topped inflorescence; petals 5, yellow or pale yellow, about $1/8$ inch long, shorter than and often hidden by the sepals, easily falling; stamens 20. *May–September*

Fruits aggregates of numerous tiny "seeds" (achenes), partly enclosed in the enlarged calyx. *June–early October*

WI range and habitats A Eurasian species, naturalized throughout, in dry disturbed habitats: abandoned fields, pastures, roadsides, lakeshores, and waste places, also dry prairies and thin woods; prefers sandy soils. *West Knoll lanes*

This is one of several cinquefoils with leaves that are whitish beneath from the dense covering of either long slender or matted hairs. The only other obvious one in our region, silver-weed (*Argentina anserina;* = *Potentilla anserina*), is generally found close to the Great Lakes shores and does not occur in the Arboretum. It has long slender stolons, pinnately compound leaves in a basal rosette, and solitary flowers on long naked stalks.

TALL CINQUEFOIL
Prairie cinquefoil, tall potentilla
Potentilla arguta

Crowded clusters (rosettes) of attractive compound leaves, tall unbranched stems, and strict narrow inflorescences distinguish this rather coarse species. **Erect perennial, 1–3 feet tall, from short rhizomes or branching crowns.**

Basal leaves with (5) 7–9 (11) toothed leaflets in two lateral rows (pinnate), usually alternating with much smaller leaflets. **Stem leaves** with a pair of leafy basal appendages and fewer smaller leaflets, these broadly elliptic to egg-shaped and toothed to shallowly cut. **Stems** and leaves softly downy with spreading brownish sticky hairs.

Flowers saucer-shaped, $3/8$–$3/4$ inch wide, arranged in small compact clusters, these together forming a narrow, sometimes ± flat-topped, rather crowded inflorescence; sepals 5, alternating with 5 bractlets, the calyx thus appearing 10-cleft; petals 5, creamy to light yellow, nearly circular, $1/4$–$3/8$ inch long, from a little shorter to a little longer than the sepals and surrounding many stamens (25–30) borne in 5 groups on a thick nectar disk. *June–August*

Fruits tiny brown achenes, obliquely inverted-egg-shaped, aggregated into a head (collected on the hemispherical receptacle) and enclosed by the enlarged calyx. *Mid July–October*

WI range and habitats Frequent throughout, in high-quality, low, medium, and dry prairies, especially common on sand prairies, high lime prairies, and cedar glades, often on bluffs, also in open woods and pastures and along roadsides and railroads, but rarely truly weedy. *Curtis Prairie (limestone knoll), Greene Prairie, Sinaiko Overlook Prairie, West Knoll*

The name *Potentilla* is derived from the Latin *potens*, powerful or potent, because of the reputed medicinal properties of one species, silver-weed (*Argentina anserina*; =*Potentilla anserina*). Tall potentilla was used by Native American groups to stop bleeding and dysentery.

The flowers attract honeybees, solitary bees including metallic green halictids, and syrphid flies.

Similar species Several other potentillas occur in Wisconsin, but none of them are ever glandular-hairy; neither do they combine pinnately compound leaves with clusters of flowers in which the terminal flower matures first. Cinquefoils can be confused with avens (*Geum* species), which differ in having pinnately-cleft leaves with enlarged terminal lobes and persistent styles that elongate after flowering. In *Potentilla* the leaflets are more uniform in shape and the styles short and inconspicuous.

Rough Cinquefoil
Norwegian cinquefoil, strawberry-weed
Potentilla norvegica

The small bright flowers of this hairy weed are reminiscent of buttercups (*Ranunculus* species) but are actually tiny roses. They are borne in rather compact, flat, very leafy clusters at the ends of slender but stiff branches, their clear yellow petals lending a touch of daintiness to the rather coarse plant. **Slender to stout, leafy annual or biennial (or short-lived perennial), 5–50 inches tall, from a shallow taproot.**

Leaves all 3-foliolate, the leaflets inverted-egg- to inverted-lance-shaped (lower leaves) or elliptic to oblong-lance-shaped (upper ones), to 3 inches long, coarsely toothed, hairy on both surfaces. **Stems** hairy with stiff spreading hairs (below) intermixed with shorter hairs (above).

Flowers $3/16$–$5/8$ or more inch wide; sepals (calyx lobes) 5, appearing double by a row of 5 small bractlets; petals 5, inverted-egg-shaped, mostly shorter than the calyx lobes; stamens 15–20; pistils many. Calyx in fruit enlarging to $1/2$ inch long. *June–September*

Fruits numerous tiny achenes collected in a head, light brown, asymmetrically kidney-shaped, with curved longitudinal ribs. *Late June–September*

WI range and habitats Very common throughout, on roadsides, fields, pastures, edges of prairies, marshes, and bogs, sandy, muddy, or rocky pond and lakeshores, sometimes in woods, on clay bluffs, sand banks, and gravel bars, also waste or disturbed places (railroad yards, gardens, empty lots); moist or dry ground. *Curtis Prairie, Sinaiko Overlook Prairie, Visitor Center parking lot*

This species is native to Europe as well as North America, and as noted by E. G. Voss, "It is quite likely that both native and European forms now occur here."

Similar species Early-season rosettes are distinguished from strawberries (*Fragaria* species) by having teeth the whole length of the margins of the leaflets, whereas in strawberries the lower $1/4$–$1/3$ of the leaflets are without teeth. Both rough cinquefoil and sulphur cinquefoil (*Potentilla recta*) are leafy herbs with palmately compound, yellowish-green leaves and yellow, saucer-shaped flowers. Rough cinquefoil has leaves with only 3 leaflets and bright yellow petals about equaling or shorter than the sepals. The main leaves of sulphur cinquefoil have 5 (–7) leaflets, and the pale yellow petals are larger. Both are very common weeds in a great number of habitats.

SULPHUR CINQUEFOIL
Rough-fruited cinquefoil,
sulphur five-fingers
Potentilla recta

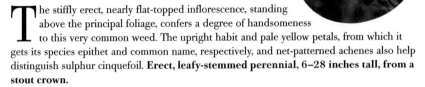

The stiffly erect, nearly flat-topped inflorescence, standing above the principal foliage, confers a degree of handsomeness to this very common weed. The upright habit and pale yellow petals, from which it gets its species epithet and common name, respectively, and net-patterned achenes also help distinguish sulphur cinquefoil. **Erect, leafy-stemmed perennial, 6–28 inches tall, from a stout crown.**

Leaves palmately compound, the basal ones on long stalks; leaflets 5–7, inverted-lance-shaped, green on both sides, $1\frac{1}{2}$–$3\frac{1}{2}$ inches long, toothed, thinly beset with straight stiffish hairs on both surfaces and (almost always) numerous glands on the lower surface.

Flowers relatively numerous, $\frac{3}{8}$–1 inch across, in a compound, nearly leafless inflorescence 3–6 inches across; sepals (calyx lobes) 5, stiffly hairy; petals inverted-heart-shaped (notch at the apex), $\frac{1}{4}$–$\frac{1}{2}$ inch long, equaling or exceeding the calyx lobes; stamens 25–30. Calyx increasing in size with age, becoming $\frac{3}{8}$–$\frac{5}{8}$ inch long. ***June–October***

Fruits numerous tiny achenes collected in a head and partly protected by the persistent calyx; achene surface dark brown with a lighter pattern of distinct branching ribs.

WI range and habitats Throughout south and central, in old fields and clearings, dry (ecologically open) to damp (grazed) prairies, pastures, beaches, sand blows, roadsides, railroads, and waste places (e.g., edges of parking lots, quarries, gravel pits, and old mine tailings); invading grassy or wooded habitats; prefers dry, sandy or gravelly soil containing lime. ***Curtis Prairie, Sinaiko Overlook Prairie, West Knoll***

Sulphur cinquefoil is naturalized from Eurasia. The exact time and place of its original introduction into North America is not known, but it was well established in Wisconsin sometime before 1920.

Despite meaning "five leaves" in French, the name cinquefoil conveniently designates the whole genus regardless of the number of leaflets. When identifying specimens, first look for the 5 sepals and below them the 5 bractlets. This "extra calyx" is sufficient to distinguish cinquefoils from buttercups. Then notice the compounding of the leaves, whether they are pinnate or palmate.

Similar species See rough cinquefoil (*Potentilla norvegica*).

Old-field Cinquefoil
Common cinquefoil,
old-field five-fingers
Potentilla simplex

The combination of palmately compound leaves, solitary flowers, and slender, usually creeping stems makes old-field cinquefoil distinctive. **Rhizomatous perennial with stems that are at first erect or ascending and 8–20 inches tall, soon spreading and arching, becoming 4 feet long and rooting at the tips.**

Leaves compound, the principal ones with 5 leaflets, these narrowly inverted-egg-shaped to narrowly elliptic or inverted-lance-shaped, $5/8$–2 inches long, green to slightly whitened beneath, with toothed margins. **Stems** and leaf stalks hairless to manifestly hairy, the hairs spreading to lying against the surfaces.

Flowers solitary, $3/8$–$5/8$ inch wide, one from each leaf axil (except the first) on a stalk several times longer than the calyx; sepals 5, triangular, spreading, alternating with 5 shorter narrower bractlets; petals 5, yellow, slightly notched at the rounded apex; stamens 20. *May–June*

Fruits aggregates of tiny, yellowish-brown achenes borne on the prolonged receptacle. *June–August*

WI range and habitats Common throughout (especially south), in dry to moist, often disturbed, sandy woodlands and prairies, oak openings, dry forests, medium woods and thickets, also on rocky ledges, often weedy in abandoned fields, eroded hillsides, and along roadsides and railroads. *Curtis Prairie, Gallistel Woods, Greene Prairie, Noe Woods*

Many of the little copper butterflies use the cinquefoil as their food plant.

This plant contains large amounts of tannins. An infusion made from the root is used in alternative medicine as an astringent, antiseptic, and tonic.

Because they have 5 yellow petals and numerous stamens and pistils, some species of *Potentilla* may be mistaken for buttercups (*Ranunculus* species). Distinguish the two genera (and families) by the fact that the leaves of the Rose Family have a pair of leafy appendages at the base (stipules), and the flowers have an expanded floral cup from which the sepals and petals arise.

White-tailed doe, Curtis Prairie edge

Roses

Native roses grow in a variety of habitats but thrive best on the prairies. **Woody shrubs to almost herbaceous perennials, 1–4 feet or taller.**

Leaves pinnately compound, typically with paired large basal appendages (stipules) that are fused to the leaf stalk for more than half their length, forming wings. **Stems** usually thorny and/or bristly.

Flowers pink (rarely white, variously colored in cultivated species, which usually have the number of petals greatly increased), large, often 1–2 inches in diameter, sweetly perfumed; sepals 5, appearing to be a continuation of the exaggerated floral tube; petals 5, inserted at the outer edge of a disk that rings the constricted opening of the spherical to urn-shaped floral tube. Stamens numerous, inserted on the disk, forming a bright yellow center. All species: *late May–late September.*

Fruits bony achenes completely enclosed by the fleshy floral tube, which at maturity becomes the so-called hip. Hips developing after the petals fall, ripening to a brilliant red or orange and simulating miniature apples. All species: *late June–winter,* the hips sometimes lasting on the plant until the next summer.

Prairie Rose
Sunshine rose
Rosa arkansana

Prairie rose is one of our most common and variable wild roses. Its stems are generally unbranched and green to reddish-green. They are densely to sparsely prickly and bristly throughout, the prickles of similar size, all short, slender, and weak. **Dwarf colonial shrub, nearly herbaceous and 7–25 inches tall or truly shrubby and becoming 3 feet tall.**

Leaflets 7–11 (mostly 9), mostly less than 1 inch long, showing all degrees of transition from hairless to commonly softly and minutely hairy (but not glandular) on the underside.

Flowers (1) 3–12 in clusters terminating the main stem of the season or rarely on short lateral branches from older stems; flower stalks and floral tubes generally without hairs or glands; sepals long-persistent, erect to spreading in fruit, glandular; petals relatively short, $5/8$–1 inch long.

WI range and habitats Frequent, especially south and northwest, in wet-medium to medium prairies, sandy fields, grassy banks, gravelly hillsides, and spreading along rights-of-way, rarely on borders of oak woods. See note under pasture rose (*Rosa carolina*). **Curtis Prairie, Grady Tract knolls, Juniper Knoll**

SMOOTH ROSE
Rosa blanda

Stems of smooth rose are branched above, purplish-red, and without prickles or very sparsely prickly; although their bases may be armed with few to many prickles, these do not extend onto the branches. **Colonial shrub, (4) 8–41 inches tall.**

Leaflets 5–7 (rarely 9), 1–1½ inches long, hairless above, finely hairy or nearly hairless and without glands on the lower surface; basal appendages nearly glandless to abundantly glandular-toothed.

Flowers usually 1 (2–5) on lateral branches from stems of the previous year; flower stalks and floral tubes generally without hairs or glands (rarely sparsely glandular); sepals long-persistent, erect in fruit, glandular; petals ¾–1¼ inches long.

WI range and habitats Throughout, in dry to moist prairies and a variety of open or brushy habitats: fields, banks and hillsides, borders of woods and thickets, fencerows, roadsides, and railroads, also frequent in transition zones near shores and marshes and on sandy lakeshores and dunes (including those of the Great Lakes). Our most common wild rose, rarely showing the acquisition of characters of swamp rose (*Rosa palustris*) and western rose (*R. woodsii*) through hybridization and back-crossing. ***Curtis Prairie, Greene Prairie, Sinaiko Overlook Prairie, Visitor Center parking lot***

Rosa arkansana   *Rosa blanda*

The taxonomy of *Rosa* is notoriously difficult. Each of our wild species is extraordinarily variable, and hybridization and polyploidy are frequent. Several variants and hybrids have been identified in our area, more than one of which may be found in a given colony. Populations of plants, rather than casually collected specimens, should be studied to arrive at accurate identifications.

Rose hips are rich in vitamins, especially A and C, and are nutritious enough to be used in soups, baby foods, jellies, and syrups. Native peoples have long used them as a supplementary food source. The word *rosary* comes from the practice in medieval Europe of using dried rose hips to make strings of prayer beads.

Multiflora rose (*Rosa multiflora*), an east Asian species with long scrambling stems, small leaflets, many-flowered racemes, and small white petals, is established as a **nuisance weed** throughout southern Wisconsin, including the Arboretum. Once planted by highway crews and landowners as a barrier hedge, it escaped to become one of our most persistent and pestiferous invasive species.

Carolina Rose
Pasture rose
Rosa carolina

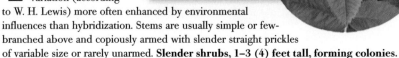

This is Wisconsin's most variable native rose, the variation (according to W. H. Lewis) more often enhanced by environmental influences than hybridization. Stems are usually simple or few-branched above and copiously armed with slender straight prickles of variable size or rarely unarmed. **Slender shrubs, 1–3 (4) feet tall, forming colonies.**

Leaves pinnately compound; leaflets (3) 5–7 (9), widely elliptic to oblong or somewhat circular, varying from inverted-egg-shaped to lance-egg-shaped but usually widest in the middle, mostly $5/8$–$1\,3/4$ inches long, dark green and hairless above, soft-hairy to hairless below; leaf stalks bristly or not, the basal appendages glandular-toothed to entire. **Flowering stems** with a pair of enlarged thorns associated with the branchlets and leaf stalks or these rarely small.

Flowers usually 1 (rarely clustered) at tips of stems of the season, $1\,1/2$–2 inches broad; flower stalks and floral tubes conspicuously stalked-glandular (rarely smooth); sepals divergent or bent downward after flowering and eventually falling, glandless or stalked-glandular; petals $3/4$–$1\,1/4$ inches long. *Late May–late September*

Fruits bony achenes inserted on the bottom of the mature floral tube or "hip," yellowish- to reddish-brown, straight on one side, long hairy on the other; hips red, spherical, $3/8$–$1/2$ inch thick. *Late July–October* (hips persist on the plant until early the next summer)

WI range and habitats Common south and west, in dry to medium prairies, embankments and hillsides, dry upland oak woods, oak openings, pine barrens, occasionally on lakeshores and stream banks, tending only slightly to spread into old fields, roadsides, railroads, and fencerows. *Curtis Prairie*

This species often co-occurs with prairie rose (*Rosa arkansana*) and freely hybridizes with it, yielding relatively frequent hybrid swarms. The most nearly intermediate individuals, those having approximately half the characters of each parent, would be considered of hybrid origin and referred to rough rose (*Rosa ×rudiuscula*).

CLIMBING PRAIRIE ROSE
Rosa setigera

Climbing prairie rose can be readily differentiated from all other North American roses by its long, wand-like stems, which do not truly climb but often recline over other vegetation, and large, dark green leaflets, which have many coarse teeth and are entirely without glands. **Leaning or trailing shrub, 4–15 feet high and long.**

Leaves pinnately compound, the leaflets 3 (occasionally 5), lance- to oblong-egg-shaped, 1–4 inches long, firm, dark green and hairless above, usually softly hairy beneath, varying to completely hairless, with a pair of very narrow, basal appendages (stipules) attached to the leaf stalk, these with a lance-shaped, free tip and glandular-hairy, marginal fringe. **Stems** armed with scattered stout, downward-curved, flattened thorns.

Flowers about 3 inches across, few to numerous in short broad, flat-topped clusters; sepals entire (never pinnate-divided), soon bent abruptly backward, falling after flowering; petals pink, fading whitish, $3/4$–$1 1/4$ inches long; styles united (but stigmas separate) into a column protruding from the mouth of the floral cup about as far as the numerous golden stamens. Flower stalks, floral tube, and sepals with stalked glands. *July*

Fruits bony achenes concealed inside the small red hips. *October–November*

WI range and habitats Known as a "wild" plant only from Dane and Milwaukee counties, where locally established, but likely to be found again in the southern part of the state; prefers evenly moist, loamy soil and partial shade. *Curtis Prairie, woods between Pond 2 and West Beltline*

This is primarily a Mideastern (or southern) species. Its northern range limit is western New York, southern Ontario, southern Michigan, and Kansas, and it is occasional to locally frequent just to our south, growing along edges of woods and in clearings, savannas, thickets, prairies, and fencerows in northern Illinois. It is locally naturalized or rarely escaped from Virginia to New England and in Wisconsin, where first collected as a spontaneous plant in 1938 and not again until 1966.

Although sometimes called "prairie rose," this species is more at home along woodland borders and clearings. It is distinguished from native roses by its vine-like growth, 3 leaflets, and narrow column of styles protruding from the orifice of the floral tube.

White Meadowsweet
Spiraea alba

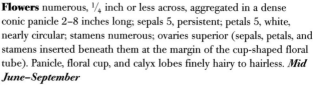

Wild spiraeas are attractive shrubs with simple stalkless leaves that unlike nearly all our native Rosaceae lack appendages at the base. Although not large, the pinkish-white flowers are conspicuous *en masse,* being aggregated into a terminal, pyramid-shaped spire. The ovaries are superior and ripen into a group of tiny, firm-textured pods that remain on the plant and create the illusion of a winter-time bloom. **Low erect shrub, 3–7 feet tall, with slender brown twigs.**

Leaves simple, lance-oblong, $1\frac{1}{4}$–$2\frac{3}{8}$ inches long, finely and sharply toothed but never lobed or compound, hairless or with sparse hairs that are hardly visible. Plants lacking spines, thorns, or prickles.

Flowers numerous, $\frac{1}{4}$ inch or less across, aggregated in a dense conic panicle 2–8 inches long; sepals 5, persistent; petals 5, white, nearly circular; stamens numerous; ovaries superior (sepals, petals, and stamens inserted beneath them at the margin of the cup-shaped floral tube). Panicle, floral cup, and calyx lobes finely hairy to hairless. *Mid June–September*

Fruits 3–5 small dry pods (follicles) per flower, yellow-brown to blackish, splitting open along one side to release the 2–several seeds. *Mid August through the winter* (seeds still present the following April)

WI range and habitats Frequent to locally common throughout, in wet to damp (rarely dry) habitats: marshy meadows, wet prairies, bogs, swamps, and thickets, as well as sandy or boggy shores, swales, and ditches. *Curtis Prairie, Greene Prairie, Pond 2*

The leaves of meadowsweet make an agreeable, albeit astringent tea that was commonly used by native peoples to prevent nausea and vomiting.

This is a notably variable species, consisting of two thoroughly intergrading varieties. True variety *latifolia* (once generally considered to be a separate species) is eastern and does not enter Wisconsin, but typical var. *alba* is widespread in the state, as are numerous plants variously intermediate in one to several characters between this relatively western taxon and var. *latifolia.* Other species of the genus are the pink-flowered native hardhack (*Spiraea tomentosa*), which grows in bogs, swamps, peaty shores, and acid sandy meadows, and the commonly planted, ornamental shrubs like bridal-wreath (*S. ✕vanhouttei*), which is rivaled in popularity only by some of the varieties of Japanese spiraea (*S. japonica*).

Other Rosaceae

Common Agrimony
Tall hairy agrimony
Agrimonia gryposepala
Native; erect perennial forb, 1–5 feet.
Flowers July–September

Yellow Avens
Geum aleppicum
Native; erect perennial
forb, 20–40 inches.
Flowers May–July

White Avens
Geum canadense
Native; erect perennial forb,
16–36 inches.
Flowers May–June

Other Rosaceae

Common Blackberry
Rubus allegheniensis
Native; perennial shrub with heavy, strong-thorned canes to 9 feet. Often found at edges of woods, spreading by rhizomes.

Black Raspberry
Rubus occidentalis
Native; perennial shrub with slender wiry, prickle-thorned canes that arch and root at the tip. Forms sprawling patches.

Northern Bedstraw
Galium boreale

Northern bedstraw is an upright herb with erect to ascending, generally smooth (occasionally hairy) stems and leaves. It is most easily recognized by the rather showy masses of snow-white or slightly creamy flowers and the 3 prominent longitudinal veins at the base of the leaves. **Slender perennial, 6–30 inches tall, from creeping rhizomes, often growing in large patches.**

Leaves in whorls of 4, linear to narrowly lance-shaped, $5/8$–$1 3/4$ inches long, with margins slightly rolled under (toward the lower surface) and 3 (rarely 5) prominent longitudinal veins. **Stems** erect, short-bearded right beneath the nodes, often with a bundle of condensed smaller leaves in the angle between the leaves and stem; neither stems (internodes) nor leaves rough to the touch.

Flowers tiny but numerous, clustered in small flattish inflorescences that are repeatedly 3-forked, forming a branched terminal panicle; calyx tube spherical (calyx teeth obsolete); corolla wheel-shaped with 4 short lobes. Ovary (and fruit) inferior, that is, fused together with the calyx and appearing to be below the point of insertion of the corolla. *Early June–mid July*

Fruits paired, about $3/32$ of an inch in diameter, when ripe separating into 2 individual dry, seed-like portions (mericarps). *Early July–late September*

WI range and habitats Ubiquitous except in the Northern Highlands, in moist to medium prairies, fens, open hardwood forests, oak openings, and a variety of other habitats, growing best in those that are sunny. *Curtis Prairie, Greene Prairie, Grady Tract, Gallistel Woods, Noe Woods, Sinaiko Overlook Prairie, Wingra Woods*

The genus name is from the Greek *gala*, milk, because a juice extracted from bedstraw curdles milk and was mixed with rennet to make cheese. Bedstraws were once used as stuffing for mattresses, because the stems remained hooked together, mat-like, and did not become brittle and turn to powder. The fruits are collected by mice and stored for the winter.

A circumpolar species, northern bedstraw is variable in stature, amount of branching, leaf characters (color, dimensions, marginal hairiness), and fruit hairiness.

Wild Madder
Blunt-leaf bedstraw
Galium obtusum

The majority of our bedstraws are weak-stemmed plants with loosely spreading stems that cling to other plants by means of minute barbs and bristles on angles and edges. Wild madder tends to have erect stems that are less intricately branched and smooth on the angles. Its sparse white flowers look like very tiny stars scattered among the bases of the taller associated plants. **Perennial, 8–30 inches tall, from hair-like rhizomes.**

Leaves mostly in 4s (rarely 5s or 6s), narrowly elliptic, the primary ones about $3/4$ inch long, $1/8$–$1/4$ inch wide, blunt at the tip; hairs on the leaf margins bristly, diverging at right angles (never directed sharply backward). **Stems** densely short-bearded at the nodes.

Flowers grouped in 2s, 3s, or 4s at the ends of stems and branches, $1/16$–$1/8$ inch wide; calyx teeth none, the 4 pointed petals (corolla lobes) and 4 stamens crowning the inferior ovary, which becomes more prominent as the fruit develops. Branches of flower clusters and individual flower stalks straight, forming little flattish inflorescences in which the terminal flower blooms earliest. *Late May–July*

Fruits paired, spherical, $1/16$ inch in diameter, smooth, separating when ripe (often only 1 develops) into 1 or 2 seed-like, non-opening, 1-seeded portions (mericarps). *August*

WI range and habitats Locally frequent throughout (except in the Northern Highlands), in sedge meadows, low prairies, low woods, swamps, and shores. ***Curtis Prairie, Greene Prairie***

The red dye in the roots gives this plant the name wild madder.

Similar species Native galiums can be difficult to distinguish. Rough bedstraw (*G. asprellum*) has mostly 5 leaves at a node; these are sharply bristle-tipped and very rough-margined. The other species have leaves mostly in 4s and blunt at the tip. Labrador marsh bedstraw (*G. labradoricum*) is similar to wild madder in having 4-lobed corollas and hairy nodes, but it has narrower leaves with down-rolled margins and even tinier fruits. Southern three-lobed bedstraw (*G. tinctorium*) and northern three-lobed bedstraw (*G. trifidum*) have smaller, predominantly 3-lobed corollas. They also differ in having hairless nodes and leaf margins with the hairs directed backward.

Bluets

Azure bluets, innocence, Quaker-ladies
Houstonia caerulea
Synonym *Hedyotis caerulea*
Wisconsin special concern

This dainty plant, conspicuous in spring in the eastern and southern U.S., was one of the colonists' favorite wildflowers. Its strikingly beautiful, blue-lavender to white flowers bloom in mid spring atop a fine tangle of very slender stems, then turn yellow with age. **Delicate low annual or weak perennial, 2–8 inches tall, from short fragile, threadlike rhizomes, forming mat-like clumps.**

Leaves mostly clustered into basal rosettes, elliptic with nearly parallel sides or spoon-shaped, on slender stalks. **Stem leaves** opposite, variable in shape, $1/8$–$3/8$ inch long, narrower than the basal leaves, at least the lower ones stalked. Short appendages connect the leaf stalks or narrowed leaf bases at their point of insertion on the threadlike stems.

Flowers single on each primary stalk of the inflorescence, pale blue to clear white with a yellow center or "eye," rather small, about $1/8$ inch across, the 4 persistent short narrow sepals (calyx lobes) and short corolla tube arising below the summit of the half-inferior ovary. Petals fused for much of their length but flaring into 4 abruptly spreading, pointed lobes. ***Late May–early June***

Fruits flattened, thin-walled capsules with 2 chambers; seeds brown, tan, or black, somewhat spherical.

WI range and habitats Locally frequent in 5 southeastern counties, in wet prairies, moist meadows, fields, and open woods. ***Greene Prairie***

Bluets flowers occur in two forms (distylous). In some individuals the slender style that surmounts the ovary is short and included within the corolla tube; in others, it is long, the end that receives the pollen therefore protruding. Pollen from a long-styled plant must reach a short-styled one for fertilization of the ovules to be successful. This mechanism ensures cross-pollination and avoids inbreeding.

Similar species Long-leaved bluets (*Houstonia longifolia*) differs in having narrow stalkless leaves all along the stem and funnel-shaped corollas that are hairy within. It flowers later and grows in drier, sandy, gravelly, or rocky soil throughout much of Wisconsin but is not found in the Arboretum.

Bastard-Toadflax
False-toadflax
Comandra umbellata

This low-growing, somewhat nondescript herb is evident in the spring but overlooked among the grasses later in the year. The foliage is fairly distinctive, the alternate, essentially stalkless, ascending leaves lining the entire stem. In the early autumn, foliage commonly turns from yellow to orange, purple, or brown, but in some clones it remains green. **Erect, light green perennial, 3–16 inches tall, forming clones.**

Leaves numerous, pale bluish-green, oblong or oval, small, $1/8$–$1/2$ inch wide, somewhat fleshy, with entire margins (not toothed or cut). **Stems** arise in clumps at intervals from very long, horizontal white rhizomes.

Flowers 3–6 in each of several compact clusters $1/4$–$1/2$ inch broad, these making up a round- or flat-topped, panicle-like inflorescence at the summit of the plant. Each whitish flower axillary on a short stalk, $1/6$ inch long, with 4 (5) petal-like segments (tepals) fused at the base into a fleshy floral cup; stamens 4 (5), opposite and adherent to the short tepal lobes. A lobed fleshy disk lines the bottom of the floral cup. *Early May–mid June*

Fruits dry, 1-seeded stone-fruits (drupes), green to yellow-green, nearly $1/4$ inch in diameter (almost like tiny fleshy nuts), slightly constricted into a short neck toward the top and crowned by the persistent tepal lobes. *Late June–August*

WI range and habitats Very common throughout, in dry to wet-medium prairies, oak barrens, open oak, aspen, or pine woods, and shores of the Great Lakes; tolerating a considerable amount of disturbance. *Curtis Prairie, Greene Prairie, Grady Tract knolls*

Bastard toad-flax is a hemiparasite; besides using photosynthesis to form carbohydrates, it parasitizes the roots of other plants to obtain some of its water and minerals. In the Great Lakes region alone, over 200 species in a number of different plant families function as its hosts. Rodents eat the fruits.

Young shoots in bud

Prairie Alumroot
Heuchera richardsonii

The luxuriant flush of long-stalked, palmately lobed, basal leaves and small, greenish-cream to brown flowers on a slender leafless stalk make this mid-spring bloomer readily identifiable. **Erect perennial, 1–3 feet tall, from a stout crown.**

Leaves circular-egg-shaped in outline, scalloped with 5–9 low rounded, coarsely toothed lobes, heavily veined, the margin fringed with hairs. **Stems** (often solitary) and leaf-stalks bearing bristly white hairs.

Flowers $3/16$–$3/8$ inch long, in a spike-like panicle 2–4 inches tall; petals 5, spoon-shaped, alternating with the 5 tiny calyx lobes; stamens 5, barely extending beyond the petals; pollen orange. The floral tube that forms each bell-shaped flower is obliquely asymmetrical, being somewhat longer on the upper side and occasionally swollen on one side at the base, and is adherent to the half-inferior ovary. *(Mid) late May–mid July*

Fruits egg-shaped, 2-beaked capsules $1/4$–$1/2$ inch long (excluding the persistent styles), erect at maturity, protruding from the persistent floral tube; seeds extremely small, dark brown. *Mid June–mid September*

WI range and habitats Frequent south and west, occasional north (especially along the Wisconsin, Chippewa, and St. Croix rivers and in the northeastern and northwestern barrens), in a variety of prairie types from dry to wet, also in open to lightly wooded, rocky or sandy ground (ledges, outcrops, cliffs, bluffs, banks) and oak and/or pine woods (including oak openings, pine barrens, and cedar glades). *Curtis Prairie (limestone knoll), Greene Prairie, Juniper Knoll*

Prairie alumroot and a related eastern species were used by Native Americans and pioneers as medicinal plants. The powdered roots and green leaves have astringent and styptic properties and were used for dressing sores and closing wounds. The dried root was chewed to stop stomach pain, and the Lakota also used the roots to make a tea for treating chronic diarrhea.

The name alumroot results from the alum-like taste, which puckers the mouth. The familiar garden flower called coral-bells (*Heuchera sanguinea*) is an alumroot. The sight of prairie alumroot always pleases prairie enthusiasts, not only because of its natural beauty, botanical interest, and role in nature, but also because as a relict of the prairie, its presence signals good-quality or recoverable habitat.

Swamp Saxifrage
Wild beet
Saxifraga pensylvanica

Swamp saxifrage is another herb of distinctive appearance, immediately identifiable by the large, almost leathery leaves in a cluster at the base of the plant, and the single erect leafless stem. The small subtle greenish flowers and swampy habitat also help identify this species. **Perennial, 1–3 feet tall.**

Leaves all basal, generally oblong with the basal end attenuated (like a druggist's spatula), varying to lance- or narrowly egg-shaped in outline, 4–8 or more inches long, the margins obscurely toothed (teeth minute, usually remote and rounded); midrib thick, conspicuous beneath. **Stems** soft-textured, ± stout, softly pubescent with gland-tipped hairs.

Flowers bisexual (having both functional, seed-bearing pistil and pollen-producing stamens), borne in an irregularly branched, open inflorescence that is at first compact (1–4 inches long), soon elongating to 4–24 inches long; calyx lobes 5, narrowly triangular, bent downward; petals 5, yellowish-white to greenish-yellow or purplish-tinged, narrowly oblong; stamens 10. Calyx lobes and petals small and inconspicuous, appearing as tiny lobes inserted on the short floral cup. *Early May–late June*

Fruits 2-chambered, deeply 2-lobed, inflated capsules about $1/8$ inch long. The two units are united to about the middle but become free above, each prolonged into a firm, slender tip (beak) that opens along the inner seam to release numerous seeds. *Early June–early August*

WI range and habitats Throughout, in damp, swampy, or marshy habitats: medium to wet prairies, sedge meadows, borders of marshes, swamps, boggy thickets, hollows, ravines, streamsides in woods, and springy or seepy places, also on dry to dripping, wooded cliffs and slopes in the Driftless Area. *Greene Prairie*

The genus name is derived from *saxum*, stone, and *frangere*, to break, and was applied to European species that produce small round tubers on the roots, which, applying the doctrine of signatures, were supposed to dissolve kidney stones.

About a dozen species of *Saxifraga* can be found in moist rocky forests and wet meadows or along streams in northeastern North America, but most members of this hardy genus prefer the cooler cliffs, mountains, and tundra of northern regions and the Andes.

Yellow False Foxglove
Large-flowered yellow false foxglove
Aureolaria grandiflora

Large, chrome-yellow flowers alone distinguish this species from anything else in its open woodland habitats, and the coarsely pinnately lobed leaves that turn blackish when bruised or picked are likewise distinctive even when flowering time is past. **Large branching perennial, 20–40 inches high.**

Leaves opposite, the lower cleft nearly to the midrib, progressively smaller and irregularly alternate upward, the upper ones often only toothed and grading into the floral bracts. Herbage ash-gray owing to the dense covering of straight soft, non-glandular hairs.

Flowers opposite or nearly opposite in an elongate raceme, on stout upcurved stalks that arise in the angle between stem and bract, to 2 inches long; sepals united into a bell-shaped, 5-lobed calyx; petals fused for most of their length into a tube, then flaring into 5 rounded lobes that are almost equal in size. Stamens 4, arranged in 2 pairs. *Late July–mid October*

Fruits egg-shaped capsules, ⅝–1 inch long; seeds irregular, the seed coat with a netlike pattern. *Late August–December*

WI range and habitats Rare to locally abundant south, in open oak woods, oak barrens, cedar glades, prairie openings, and Driftless Area pine relicts, occasionally on the edges of hardwood stands; prefers dry, sandy or rocky ground. *Grady Tract savannas*

All *Aureolaria* species are root parasites on oaks and perhaps other woody plants. This hemiparasitic property may explain why populations are erratic in their presence at a given spot; they seem not to persist long but appear to shift a few yards away from where they were located the previous year.

Analyses of DNA sequence data have shown that parasitism evolved once in the Figwort/Broomrape Family pair; in other words, all the parasitic plants in the old Figwort Family and traditional Broomrape Family (Orobanchaceae) are descended from a single ancestor. Therefore, it has been proposed that *Agalinis, Aureolaria, Castilleja, Pedicularis,* and the other parasitic scrophs be included in a redefined Broomrape Family.

Similar species Annual false foxglove (*Aureolaria pedicularia*) differs in having all leaves lobed and the upper part of the stem and flower stalks glandular-sticky. It has much the same geographical distribution in Wisconsin as yellow false foxglove but is not found in the Arboretum.

Indian Paintbrush
Scarlet painted-cup
Castilleja coccinea

Brightly colored tips of the bracts and calyx are reminiscent of a painter's brush, making this an easy plant to spot despite its short stature. The prominent leafy, 3-lobed bracts are often mistaken for the flowers; the actual flowers are hidden. **Single-stemmed annual or biennial, 5–24 inches tall.**

Basal leaves entire. **Stem leaves** alternate, stalkless, entire or more typically 3- to 5-cleft into linear lobes; upper leaves and floral bracts with variously colored tips (red, scarlet, orange, or yellow).

Flowers in the angle between bract and stem, forming a spike, yellowish-green; calyx tubular, each half widened and entire at its tip; corolla tubular, long and slender, 2-lipped and extremely asymmetrical, the upper lip hooded and beaklike, its united lobes enfolding the 4 stamens. *May–September*

Fruits asymmetrical, egg-shaped capsules containing many wingless seeds. *Late May–August or later*

WI range and habitats Infrequent to locally common throughout much of the state (absent from most of the Northern Highlands), in low open ground in full sun, especially damp sandy prairies, flats, and swales, also lakeshores, meadows, fens, and old marshes. *Greene Prairie*

This is one of our native plants pollinated by hummingbirds. It is a hemiparasite; its roots must attach to nearby roots of another species (not host specific) in order for the plant to mature beyond the seedling stage. All the parasitic figworts (old Scrophulariaceae) and broomrapes (traditional Orobanchaceae) are now classified together in a newly defined Orobanchaceae.

A Native American legend tells of a brave who, having difficulty while trying to paint a sunset, asked the Great Spirit for help. He received paint brushes dripping in bright colors, which he placed on the ground after the painting was finished. The discarded brushes sprang up into these delightful plants.

Similar species Downy paintbrush or downy painted-cup (*Castilleja sessiliflora*) is distinguished by its doubly cleft calyx, green or sometimes pink-tipped bracteal leaves, and purplish to yellow or white corolla. This rather rare, dry-prairie species is not found in the Arboretum.

Scrophulariaceae • Figwort Family

TURTLEHEAD
Chelone glabra

Tall stems and opposite narrow leaves help distinguish the handsome white turtlehead even when it is not flowering. In bloom, the large, creamy-white or greenish-tinged flowers are unmistakable. **Smooth perennial, 2–3 feet tall.**

Leaves opposite, stalkless or nearly so, narrow, 2–7 inches long by $3/8$–$1 3/8$ inches wide, sharply toothed. **Stem** upright, unbranched or sparingly branched.

Flowers closely overlapped by broad concave bracts and bractlets, forming a dense spike that becomes looser in fruit; sepals nearly circular, overlapping; corolla tubular, strongly bilaterally symmetrical (bisected by only one plane into similar halves), 1–$1 1/2$ inches long, the upper lip broad and arched, the lower 3-lobed and bearded in the throat; stamens 4 fertile plus 1 reduced to a sterile greenish filament. *Early July–early October*

Fruits 2-chambered, spherical capsules that enclose flat winged seeds. *Mid August–early October*

WI range and habitats Common throughout, in sedge meadows, wet prairies, fens, edges of bogs and swamps, wet thickets, shores, and stream banks. *Curtis Prairie, Greene Prairie, wetlands throughout the Arboretum*

The genus name (Greek *chelone,* tortoise) and common name refer to the corolla, which is shaped like a turtle's head with its mouth open. The lower lip is a landing platform for pollinating bumblebees, which must push hard to squeeze between the closed lips. Only long-tongued bumblebees can reach the nectar. This plant has horticultural promise (pink-flowered relatives have long been in the nursery trade), the large flowers making a show from midsummer until the first hard frost.

Turtlehead is highly variable in leaf characters (shape, dimensions, hairiness) and was subdivided by old-time botanists into a series of varieties and forms. Several recent studies using DNA sequence data indicate clearly that traditional views of the Figwort Family are untenable and support the inclusion of the majority of scrophs, including *Chelone,* in an expanded Plantain Family (Plantaginaceae).

Pink turtlehead from the UW-Botany Garden. Note bumblebee.

WOOD-BETONY
Lousewort
Pedicularis canadensis

Low stature, mostly basal, deeply divided leaves, and leafy-bracted spikes make wood-betony readily identifiable. **Perennial, 4–16 inches tall, from short rhizomes, often forming large clumps or colonies.**

Leaves scattered on the stem, pinnately cleft into coarse segments, toothed with broad shallow blunt teeth. **Stems** unbranched, loosely clustered, hairy.

Flowers clustered in a dense short, spike-like raceme that elongates in fruit; sepals 5, united into an oblique calyx that is split in front; petals 5, pale yellow (upper lip sometimes purplish), forming a strongly 2-lipped, tubular corolla that curves to the side. The upper corolla lip is arched and hooded and enfolds the 4 stamens. *May–June*

Fruits oblong, asymmetrical (slightly wider near the base), somewhat flattened capsules twice as long as the calyx. *Mid May–mid August*

WI range and habitats Locally common throughout (absent from the north-central, yellow silt-loams), in dry to medium prairies, open oak or pine woods, oak savannas, cedar glades, and bracken grasslands, also on clay bluffs along Lake Michigan. *Curtis Prairie, Greene Prairie, Grady Tract knolls, Juniper Knoll, Sinaiko Overlook Prairie*

This is one of the few eastern North American representatives of an enormous genus containing about 500 species. The genus name is derived from the Latin word *pediculus,* louse, because there is an old superstition in Europe that livestock, feeding in marshes and wet meadows containing *Pedicularis palustris,* would become covered with lice. Like castillejas, these plants are hemiparasites living on a wide range of host species. In the Angiosperm Phylogeny Group (APG) system, these and the other hemiparasitic figworts are included with the holoparasitic broomrapes in a redefined Broomrape Family (Orobanchaceae).

Similar species Fen- or swamp-betony (*Pedicularis lanceolata*) is distinguished by its opposite to nearly opposite, stalkless leaves, hairless raceme axis, and fall blooming time. It is not found in the Arboretum.

Gone to seed

Tall Beard-tongue
False foxglove
Penstemon digitalis

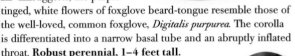

As one of its common names suggests, the purple-tinged, white flowers of foxglove beard-tongue resemble those of the well-loved, common foxglove, *Digitalis purpurea*. The corolla is differentiated into a narrow basal tube and an abruptly inflated throat. **Robust perennial, 1–4 feet tall.**

Stem leaves dark green, narrowly oblong or narrowly lance-shaped, relatively large, the upper ones widely spaced and reduced in size, entire to finely toothed, hairless. **Stems** hairless, slightly tinted whitish-green and somewhat shiny.

Flowers strongly bilaterally symmetrical but only slightly 2-lipped, the 3 lower corolla lobes almost equaling the 2 upper ones; calyx 5-parted; corolla $5/8$–$1 1/4$ inches long, the lower inner surface marked with faint purple lines. Anthers of the 4 fertile stamens with few coarse white hairs on the back. *Mid June–mid July*

Fruits egg-shaped, many-seeded capsules, $3/8$–$1/2$ inch long. *Mid July–early September*

WI range and habitats Occasional south and central, sometimes locally abundant, on roadsides, low pastures, old fields, clearings in medium woods, and recovering, disturbed or weedy, dry to medium prairies. *Curtis Prairie, East Knoll, Greene Prairie, Sinaiko Overlook Prairie*

In habit, leaves, and flowers, this beautiful wildflower resembles our other penstemon species; a close look is usually required to see the subtle differences between them. All are erect perennials that arise from a rosette of stalked basal leaves, bear opposite stalkless leaves, and produce flowers in a compact panicle. Tall beard-tongue was originally indigenous to the Mississippi Basin, but spread north and east in the later nineteenth century and is now part of our floristic landscape. It is one of many species of this large and complex genus that are in cultivation.

Modern systematic studies put most of the non-parasitic plants earlier assigned to the Figwort Family, including *Digitalis*, *Penstemon*, and *Veronica*, into an expanded Plantain Family (Plantaginaceae).

SLENDER BEARD-TONGUE
Penstemon gracilis

This is Wisconsin's smallest, most slender-stemmed, latest-flowering beard-tongue species. It is most easily identified by corolla shape and color. **Slender erect perennial, 8–24 inches tall, from a short subterranean crown.**

Stem leaves opposite, linear to lance-shaped, rarely more than $3/8$ inch wide, the margins almost untoothed to minutely toothed, hairless. **Stems** minutely hairy, the hairs toward the base often down-curved and non-glandular but glandular near the inflorescence.

Flowers strongly bilaterally symmetrical (bisected by only one plane into similar halves), in several, few-flowered, false whorls; corolla pale lavender to mauve, tubular and slender, $1/2$–$7/8$ inch long, hairy externally and internally, with a flattened throat that is not distinctly differentiated from the narrower tube. Lobes of the lower lip surpass the upper lobes, creating a 2-lipped appearance. Stamens 4 fertile plus 1 sterile, the anthers hairless. *Late May–July*

Fruits egg-shaped, many-seeded capsules, $1/4$ inch long. *Late June–early September*

WI range and habitats Locally frequent west and central (rarely beyond), in sand prairies, barrens, dry open woods, and disturbed ground (fields, roadsides, pine plantations); dry sandy soil, including cliffs and outcrops of the Driftless Area and the acid sandy soils of the Central Plain. *West Knoll*

The names penstemon and beard-tongue both refer to the conspicuous infertile stamen, which lacks the anther found on the other four stamens but in many species sports a thick beard. See notes under tall beard-tongue (*Penstemon digitalis*).

Slender beard-tongue is primarily pollinated by bees, but wasps, flies, and beetles can also pollinate it. It is divisible into two subspecies; the total geographical range of *Penstemon gracilis* subspecies *wisconsinensis* is limited to central Wisconsin, where it apparently originated by hybridization of pale beard-tongue (*P. pallidus*) into the typical race of slender beard-tongue (*P. gracilis* subspecies *gracilis*) and formed a stable population.

LARGE-FLOWERED BEARD-TONGUE
Penstemon grandiflorus

This strikingly handsome wildflower is unmistakable owing to the bluish waxy appearance of the stems and leaves and especially the large lavender flowers, from which it gets both its common name and species epithet. They are the largest flowers of all beard-tongues. **Stout biennial or short-lived perennial, 12–40 inches tall, somewhat taprooted.**

Stem leaves opposite, spoon-shaped to mostly broadly egg-shaped, 5/8–2 inches wide, the upper and bracteal leaves (those subtending the flowers) becoming somewhat heart-shaped and clasping at the base, firm-textured, entire. **Stems** and leaves hairless and with a whitish powdery covering.

Flowers 2–4 together in each angle between leaf and stem, forming a raceme-like panicle, strongly bilaterally symmetrical and distinctly 2-lipped; calyx deeply 5-parted, 1/4–7/8 inch long; corolla tubular, well-inflated, 1 1/4–1 3/4 inches long, hairless inside and out, the throat lined on the inner lower surface with magenta nectar guides. Stamens 4 fertile plus 1 sterile, the latter bearded and lying close to the lower lip of the corolla. *Early June–early July*

Fruits egg-shaped, many-seeded capsules, 5/8–7/8 inch long. *Mid June–mid August*

WI range and habitats
Infrequent southwest, in dry, sandy or gravelly prairies and barrens, now mostly adventive on roadsides and in gravel pits. *Upper Greene Prairie, West Knoll*

All beard-tongues have tubular flowers, in which the 2 lobes of the upper lip project like a roof over the 3 lobes of the lower lip, which serves as a landing platform for bees and sometimes wasps and flies. The variation in flower size, shape, and color is the product of co-evolutionary interplay between the flowers and their pollinating insects. See notes under tall beard-tongue (*Penstemon digitalis*).

Culver's-root
Veronicastrum virginicum

Interesting whorled leaves and densely flowered, terminal spikes, 1 or 3–5 per plant, make Culver's-root a particularly elegant plant. Even though the foliage is olive-green and may turn partly blackish, it is a non-parasitic plant. **Erect perennial, 3–5 feet tall.**

Leaves mostly whorled (some of the uppermost opposite), 3–7 at each node, lance- or narrowly egg-shaped, 2–6 inches long, narrowed to a short-stalked base, long-pointed, sharply toothed. **Stems** and leaves hairless or covered with long soft hairs.

Flowers in spike-like racemes 2–6 inches long, white to pale purplish, tubular, nearly symmetrical, $\frac{1}{4}$ inch long or a little less; sepals 5 (4), short, separate nearly to the base; petals 4, united for most of their length, the lobes much shorter than the tube. The 2 fertile stamens and equally long style project well beyond the corolla. *Late June–August*

Fruits narrowly egg-shaped capsules, opening by 4 short apical slits. *Mid August–early October*

WI range and habitats Common throughout, but conspicuously absent from the Northern Highlands, in wet-medium to medium prairies, fens, oak openings, edges of woods, occasionally on lakeshores and ditches; prefers deep loamy soil, in sun or light shade. ***Curtis Prairie, Greene Prairie, Juniper Knoll, Marion Dunn Prairie, Sinaiko Overlook Prairie***

The genus name is derived from that of the related genus *Veronica,* out of which it was segregated, and the suffix *astrum,* i.e., false Veronica. At one time the American species was called Culver's-physic in reference to its use in home remedies; its roots are toxic, their sap containing a strong emetic and laxative.

The majority of genera assigned to the Figwort Family in a traditional sense, including the speedwells (*Veronica* species) and Culver's-root, as well as snapdragons (*Antirrhinum*), water-hyssops (*Bacopa*), kitten's-tails (*Besseya*), toadflaxes (*Linaria*), etc., have been placed in the newly reconstituted Plantain Family in the Angiosperm Phylogeny Group (APG) system.

Other Scrophulariaceae

Butter-and-Eggs
Linaria vulgaris
Introduced, naturalized; erect perennial forb, 1–3 feet. Common in disturbed and waste places.
Flowers June–October

Monkey-Flower
Mimulus ringens
Native; erect perennial forb of wetlands, 6–40 inches.
Flowers July–September

Mullein
Flannel plant
Verbascum thapsus
Introduced, naturalized; potentially invasive erect biennial forb, 1–6 feet. Often seen towering in pastures because cows dislike the fuzzy leaves.
Flowers June–October

Clammy Ground-cherry
Physalis heterophylla

Whitish to yellowish or yellow-green corollas, usually marked with brownish or bluish spots, and inflated fruiting calyces make ground-cherries easily recognizable. Relatively broad leaves and spreading, often sticky stem hairs distinguish clammy ground-cherry from our other native species. **Perennial, 6–28 inches, tall, from deeply buried rhizomes.**

Leaves alternate, simple, broadly egg- or egg-diamond-shaped, the main ones 2–4 inches long and $1\frac{1}{2}$–$2\frac{1}{2}$ inches wide, broadly rounded to straight across or heart-shaped at the base, irregularly wavy-toothed. **Stem** erect, spreading and often branched at top; herbage (including calyx) covered with varying proportions of short sticky, long gland-tipped, and long jointed hairs.

Flowers solitary in the axils of leaves or bracts, nodding, $\frac{3}{8}$–$\frac{3}{4}$ inch long; calyx 5-cleft, $\frac{1}{4}$–$\frac{1}{2}$ inch long in flower, the venation forming an interconnected network, enlarging as the fruit matures and becoming much larger than the fruit inside; corolla wheel- to short-funnel-shaped, $\frac{5}{8}$–1 inch broad, 5-angled and minutely 5-toothed; stamens 5, the anthers yellow or blue-tinged. *Early June–end of August*

Fruits spherical, many-seeded berries, turning yellow, pulpy, loosely enclosed within an enlarged calyx $1\frac{1}{4}$–$1\frac{1}{2}$ inches long and equally as thick. *Early September–early October*

WI range and habitats Infrequent to locally common nearly throughout except the Northern Highlands, in dry to medium prairies, open jack pine or white or black oak woods, and fields (cultivated and fallow), less common on cleared or pastured hillsides, embankments, fencerows, roadsides, railroads, and other disturbed habitats, e.g., gravel pits and dumps. *Curtis Prairie, Greene Prairie, West Knoll*

A few ground-cherries like Chinese lantern-plant (*Physalis alkekengi*) are cultivated as ornamentals. In large quantities, the leaves and unripe fruits are poisonous to livestock, but the ripe berries are non-poisonous (at least to humans). Though rather tasteless, they are used for jelly and preserves, and in Latin America, especially Mexico and the West Indies, in the preparation of savory salsas.

Virginia Ground-cherry
Lance-leaved ground-cherry
Physalis virginiana

Drooping, broadly funnel-shaped flowers of this common wildflower are promptly followed by inflated papery husks that completely enclose the berry, which resembles a miniature tomato. **Perennial, 4–21 inches tall, from deeply buried rhizomes.**

Leaves alternate, simple, egg- to lance-shaped or elliptic, 1–2¼ inches long, narrowly to broadly wedge-shaped at the base, entire to wavy or even slightly toothed. **Stem** erect, spreading at the top, often forked; hairs of the younger stems, leaf stalks, and calyces (under magnification) short, stiff, abruptly down-curved (occasionally mixed with long spreading hairs).

Flowers arising in the angles formed by the leaf and stem or the stem and branch, mostly solitary, nodding, on stalks ¼–¾ inch long; sepals 5, fused, forming a calyx ½–⅔ the length of the corolla, in fruit egg-shaped and inflated, minutely hairy over the entire surface; corolla of 5 fused petals, ⅜–¾ inch long, yellowish or yellow-green with a purple blotch toward the center; anthers yellow or sometimes with a blue tinge, ⅛ inch long. *Late May–early September*

Fruits spherical juicy berries, turning orange, enclosed within a 5-angled, inflated papery husk (the enlarged calyx) 1–1⅜ inch long. *Early July–mid November*

WI range and habitats Occasional to locally abundant throughout but sporadic north-central, in dry, sandy, gravelly, or rocky prairies, jack pine-black oak barrens, steep hillsides, open ridge tops, and sandstone outcrops, also medium prairies and beaches and dunes along Lake Michigan, often weedy in abandoned sandy fields and along sandy or gravelly roadsides and railroads. *Curtis Prairie, Greene Prairie, Grady Tract knolls*

Similar species Smooth long-leaved ground-cherry (*Physalis longifolia*) is sometimes found in prairies but more often in pastures, fields, and cultivated or disturbed ground adjoining prairies. Except for sparse short hairs on younger parts and 10 lines of short hairs on the calyx tube, it is otherwise hairless. Carolina horse-nettle (*Solanum carolinense*) belongs to a related genus that differs from ground-cherries in having the corolla widely spreading and anthers opening by pores instead of slits. It is a low spiny perennial with purple to pale violet or white flowers and yellow berries, weedy in fields, pastures, and degraded prairies.

Other Solanaceae

Deadly Nightshade
Bittersweet nightshade, climbing nightshade
Solanum dulcamara
Introduced, naturalized; potentially invasive trailing/twining perennial woody vine, to 8 feet. Flowers May–September

Both poisonous

purple flowers, red fruits

white flowers, black fruits

Nightshade
Solanum ptychanthum
Synonyms *Solanum americanum, S. nigrum*

Broad-leaved Cat-tail
Common cat-tail
Typha latifolia

Almost everyone is acquainted with the familiar sight of large stands of cat-tails, the common name for *Typha,* of which we have three taxa. This small but easily recognized genus has long, grasslike leaves that sheathe the solitary unbranched stem, on top of which is perched an extremely dense, club-like, cylindrical spike of minute flowers. **Coarse perennial, 3–9 feet tall, strongly rhizomatous, forming tremendous colonies.**

Leaves (larger ones) almost always more than $5/16$ inch wide, flat below the middle, the sheaths produced into a tapered or rounded flange at the summit.

Flowers in long dense spikes, the dark brown female portion $5/8$–$1\,3/8$ inch thick in fruit, contiguous with the thinner yellowish male portion above or separated by an interval of stem less than $3/4$ inch long. Female flowers reduced to stalked, 1-chambered, fertile or abortive ovaries, each with a broadened, tongue-shaped stigma and copious, thread-like bristles. Male flowers reduced to stamens, covered with yellow pollen at maturity, then disintegrating.

Fruits minute, long-stalked nutlets equipped with a delicate parachute of silky hairs.

WI range and habitats Throughout in shallow water or intermittently wet habitats: borders of lakes, ponds, and rivers, swamps, bogs, sedge meadows, seeps, swales, and roadside ditches. *East Curtis Prairie, Gardner and Redwing marshes and other Arboretum wetlands*

It has been estimated that a single well-developed cat-tail spike may produce an average of more than 220,000 seeds. The leaves of cat-tails are made up mostly of air spaces separated by very thin, pithy partitions. This structure confers great buoyancy to the long leaves, which rarely break no matter how violently they may be blown about.

Similar species Two well-marked species, broad-leaved cat-tail and narrow-leaved cat-tail *(Typha angustifolia),* and hybrids intermediate between them (hybrid cat-tail, *T.* ×*glauca*), are abundant in our area. Narrow-leaved cat-tail has linear stigmas and well-developed, brown-tipped bracts among the bristles, whereas hybrid cat-tail has slenderly broadened (lance-shaped) stigmas and poorly developed, often pale-tipped bracts. Unfortunately, more readily observed characters are not all that helpful because of overlap in the measurements.

EDIBLE VALERIAN
Tap-rooted valerian
Valeriana edulis

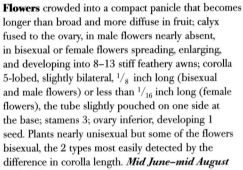

Edible valerian is recognized by its predominantly basal leaves, which vary from entire to pinnate-compound or -cleft, and numerous tiny, white to cream-colored flowers in a terminal compound panicle. A calyx is present but develops late; its divisions, inrolled during flowering, unroll as the fruit develops, forming a conspicuous feathery pappus. **Perennial, 9–32 inches tall, from a short crown atop a long stout, strong-scented taproot.**

Basal leaves inverted-lance-shaped, to 15 inches long, entire or some or all with basal divisions. **Stem leaves** 2–6 pairs, smaller, deeply pinnate-lobed or parted (rarely simple), the axis broader than the lateral segments; leaf margins with a very dense fringe of short hairs.

Flowers crowded into a compact panicle that becomes longer than broad and more diffuse in fruit; calyx fused to the ovary, in male flowers nearly absent, in bisexual or female flowers spreading, enlarging, and developing into 8–13 stiff feathery awns; corolla 5-lobed, slightly bilateral, $1/8$ inch long (bisexual and male flowers) or less than $1/16$ inch long (female flowers), the tube slightly pouched on one side at the base; stamens 3; ovary inferior, developing 1 seed. Plants nearly unisexual but some of the flowers bisexual, the 2 types most easily detected by the difference in corolla length. *Mid June–mid August*

Fruit a compressed, oval to oblong-egg-shaped achene with 3 veins on the outer face and 1 on the inner. *Mid August–early October*

WI range and habitats Rare to locally common south, confined to calcareous prairies (wet or dry) and fens, from rich, medium to moist prairies on beds of old glacial lakes to dry-medium to dry prairies on knolls, bluffs, and gravelly hills. *Greene Prairie*

The edible root was used as food by native peoples in the West and probably in our region as well. Members of the family often have a very characteristic, slightly unpleasant odor that has been compared with that of a wet dog, owing to volatile oils occurring mainly in the root and rhizome. The genus includes the old-fashioned garden flower, garden-heliotrope (*Valeriana officinalis*).

Blue Vervain
Simpler's-joy
Verbena hastata

Usually a tall conspicuous plant, this vervain has square stems, paired leaves, and numerous small flowers borne in a candelabra-like panicle of several pencil-thin spikes. Blooming from the bottom up, the flowers form a bright blue-violet wreath that ascends the spike until reaching the top, at which time, according to old-timers, the first frost will occur. **Short-lived perennial, 1–5 feet tall.**

Leaves opposite, narrowly egg-shaped to elliptic, 4–6 inches long, the lower sometimes with 1 or 2 lobes at the base, sharply toothed. **Stem** somewhat 4-sided; whole plant thinly hairy.

Flowers (and fruits) densely overlapping, $1/_8$ inch long and broad; spikes 2–6 inches long; calyx tubular, 5-toothed; petals fused into a short tube, the 5 abruptly flaring lobes weakly 2-lipped; stamens 4, of two lengths, inserted on and included within the corolla tube. ***Late June–September***

Fruits 4 tiny pink nutlets ("seeds"), faintly lined or smooth, enclosed by the persistent calyx. ***Early July–mid October***

WI range and habitats Common throughout, in moist sunny habitats: sedge meadows, low prairies, marshes, wet thickets, openings in swamps, stream edges, and lakeshores; tolerant of disturbance, hence also in pastures, roadsides, railroad rights-of-way, and occasionally abandoned fields. ***Curtis Prairie, Lower Greene Prairie, Marion Dunn Prairie, Visitor Center parking lot***

Although odorless, the flowers secrete a large amount of nectar, attracting many bees and butterflies, including skippers and swallowtails.

The genus name means "sacred plant" in Latin, recalling ancient times when vervains were thought to have medicinal properties or spiritual or magical powers. In the Midwest the roots were used medicinally, the Menominee making a tea for urinary disorders and the Meskwaki preparing a remedy for eczema. Actually, vervains have no therapeutic value, but we should take the meaning of their name to heart and treat all wild plants as sacred, respecting their existence and protecting them from destruction.

Similar species Hairy white vervain (*Verbena urticifolia*), common in savannas and degraded woodlands, has graceful panicled spikes with white flowers that are well separated along the very slender spike axis. The tiny nutlets are smooth or corrugated on the back. See also hoary vervain (*Verbena stricta*). Hairiness, spike number, corolla color, or nutlet surface details—any will serve to distinguish between these three species.

Hoary Vervain
Verbena stricta

Hoary vervain is usually quite tall, and its stems, leaves, and spikes are gray-hairy. The spikes, usually fairly elongate by blooming time, are fewer and relatively stout compared to those of blue vervain. The flowers are larger and paler, opening a few at a time and forming a lavender wreath somewhere along the length of the spikes. **Erect perennial, 1–4 feet tall.**

Leaves egg-shaped, elliptic, or almost circular, $1\frac{1}{4}$–$3\frac{3}{4}$ inches long, wedge-shaped at the base and stalkless or the larger somewhat stalked, sharply and mostly double-toothed, wrinkled above, prominently veined beneath. Whole plant densely covered with spreading, stiffish (spikes) to soft (stems and leaves) hairs.

Flowers and fruits overlapping in stiff, short-stalked spikes terminating the stem or each branch, each subtended by a lance-awl-shaped bract about as long as the calyx; sepals united to form a tubular calyx $\frac{1}{8}$ inch long, the 5 lobes tapering gradually to a point; corolla protruding slightly beyond the calyx, lavender or rarely pink or white, the expanded flat part $\frac{3}{8}$ inch broad, hairy outside. *Late June–early October*

Fruits 4 ellipsoid nutlets, each a little less than $\frac{1}{8}$ inch long, marked with a network-like pattern above and fine lines below. *Early July–October*

WI range and habitats Frequent southwest, sporadic (and probably adventive) northward, in dry, sandy or limy remnant and degraded prairies, sandy or gravelly abandoned fields and pastures, and sometimes open oak or oak-pine woods, spreading along railroads and roadsides. *Curtis Prairie (limestone knoll), Marion Dunn Prairie*

The flowers of hoary vervain are capable of being fertilized by their own pollen but are partly self-incompatible. They are visited by flies, bees, and butterflies in equal numbers.

Similar species Flowers and often leaves are superficially similar among all 5 native and 2 introduced verbenas known from Wisconsin. These species are all wide-ranging and interfertile, and occasional specimens clearly represent first-generation hybrids, of which six types occur in the state. Creeping vervain (*Verbena bracteata*) has a prostrate habit and thick inflorescences, and narrow-leaved vervain (*V. simplex*) has leaves no more than $\frac{3}{4}$ inch wide. Neither is found in the Arboretum. See blue vervain (*V. hastata*).

Other Verbenaceae

White Vervain
Nettle-leaved vervain
Verbena urticifolia
Native; erect perennial forb, 16-40 inches. Dry to medium and floodplain woods (especially along edges and trails), thickets, pastures, grazed or drained sedge meadows, fencerows, roadsides, and other degraded habitats; probably increasing as good-quality habitats decline.
Flowers July–October

Bird's-foot Violet
Viola pedata

Perhaps the most beautiful violet of all, this species ranges throughout much of the continent east of the Rockies. The namesake leaves (*pedata* means foot-like) are deeply divided like the clawed toes of a bird. The large flowers have a flattened appearance, revealing their kinship to the annual pansy. **Perennial, 3–7 inches tall, from a fleshy vertical rhizome.**

Leaves all basal, characteristically deeply 3-divided, the divisions further 3–5-cleft, the ultimate lobes linear or spatula-shaped and often with 2–4 teeth or lobes near the tip.

Flowers solitary on stalks that emerge directly from the plant base, bilaterally symmetrical, $3/4$–$1 1/2$ inches wide; petals 5, uniformly light violet or lilac-purple (the upper pair very saturated purple in a strikingly beautiful, rare form), the lowest larger than the others and projecting backward as a spur; all petals beardless. Stamens 5, standing together and closely surrounding the style, holding the pollen, in effect, in a basket, their orange tips conspicuously protruding from the center of the flower. *Late April–mid June* (irregularly into October)

Fruits yellow-brown, ellipsoid capsules $1/4$–$3/8$ inch long, splitting when ripe into 3 little canoe-shaped sections, each filled with a row of shiny brown seeds. *Mid May–September*

WI range and habitats Locally common south and central, lacking from the Northern Highlands and Niagara Uplands, in dry, sandy, gravelly, or rocky prairies, open oak and/or pine woodlands, black oak savannas, jack pine barrens, and cedar glades, also pine plantations, abandoned fields, embankments, roadsides, and sand blows. ***Grady Tract knolls***

Most of the 20 species of violets native to Wisconsin (31 total taxa, including introduced species, varieties, and hybrids) favor wooded habitats and bloom early in spring. This species is distinctive not only in the shape of the leaves, which are not the familiar heart shape common to other violet species, but also in being the only American violet that does not produce inconspicuous, self-fertilizing flowers in addition to the ordinary flowers.

In normal violet flowers, the 2 lower stamens bear nectaries that extend into the spur; when these are touched by a probing insect, the anthers are disturbed and shed their pollen on the visitor.

Prairie Violet
Viola pedatifida

Like bird's-foot violet, this species has dissected leaves and does not produce runners, but the flowers have bearded lateral petals and non-protruding stamens, making it readily distinguishable. **Perennial, 3–7 inches tall, from a slender to stout, ascending rhizome.**

Early leaves palmately cleft nearly to the base into usually 3 parts, each of which is cleft into linear or spatula-shaped segments, these again usually cut into 2–4 lobes. **Late spring and summer leaves** are often only incised or shallowly lobed and can be mistaken for hybrids involving various species.

Ordinary flowers borne on erect stalks a little shorter to usually longer than the stalks of the leaves; sepals 5, separate, persistent; petals 5, unequal, light to dark blue-violet, the lower with dark veins and a dense beard; stamens included within the throat. **Self-fertilizing flowers** are also produced; these remain small, not developing petals and never opening, but are nonetheless very fruitful. *Late April–early June*

Fruits yellow-green capsules $5/16$–$1/2$ inch long, with 1 chamber and 3 rows of seeds. When the capsule dries to a certain point, its 3 sections explode, shooting the seeds in all directions. *Mid June–mid October*

WI range and habitats Locally frequent south and west, from steep rocky dry prairies to rich, black-soil medium prairies, oak savannas, and cedar glades, also pine plantations, grazed oak and red cedar stands, fields, hillsides, and banks; prefers soil rich in humus. *Curtis Prairie, Greene Prairie*

Prairie violet is one of the earliest spring bloomers. The purple veins serve the purpose of guiding pollinating insects to the pollen and nectar.

Similar species In Wisconsin this species is easily confused only with the more common bird's-foot violet (*Viola pedata*), with which it may grow, but it produces fewer, darker blue flowers that are not flat-faced and have a less pronounced orange center. Lobed blue violet (*V. ×subsinuata*), occasional in prairies, savannas, and oak woods, usually occurs with prairie violet and near common wood violet (*V. sororia*), its putative parental species. True *V. ×palmata* (*V. sagittata × V. sororia*) is exceedingly rare in Wisconsin.

Arrow-leaved Violet
Viola sagittata

Many of the leaf blades of this plant are somewhat narrower ($1\frac{1}{2}$–4 times as long as wide) than in our other "stemless" blue violets. Furthermore, the shape of the leaf changes so much as the season progresses it is difficult to believe you are looking at the same plant. The flowers are the familiar, white-centered, blue-violet, bilaterally symmetrical flowers of the genus. **Perennial, 2–6 inches tall, from a short, horizontal to vertical rhizome.**

Leaves narrowly egg- (early ones) to lance-egg- or lance-oblong-shaped, $1\frac{1}{2}$–4 inches long (often smaller at flowering time), straight across to ± heart-shaped at the base, evenly or shallowly toothed near the base and scalloped above in early leaves, with 4–6 unusually large teeth or short diverging lobes near the base in later leaves. Stalks of early leaves spreading, those of mid-season leaves erect. Leaves prostrate to weakly ascending and the blades elliptic to egg-shaped and heavily hairy in a very rare variety.

Flowers $\frac{3}{8}$–$\frac{3}{4}$ inch across, solitary on stalks shorter to longer than the leaves; sepals lance-shaped, narrowly pointed at the apex, the basal lobes short (out-crossing flowers) to prominent (at least $\frac{1}{2}$ as long as the sepals in self-pollinating flowers); lower 3 petals bearded. Small, self-fertilizing flowers abundantly produced during summer. *Late April–beginning of June (mid June in Door County)*

Fruits yellowish to green, ellipsoid capsules $\frac{3}{8}$–$\frac{1}{2}$ inch long; seeds numerous, bronze-brown. *Late June–early October*

WI range and habitats Infrequent to locally common south, sporadic northeast, in oak barrens, savannas, dry woodlands (oak with jack pine, hickory, or white birch), dry to moist prairies, clearings, edges of fields, slopes, banks, also bluffs, cliffs, and outcrops and moist, sandy to peaty meadows, swales, shores, and ditches, including previously cultivated or excavated ground. *Greene Prairie*

Violets form a well-defined genus, but many of the species are variable and difficult to distinguish, especially because the characteristics are often taken from the shape of the leaves and the nature of the hairs. Characters are open to ambiguity, and hybrids are frequent, so one must exercise good judgment and intuition when making identifications.

Violaceae • Violet Family

Common Blue Violet
Door-yard violet, hairy wood violet
Viola sororia
Synonym *Viola papilionacea?*

Most violets encountered, including in urban areas, are this common unassuming species, the state flower of Wisconsin, whether the flowers are light lilac, red- or blue-violet, white or gray with purple veins, or white with blue speckles. **Perennial, $1\frac{1}{2}$–6 inches tall, from short stout rhizomes.**

Leaves egg- to kidney- or heart-shaped (never triangular), the larger ones $1\frac{1}{2}$–5 inches broad, usually at least slightly hairy (use a hand lens), varying from almost shaggy to completely hairless on either or both surfaces.

Flowers $\frac{5}{8}$–1 inch across, solitary on stalks that emerge directly from the rhizomes, the stalks at first longer than but soon only equaling or shorter than the leaves; ear-shaped sepal-lobes less than $\frac{1}{16}$ inch long; petals 2 upper, 2 side, and 1 lower, 3 of which are strongly veined, the central one of these extending backward into a small spur, the side petals bearded with slender hairs that are barely or not expanded at the tip. *April throughout summer*

Fruits purple or purple-spotted, ellipsoid capsules, their short stalks arching or lying upon the ground; seeds buff to brown. *May throughout summer*

WI range and habitats Common throughout, in dry to moist forests, woodlands, and thickets, wooded ravines, shaded cliffs, riverbanks, prairie edges, fencerows, fields, and disturbed sites (roadsides, waste areas, lawns, gardens); native but somewhat weedy.
Throughout the Arboretum

Violets attract many butterflies, including fritillaries and the spring azure. The leaves, flowers, and seeds have been used for medicinal purposes since ancient times. The springtime leaves are rich in vitamins A and C.

Besides ordinary out-crossing or sterile flowers, this species produces inconspicuous, at first partly subterranean, bud-like, self-pollinating flowers. These normally do not have petals, nor do they open, but they produce more seeds than the normal flowers.

Similar species Blue marsh violet (*Viola cucullata*) and northern bog violet (*V. nephrophylla*) are much like common wood violet but are quite hairless and grow in wet places. The former produces pale to medium violet flowers with a dark eye (varying to white) and short stout beard-hairs that are club-shaped or knob-tipped. The latter has relatively small leaves that are commonly tinged bluish on the lower surface and more elongate flowers with a bearded spurred petal and blunt sepals with very short, basal lobes.

SELECTED BIBLIOGRAPHY

Ahrenhoester, R. and T. Wilson. 1981. *Prairie Restoration for the Beginner.* Prairie Seed Source, North Lake, WI. 32 pp.

Allen, D. L. 1967. *The Life of Prairies and Plains.* McGraw-Hill, New York, NY. 232 pp.

Anderson, R. C. 1972. *The Use of Fire as a Management Tool on the Curtis Prairie.* Univ. of Wisconsin Arboretum, Madison, WI. 8 pp.

Anderson, R. C. 1983 ["1982"]. The eastern prairie-forest transition—an overview. Pp. 86–92. *In:* R. Brewer, ed. *Proc. Eighth N. Amer. Prairie Conf.,* August 1–4, 1982. Dept. Biology, Western Michigan Univ., Kalamazoo, MI.

Anderson, R. C., J. S. Fralish and J. M. Baskin, eds. 1999. *Savannas, Barrens, and Rock Outcrop Plant Communities of North America.* Cambridge Univ. Press, Cambridge, U.K. 470 pp.

Angiosperm Phylogeny Group. 2003. An update of the Angiosperm Phylogeny Group classification for the orders and families of flowering plants: APG II. J. Linn. Soc. Bot. 141: 399-436. www.mobot.org/mobot/research/APweb/

Axelrod, D. I. 1985. Rise of the grassland biome, central North America. Bot. Rev. 51:163–201.

Ballard, H. E., Jr. 1994. Violets of Michigan. Michigan Bot. 33:131–199.

Bayer, R. J. and G. L. Stebbins. 1982. A revised classification of *Antennaria* (Asteraceae: Inuleae) of the eastern United States. Syst. Bot. 7:300–313.

Blewett, T. J. and G. Cottam. 1984. History of the University of Wisconsin Arboretum Prairies. Trans. Wisconsin Acad. Sci. 72:130–144.

Borchert, J. R. 1950. The climate of the central North American grassland. Ann. Assoc. Amer. Geogr. 40:1–39.

Bragg, T. B. 1995. The physical environment of the Great Plains grasslands. Pp. 49–81. *In:* A. Joern and K. H. Keeler, eds. *The Changing Prairie.* Oxford Univ. Press, New York, NY.

Bureau of Endangered Resources. 1993. *Guide to Wisconsin's Endangered and Threatened Plants.* Bureau of Endangered Resources. Wisconsin DNR PUBL-ER-067. Madison, WI. 128 pp.

Carpenter, J. R. 1940. The grassland biome. Ecol. Monogr. 10:617–684.

Case, F. W., Jr. 1987. *Orchids of the Western Great Lakes Region.* Rev. ed. Cranbrook Inst. Sci. Bull. 48. Bloomfield Hills, MI. 251 pp.

Christiansen, P. and M. Müller. 1999. *An Illustrated Guide to Iowa Prairie Plants.* Univ. of Iowa Press, Iowa City, IA. 237 pp.

Cochrane, T. S. and H. H. Iltis. 2000. *Atlas of the Wisconsin Prairie and Savanna Flora.* Wisconsin Dept. Nat. Res. Techn. Bull. No. 191. Madison, WI. 226 pp.

Collins, S. L. and L. L. Wallace, eds. 1990. *Fire in North American Tallgrass Prairies.* Univ. of Oklahoma Press, Norman, OK. 175 pp.

Costello, D. F. 1969. *The Prairie World.* Thomas Y. Crowell, New York, NY. 242 pp. [Reprinted, 1980, Univ. of Minnesota Press, Minneapolis, MN. 244 pp.]

Cottam, G. and H. C. Wilson. 1966. Community dynamics on an artificial prairie. Ecology 47:88–96.

Courtenay, B. and J. H. Zimmerman. 1978. *Wildflowers and Weeds.* Prentice Hall Press, New York, NY. 144 pp.

Curtis, J. T. 1959. *The Vegetation of Wisconsin: an Ordination of Plant Communities.* Univ. of Wisconsin Press, Madison, WI. 657 pp.

Egan, D. J. 2002. *Arboretum Prairies.* Univ. of Wisconsin Arboretum, Madison, WI. 56 pp.

Davis, A. 1977. The prairie-deciduous forest ecotone in the upper Middle West. Ann. Assoc. Amer. Geogr. 67:204–213.

Deevey, E. S. and R. F. Flint. 1957. Post-glacial hypsithermal interval. Science 125:182–184.

Dietrich, W., W. L. Wagner and P. H. Raven. 1997. Systematics of *Oenothera* sect. *Oenothera* subsect. *Oenothera* (Onagraceae). Syst. Bot. Monogr. 50:1–234.

Fassett, N. C. 1951. *Grasses of Wisconsin*. Univ. of Wisconsin Press, Madison, WI. 173 pp.

Fassett, N. C. 1976. *Spring Flora of Wisconsin*, 4th ed., rev. and enlarged by O. S. Thomson. Univ. of Wisconsin Press, Madison, WI. 413 pp.

Fassett, N. C., H. H. Iltis, et al. 1929–1988 ["1987"]. Preliminary reports on the flora of Wisconsin. Trans. Wisconsin Acad. Sci. An irregularly appearing series, some 70 studies to date, of selected families and genera. A bibliography of these reports is available electronically at www.botany.wisc.edu/herbarium/info/pdf/appendex_index.pdf

Fernald, M. L. 1950. *Gray's Manual of Botany*, 8th ed. American Book Co., New York, NY. 1632 pp.

Flora of North America Editorial Committee, eds. 1993–2006+. *Flora of North America North of Mexico*. 12+ vols. Oxford Univ. Press, New York, NY.

Foster, S. and J. A. Duke. 1990. *A Field Guide to Medicinal Plants: Eastern and Central North America*. Houghton Mifflin Co., Boston, MA. 366 pp.

Geiss, J. W. and W. R. Boggess. 1968. The Prairie Peninsula: its origin and significance in the vegetational history of central Illinois. Pp. 89–95. *In:* R. E. Bergstrom, ed. *The Quaternary of Illinois*. Univ. of Illinois, Coll. Agric. Special Publ. No. 14. Urbana, IL.

Gleason, H. A. and A. Cronquist. 1991. *Manual of Vascular Plants of Northeastern United States and Adjacent Canada*, 2nd ed. New York Botanical Garden, New York, NY. 910 pp.

Great Plains Flora Association, eds. 1986. *Flora of the Great Plains*. Univ. Press of Kansas, Lawrence, KS. 1392 pp.

Greene, H. C. and J. T. Curtis. 1953. The re-establishment of an artificial prairie in the University of Wisconsin Arboretum. Wild Fl. 29:77–88.

Hanson, H. C. 1950. Ecology of the grassland. II. Bot. Rev. 16:283–360.

Heiser, C. B., Jr., D. M. Smith, S. B. Clevenger and W. C. Martin, Jr. 1969. *The North American Sunflowers (*Helianthus*)*. Mem. Torrey Bot. Club 22(3). 218 pp.

Henderson, R. A. 1982. Vegetation: fire ecology of tallgrass prairie. Nat. Areas J. 2:17–26.

Henderson, R. A. 1995. *Plant Species Composition of Wisconsin Prairies: an Aid to Selecting Species for Plantings and Restorations Based upon University of Wisconsin–Madison Plant Ecology Laboratory Data*. Wisconsin Dept. Nat. Res. Techn. Bull. No. 188. Madison, WI. 58 pp.

Henderson, R. A. and D. W. Sample. 1995. Grassland communities. Pp. 116–129. *In:* Wisconsin Dept. Nat. Resources. *Wisconsin's Biodiversity as a Management Issue: a Report to Department of Natural Resources Managers*. Wisconsin Dept. Nat. Res. Publ.-RS-915 95. Madison, WI.

Henderson, R. A. and S. H. Statz. 1995. *Bibliography of Fire Effects and Related Literature Applicable to the Ecosystems and Species of Wisconsin*. Wisconsin Dept. Nat. Res. Techn. Bull. No. 187. Madison, WI. 56 pp.

Hitchcock, A. S. 1951. *Manual of Grasses of the United States*. 2nd ed., rev. by A. Chase. U.S. Dept. Agric. Misc. Publ. 200. 1051 pp.

Illinois Natural History Survey. [Undated.] *Tallgrass Prairies of Illinois* (website). www.inhs.uiuc.edu/~kenr/prairieintroduction.html

Iltis, H. H. 1969. A requiem for the prairie. The Prairie Naturalist 1:51–57. [Reprinted, pp. 219–224. *In:* J. H. Zimmerman, ed. 1972. Proc. Second Midwest Prairie Conf. J. H. Zimmerman, Madison, WI.]

Kilburn, P. D. 1959. The prairie-forest ecotone in northeastern Illinois. Amer. Midl. Naturalist 62:206–217.

Kindscher, K. 1987. *Edible Wild Plants of the Prairie: an Ethnobotanical Guide*. Univ. Press of Kansas, Lawrence, KS. 276 pp.

Kindscher, K. 1992. *Medicinal Wild Plants of the Prairie: an Ethnobotanical Guide.* Univ. Press of Kansas, Lawrence, KS. 340 pp.

Kline, V. M. 1992. Long range management plan: Arboretum ecological communities. Madison, WI: Univ. of Wisconsin Arboretum. digital.library.wisc.edu/1711.dl/EcoNatRes.ArbMgmtPlan

Leach, M. and T. Givnish. 1996. Ecological determinants of species loss in remnant prairies. Science 273:1555–1558.

Madson, J. 1993. *Tallgrass Prairie.* The Nature Conservancy and Falcon Press, Helena, MT. 112 pp,

Madson, J. 1995. *Where the Sky Began; Land of the Tallgrass Prairie.* Rev. ed. Iowa State Univ. Press, Ames, IA. 326 pp.

Moerman, D. E. 1998. *Native American Ethnobotany.* Timber Press, Portland, OR. 927 pp.

Newcomb, L. 1977. *Newcomb's Wildflower Guide.* Little, Brown, & Co., Boston, MA. 490 pp.

Ownbey, G. B. and T. Morley. 1991. *Vascular Plants of Minnesota; A Checklist and Atlas.* Univ. of Minnesota Press, Minneapolis, MN. 307 pp.

Packard, S. and C. F. Mutel, eds. 1997. *The Tallgrass Restoration Handbook for Prairies, Savannas, and Woodlands.* Island Press, Washington, DC. 463 pp.

Pemble, R. H., R. L. Stuckey and L. E. Elfner. [1975.] *Native Grassland Ecosystems East of the Rocky Mountains in North America: a Preliminary Bibliography.* A Supplement to [Wali, M. K., ed.] *Prairie: a Multiple View.* [Univ. of North Dakota Press, Grand Forks, ND.] 466 pp.

Pyne, S. J., P. L. Andrews and R. D. Laven. 1996. *Introduction to Wildland Fire.* 2nd ed. Wiley, New York, NY. 769 pp.

Read, R. H. (with the collaboration of the University of Wisconsin Herbarium). 1976. *Endangered and Threatened Vascular Plants of Wisconsin.* Wisconsin Dept. Nat. Res. Techn. Bull. No. 92. Madison, WI. 58 pp.

Reich, B. 1971. *Guide to the Arboretum Prairies: University of Wisconsin-Madison.* Univ. of Wisconsin Arboretum, Madison, WI. 62 pp.

Robertson, K. R., R. C. Anderson and M. W. Schwartz. 1997. The tallgrass prairie mosaic. Pp. 55–87. *In:* M. W. Schwartz, ed. *Conservation in Highly Fragmented Landscapes.* Chapman & Hall, New York, NY.

Rock, H. W. 1977. *Prairie Propagation Handbook: a List of Prairie Plants of Wisconsin and Suggested Techniques for Growing Them as Part of a Prairie or Wild Garden to Preserve Them for Our Joy and for Future Generations.* 5th ed. Wehr Nature Center, Hales Corners, Wisconsin. 75 pp.

Runkel, S. T. and D. M. Roosa. 1989. *Wildflowers of the Tallgrass Prairie: The Upper Midwest.* Iowa State Univ. Press, Ames, IA. 279 pp.

Sachse, N. D. 1965. *A Thousand Ages.* Univ. of Wisconsin Arboretum, Madison WI. 149 pp.

Samson, F. B. and F. L. Knopf. 1994. Prairie conservation in North America. Bioscience 44:418–422.

Samson, F. B. and F. L. Knopf, eds. 1996. *Prairie Conservation: Preserving North America's Most Endangered Ecosystem.* Island Press, New York, NY. 339 pp.

Smith, J. R. and B. S. Smith. 1980. *The Prairie Garden. Seventy Native Plants You Can Grow in Town or Country.* Univ. of Wisconsin Press, Madison, WI. 219 pp.

Snyder, T. A., III. 2004. A spatial analysis of grassland species richness in Curtis Prairie. M.S. Thesis, Univ. of Wisconsin-Madison, WI.

Sperry, T. M. 1983. Analysis of the University of Wisconsin-Madison prairie restoration project. Pp. 140–147. *In* R. Brewer, ed., *Proc. N. Amer. Prairie Conference.* Dept. of Biology, Western Michigan Univ., Kalamazoo, MI.

Swink, F. and G. Wilhelm. 1994. *Plants of the Chicago Region,* 4th ed. Indiana Acad. Science, Indianapolis, IN. 921 pp.

Thomson, J. W. 1940. Relic prairie areas in central Wisconsin. Ecol. Monogr. 10:685–717.

Transeau, E. N. 1935. The prairie peninsula. Ecology 16:423–437.

Tryon, R. 1980. *Ferns of Minnesota,* 2nd ed. Univ. of Minnesota Press, Minneapolis, MN. 165 pp.

Utech, F. H., and H. H. Iltis. 1970. Preliminary reports on the Flora of Wisconsin No. 61. Hypericaceae—St. John's-wort family. Trans. Wisconsin Acad. Sci. 58:325–351.

Umbanhowar, C. E., Jr. 1992. Reanalysis of the Wisconsin prairie continuum. Amer. Midl. Naturalist 127:268–275.

Umbanhowar, C. E., Jr. 1993. Classification of Wisconsin prairies: reanalysis and comparison of classification methods. Pp. 289–303. *In:* J. S. Fralish, R. P. McIntosh and O. L. Loucks, eds. *John T. Curtis: Fifty Years of Wisconsin Plant Ecology.* Wisconsin Acad. Sci., Arts & Letters, Madison, WI.

Vogl, R. 1974. Effects of fire on grasslands. Pp. 139–194. *In:* T. T. Kozlowski and A. C. Ahlgren, eds. *Fire and Ecosystems.* Academic Press, New York, NY.

Voss, E. G. 1972. *Michigan Flora,* Part I: Gymnosperms and Monocots. Cranbrook Inst. Sci. and Univ. of Michigan Herbarium Bull. 55. Bloomfield Hills, MI. 488 pp.

Voss, E. G. 1985. *Michigan Flora,* Part II: Dicots (Saururaceae–Cornaceae). Cranbrook Inst. Sci. and Univ. of Michigan Herbarium Bull. 59. Ann Arbor, MI. 724 pp.

Voss, E. G. 1996. *Michigan Flora,* Part III: Dicots (Pyrolaceae–Compositate). Cranbrook Inst. Sci. and Univ. of Michigan Herbarium Bull. 61. Ann Arbor, MI. 622 pp.

Weaver, J. E. 1954. *North American Prairie.* Johnsen Publishing Co., Lincoln, NE. 348 pp.

Weaver, J. E. 1968. *Prairie Plants and Their Environment: a 50-year Study in the Midwest.* Univ. of Nebraska Press, Lincoln, NE. 276 pp.

Wetter, M., T. S. Cochrane, M. R. Black, H. H. Iltis, and P. E. Berry. 2001. *Checklist of the Vascular Plants of Wisconsin.* Wisconsin Dept. Nat. Res. Techn. Bull. No. 192. Madison, WI. 258 pp. See also: www.botany.wisc.edu/wisflora/

Wisconsin Natural Heritage Program. 2004. *Wisconsin Natural Heritage Working List* [of Rare Species]. Bureau of Endangered Resources, Wisconsin DNR, Madison, WI. www.dnr.state.wi.us/org/land/er/working_list/taxalists/

Curtis Prairie checkerspot

Acknowledgments

We wish initially to acknowledge Dr. Paul E. Berry for encouraging the project and for unflagging advice and support during the past three years, for calling meetings, arranging funding, and allowing the authors time to make the sustained effort required to complete the book. We are thankful for the very significant contributions of Arboretum staff, especially Molly Fifield Murray, Outreach and Education Manager, who initiated the Flora of the UW-Madison Arboretum series, of which this is the second book, and obtained funding for various phases of the project; Kathy D. Miner, Arboretum tour scheduler and coordinator, who helped with the difficult job of proofreading; and David J. Egan, Senior Editor, who guided the manuscript through press. Other contributors include Dr. Andrew L. Hipp, who provided a copy of his checklist of the Arboretum's prairie flora; Sylvia Marek, Arboretum Naturalist, who accompanied UW–Botany Senior Artist Claudia S. Lipke on several field trips to locate and identify species, and Dr. Mark K. Leach, Arboretum Ecologist, and Stephen B. Glass, Land Care Manager, who provided information about the presence, abundance, and distribution of given plant species in the Arboretum; and Dr. Berry, Ms. Murray, Ms. Barbara A. Cochrane, and UW-Botany Professor Emeritus Dr. Hugh H. Iltis, who read segments of the manuscript. Special thanks go to Dr. Henry L. Cuthbert, Senior University Legal Counsel, for advice and guidance. The facilities, grounds, and holdings of the UW–Madison's Arboretum and the Wisconsin State Herbarium provided the specimens, living and preserved; and the Botany Department and Multimedia Facility provided the studio, equipment, and libraries that made this book possible.

This work was funded in part by the Department of Botany, the Wisconsin State Herbarium, the Arboretum, and the Evjue Foundation.

Wisconsin State Herbarium Senior Academic Curator Theodore S. Cochrane is responsible for the text and the botanical sins that inevitably will be found in it, as well as for the taxonomic, ecological, and geographical views expressed. Kandis Elliot served as Project Manager, graphic designer, illustrator, and compositor for the book. Claudia Lipke made innumerable trips to the field and took thousands of photographs from which book figures were selected. For a few desired species, we relied on additional photographs taken by and used with permission from Mr. Cochrane, Ms. Elliot, Dr. Robert W. Freckmann (RWF), Dr. Hugh H. Iltis (HHI), Ms. Kim L. Karow (KLK), Dr. Virginia M. Kline (VMK), Mr. James R. Sime (JRS), Dr. Kenneth J. Sytsma (KJS), and Mr. Kenneth W. Wood (KWW).

Theodore S. Cochrane
Kandis Elliot
Claudia S. Lipke

Claudia, Ted, Kandis
Photo by Jeff Vogtschaller

Index to Scientific and Common Names

Acer negundo • 41
Achillea millefolium • 60
Agalinis • 323
Agastache scrophulariaefolia • 225
Ageratina altissima • 61
Agrimonia gryposepala • 315
Agropyron repens • 273
Agrostis
   gigantea • 273
   hyemalis • 273
   scabra • 273
Ague-weed • 205
Allegheny blackberry • 7, 316
Allium
   canadense • 226
   cernuum • 227
   stellatum • 227
Alsike clover • 197
AMARANTH FAMILY • 36
AMARANTHACEAE • 36
Ambrosia
   artemisiifolia • 62
   psilostachya • 62
   trifida • 62, 131
American bittersweet • 150-151
American burn-weed • 78, 133
American feverfew • 102
American germander • 223-224
American hazelnut • 7
American pasqueflower • 9, 293
American plantain • 255
American vetch • 194
American water-horehound • 216
Amethyst aster • 120
Ammophila breviligulata • 261
Amorpha canescens • 178
Amphicarpaea bracteata • 41, 179
ANACARDIACEAE • 38-41
Anaphalis margaritacea • 104
Andropogon
   gerardii • 258
   scoparius • 269
Anemone • 19
   canadensis • 291
   cylindrica • 292, 294
   patens • 293
   riparia • 294
   virginiana • 292, 294
Angelica
   archangelica • 42
   atropurpurea • 42
Annual bur-sage • 62
Annual false foxglove • 323
Annual fleabane • 76, 78

Antennaria • 104, 344
   howellii • 63
   neglecta • 63
   parlinii • 64
   plantaginifolia • 63-64
Anthemis cotula • 94, 131
Anticlea elegans • 232
APIACEAE • 42-51
APOCYNACEAE • 52-53, 55
Apocynum
   androsaemifolium • 52
   cannabinum • 53
   sibiricum • 53
Aquilegia • 295-297
   canadensis • 295
Arabis lyrata • 140
Arctic brome • 275
Arctium minus • 131
Arethusa bulbosa • 246
Argentina anserina • 304-305
Arkansas rose • 7
Arnoglossum
   atriplicifolium • 65
   plantagineum • 65-66
   reniforme • 65
Aromatic aster • 90, 126
Arrow-leaved tear-thumb • 285
Arrow-leaved violet • 342
Artemisia
   campestris • 67
   caudata • 67
   frigida • 68
   ludoviciana • 68
   serrata • 68
ASCLEPIADACEAE • 54-59
Asclepias
   amplexicaulis • 54
   incarnata • 55
   purpurascens • 56-57
   sullivantii • 54, 56
   syriaca • 57, 59
   tuberosa • 58-59
   verticillata • 59
   viridiflora • 14, 59
Asian bittersweet • 150
Asparagus • 233
Asparagus officinalis • 233
Aster • 6-7, 15
   azureus • 127
   borealis • 67
   ericoides • 120
   firmus • 121
   laevis • 122
   lanceolatus • 123

   lateriflorus • 124
   linariifolius • 90
   lucidulus • 121
   novae-angliae • 125
   oblongifolius • 126
   oolentangiensis • 127
   paniculatus • 123
   pilosus • 128
   puniceus • 135
   sericeus • 129
   simplex • 123
   umbellatus • 133
ASTERACEAE • 60-135, 344
Astragalus • 14, 195, 297
Aureolaria
   grandiflora • 323
   pedicularia • 323
Avens • 302, 305, 315
Awl aster • 128
Azure aster • 127
Azure bluets • 319
Bald spike-rush • 7, 170
Balsam groundsel • 100
Balsam ragwort • 100
BALSAMINACEAE • 136
Baptisia
   alba • 180
   bracteata • 181
   leucantha • 180
   leucophaea • 181
Barbarea vulgaris • 141
Basal-leaved rosinweed • 112
Bastard evening-primrose • 242
Bastard-toadflax • 320
Beach pea • 186
Beaked willow • 7
BEAN FAMILY • 42-51, 183, 280
Beard-tongue • 23, 327-329
Bee balm • 7, 218-219
Beggar's-lice • 139
Beggar-ticks • 7, 69
BELLFLOWER FAMILY • 145
Bent grass • 273
Berteroa incana • 142
Bicknell's oval sedge • 7, 160
Bicknell's rock-rose • 152-153
Bicknell's sedge • 160
Bicknell's crane's-bill • 208
Bidens cernuus • 69
Biennial bee-blossom • 241
Biennial gaura • 241
Big bluestem • 3, 5-7, 9, 11, 258, 269-270
Bird's-foot deer-vetch • 196

Bird's-foot trefoil • 196
Bird's-foot violet • 9, 340-341
Bitter milkwort • 279
Bittersweet • 150
BITTERSWEET FAMILY • 150-151
Bittersweet nightshade • 334
Black-eyed Susan • 6-7, 105-108
Black locust • 7, 11
Black medick • 196
Black oak • 6-7
Black raspberry • 41, 316
Black-seeded plantain • 255
Blackberry • 7, 316
Bladder campion • 147
Blood milkwort • 280
Blue cardinal-flower • 234
Blue false indigo • 23
Blue flag • 23, 212
Blue-fruited dogwood • 158
Blue grama • 260
Blue-jacket • 155
Blue-joint grass • 6-7, 261
Blue marsh violet • 343
Blue-sailors • 132
Blue skullcap • 222
Blue vervain • 337-338
Bluebell bellflower • 145
Bluets • 319
Blunt-leaf bedstraw • 318
Blunt-leaf milkweed • 54
Boneset • 79, 92
BORAGE FAMILY • 137-139
BORAGINACEAE • 137-39
Bottle gentian • 6, 10, 202-203
Bottlebrush grass • 273
Bouncing-Bet • 149
Bouteloua
   curtipendula • 259-260
   gracilis • 260
   hirsuta • 260
Box elder • 41
Bracken fern • 200
Brassica • 141
BRASSICACEAE • 140-142
Breadroot scurf-pea • 192
Brickellia eupatorioides • 92
Bridal-wreath • 314
Bristly aster • 135
Bristly buttercup • 299
Bristly crowfoot • 299
Brittle prickly-pear • 143
Broad-leaved cat-tail • 335
Broad-leaved plantain • 256
Broad-leaved purple coneflower • 75
Broad-leaved woolly sedge • 7, 161
Brome grass • 275
Bromus
   ciliatus • 275
   inermis • 275
   kalmii • 275

BROOMRAPE FAMILY • 323, 326
Broom sedge • 7
Brown fox sedge • 169
Brown-eyed Susan • 106-108
Buckthorn • 7
BUCKTHORN FAMILY • 300
Bulblet water-hemlock • 43
Bull thistle • 70, 133
Bur oak • 7, 9, 14
Burdock • 131
Butter-and-eggs • 23, 331
BUTTERCUP FAMILY • 291-299
Buttercup • 14, 291, 295, 299, 306-308
Butterfly-weed • 24, 58-59
Buxbaum's sedge • 7, 25, 161
Cacalia
   atriplicifolia • 65
   tuberosa • 66
CACTACEAE • 143
CACTUS FAMILY • 143
CAESALPINIACEAE • 144
Calamagrostis
   canadensis • 261
   stricta • 261
Calamovilfa longifolia • 261
Calico aster • 123-124
Calopogon
   oklahomensis • 246
   tuberosus • 246
Caltha palustris • 299
Calystegia sepium • 156
Camassia scilloides • 228
Campanula
   aparinoides • 145
   rotundifolia • 145
CAMPANULACEAE • 145
Canada anemone • 291
Canada bluegrass • 257, 274
Canada lettuce • 93
Canada lily • 230
Canada thistle • 70, 132
Canada wild-rye • 6-7, 264
Canadian columbine • 295
Canadian germander • 224
Canadian goldenrod • 113-115
Canadian horseweed • 72
Canadian tick-trefoil • 9, 184
Candle anemone • 292, 294
Caraway • 44
Cardinal-flower • 234-236
Carduus acanthoides • 71, 132
Carex • 18, 159
   annectens • 169, 172
   bicknellii • 160
   blanda • 172, 311
   brevior • 160, 163, 172
   buxbaumii • 161
   communis • 164
   conoidea • 168, 172
   foenea • 165
   haydenii • 162, 166

   interior • 173
   lacustris • 161
   meadii • 168
   molesta • 160, 163
   normalis • 163, 167
   pellita • 161
   pensylvanica • 164
   siccata • 165
   stipata • 169
   stricta • 161, 166
   tenera • 167
   tetanica • 168
   umbellata • 164
   vulpinoidea • 169
Carolina horse-nettle • 333
Carolina larkspur • 297
Carolina puccoon • 138
Carolina rose • 7, 312
Carrion-flower • 233
CARROT FAMILY • 42-51
Carum carvi • 44
CARYOPHYLLACEAE • 146-149, 276
CASHEW FAMILY • 38-41
Cassia fasciculata • 144
Castilleja
   coccinea • 324
   sessiliflora • 14, 324
Cat's-foot • 63, 104
CAT-BRIER FAMILY • 15, 233
Catnip • 225
Ceanothus
   americanus • 300
   herbaceus • 300
CELASTRACEAE • 150-151
Celastrus
   orbiculata • 150
   scandens • 151
Centaurea
   biebersteinii • 132
   maculosa • 132
   stoebe • 71, 132
Cerastium fontanum • 149
Chamaecrista fasciculata • 144
Cheeses • 239
Chelone glabra • 325
Chicory • 91, 132
Chinese lantern-plant • 332
Chrysanthemum leucanthemum • 94
Cichorium intybus • 132
Cicuta
   bulbifera • 43
   maculata • 43
Cinnamon fern • 199
Cinnamon willow-herb • 245
Cinquefoil • 304-305, 307
Cirsium
   altissimum • 70
   arvense • 132
   discolor • 70
   muticum • 71
   vulgare • 70, 133

CISTACEAE • 152-154
Clammy ground-cherry • 332
Clasping dogbane • 53
Clasping milkweed • 54
Cleland's evening-primrose • 243
Clematis
   occidentalis • 296
   virginiana • 41, 293, 296
Cliff cudweed • 104
Climbing bittersweet • 151
Climbing nightshade • 334
Climbing poison-ivy • 40
Climbing prairie rose • 313
Clover • 182, 190
Coeloglossum viride • 249
Colonial oak sedge • 164
Columbine • 24, 295, 297
Comandra umbellata • 320
COMMELINACEAE • 155
Common agrimony • 315
Common asparagus • 233
Common blackberry • 316
Common blue violet • 9, 343
Common blue-eyed-grass • 213
Common boneset • 79
Common cat-tail • 335
Common cinquefoil • 308
Common daisy • 94
Common evening-primrose • 7, 242-243
Common flat-topped goldenrod • 80
Common fox sedge • 169
Common foxglove • 327
Common goldenrod • 113
Common gold-star • 229
Common great angelica • 42
Common horsetail • 174
Common ironweed • 130
Common lake sedge • 161
Common mallow • 239
Common milkweed • 6-7, 56-57, 59
Common oak sedge • 164
Common plantain • 255-256
Common ragweed • 7, 62, 210
Common reed • 261
Common rock-rose • 152-153
Common scouring rush • 175
Common sneezeweed • 82
Common spiderwort • 9, 155
Common St. John's-wort • 210
Common star-grass • 229
Common stiff sedge • 7, 168
Common tussock sedge • 161-162, 166
Common vetch • 195
Common water-hemlock • 43
Common water-horehound • 216
Common wood sedge • 172
Common wood violet • 341, 343
Common yarrow • 60
Common yellow oxalis • 253
Compass-plant • 6-7, 10, 110, 112

Coneflower • 9, 11, 74, 82, 106
Conioselinum chinense • 43
Conium maculatum • 43
CONVOLVULACEAE • 156
Convolvulus
   arvensis • 156
   sepium • 156
Conyza canadensis • 72
Cord grass • 271
Coreopsis • 110
   lanceolata • 73
   palmata • 73
CORNACEAE • 157-158, 347
Cornus
   amomum • 157-158
   foemina • 157
   racemosa • 157
   stolonifera • 157-158
Coronilla varia • 196
Cotton-weed • 36
Cow vetch • 195
Cowslip • 299
Crane's-bill • 208
Cream gentian • 202-203
Cream wild indigo • 6-7, 180-181
Creeping thistle • 132
Creeping vervain • 338
Creeping-Charlie • 225
Crepis • 78
Crown-vetch • 196
CRUCIFERAE • 140-142
Culver's-physic • 330
Culver's-root • 330
Cup-plant • 109, 111
Curly dock • 287
Cylindrical blazing-star • 96
CYPERACEAE • 159-173
Cyperus strigosus • 173
Cypress spurge • 177
Cypripedium
   calceolus • 248
   candidum • 247-248
   parviflorum • 247-248
Daisy • 94, 106, 210
Daisy fleabane • 7, 76, 78
Dalea
   candida • 182
   purpurea • 183
Dame's rocket • 142
Dandelion • 135
Daucus carota • 44
Daylily • 230
Deadly nightshade • 334
Death camas • 228, 232
Delphinium • 24
Delphinium carolinianum • 297
DENNSTAEDTIACEAE • 200
Dense gay-feather • 99
Deptford pink • 146
Desmodium • 188
   canadense • 184
   illinoense • 185
Devil's-darning-needle • 296

Devil's plague • 44
Dianthus armeria • 146
Dichanthelium
   leibergii • 262
   oligosanthes • 263
   ovale • 263
   praecocius • 263
Digitalis purpurea • 327
Dillenius's oxalis • 253
Divaricate woodland sunflower • 85
Dodecatheon meadia • 288
Doellingeria umbellata • 133
DOGBANE FAMILY • 15, 52-53, 55
Dog-fennel • 94, 131
DOGWOOD FAMILY • 157-158
Door-yard violet • 343
Dotted horsemint • 219
Dotted monarda • 219
Dotted smartweed • 284
Downy gentian • 203-204
Downy paintbrush • 324
Downy painted-cup • 14, 324
Downy phlox • 276-277
Dragon's-mouth • 246
Dropseed • 5-6, 272
Dry-spiked sedge • 165
Dudley's rush • 214-215
Dune grass • 261
Dwarf fleabane • 76, 78
Dyer's-weed goldenrod • 116
Early goldenrod • 115
Early oak sedge • 7, 164
Eastern camas • 228
Eastern daisy fleabane • 76
Eastern feverfew • 102
Eastern parthenium • 102
Eastern prairie fringed orchid • 250
Eastern prickly-pear • 143
Eastern purple coneflower • 75
Eastern red-cedar • 7
Eastern shooting-star • 6-7, 288
Eastern willow-herb • 245
Eastern woodland sedge • 172
Echinacea • 82, 106
   angustifolia • 75
   pallida • 74
   purpurea • 75
Edible valerian • 336
Elderberry • 7
Eleocharis
   compressa • 170
   elliptica • 170
   erythropoda • 170
Elliptic spike-rush • 170
Elm-leaved goldenrod • 134
Elymus
   canadensis • 264
   hystrix • 273
   virginicus • 264
Elytrigia repens • 273
English plantain • 256
Epilobium coloratum • 245

EQUISETACEAE • 174-175
Equisetum
    arvense • 174
    hyemale • 175
    laevigatum • 175
    pratense • 174
    X*ferrissii* • 175
Eragrostis spectabilis • 274
Erechtites hieracifolius • 78, 133
Erigeron • 122
    annuus • 76, 78
    canadensis • 72
    pulchellus • 77
    strigosus • 76, 78
Eryngium yuccifolium • 45, 66
Eupatorium
    altissimum • 92
    maculatum • 81
    perfoliatum • 79
    rugosum • 61
Euphorbia • 23
    corollata • 176
    cyparissias • 177
    esula • 177
    waldsteinii • 177
    X*pseudovirgata* • 177
EUPHORBIACEAE • 176-177
Euthamia
    graminifolia • 80, 118
    gymnospermoides • 80
    tenuifolia • 80
Eutrochium
    maculatum • 81
    purpureum • 81
EVENING-PRIMROSE FAMILY • 241-245
FABACEAE • 178-197
False boneset • 92
False-dandelion • 91
False foxglove • 323, 327
False nut sedge • 173
False Solomon's-seal • 233
False spikenard • 233
False sunflower • 88
False water-pepper • 284
False-toadflax • 320
Fen-betony • 326
Fern • 14, 17, 19, 198-201
Ferriss' horsetail • 175
Fescue sedge • 160, 163, 172
Few-flowered panic grass • 7
Few-headed blazing-star • 96
Few-leaved sunflower • 86-87
Field bindweed • 156
Field hawkweed • 133
Field horsetail • 7, 174
Field milkwort • 280
Field mint • 217
Field oval sedge • 7, 160, 163
Field pepper-weed • 142
Field pussy-toes • 7, 63

Field sage-wort • 7, 67
Field sorrel • 286
Field thistle • 70, 132
Field wormwood • 67
Fieldcress • 142
FIGWORT FAMILY • 15, 323-331
Figwort giant hyssop • 225
Finger tickseed • 73
Fireweed • 133
Fistulous goat's-beard • 135
Flannel plant • 331
Flat-stemmed spike-rush • 7, 170
Flat-top aster • 133
Flax-leaved aster • 90
Fleabane • 72, 76, 78, 122
Florida snake-cotton • 36
Flowering spurge • 6-7, 9, 53, 176
Forest pea • 187
Fowl manna grass • 7
Fox sedge • 169, 172
Fragaria • 306
    chiloense • 301
    vesca • 301
    virginiana • 41
Fragrant cudweed • 104
Fringed brome • 275
Fringed gentian • 206
Fringed loosestrife • 290
Froelichia
    floridana • 36
    gracilis • 36
Frost aster • 6, 120, 124, 128
Frostweed • 152-153
Galium
    asprellum • 318
    boreale • 317
    labradoricum • 318
    obtusum • 318
    tinctorium • 318
    trifidum • 318
Garden asparagus • 233
Garden-heliotrope • 336
Garden lettuce • 93
Gaura
    biennis • 241-242
    coccinea • 241
GENTIAN FAMILY • 202-206
Gentiana
    alba • 202-203
    andrewsii • 203
    crinita • 206
    flavida • 202
    puberula • 204
    puberulenta • 203-204
    quinquefolia • 205
GENTIANACEAE • 202-206
Gentianella quinquefolia • 205
Gentianopsis
    crinita • 206
    procera • 206
GERANIACEAE • 208

Geranium
    bicknellii • 208
    maculatum • 208
    robertianum • 208
GERANIUM FAMILY • 208
Germander • 223-224
Geum • 305
    aleppicum • 302, 315
    canadense • 315
    laciniatum • 302
    triflorum • 293, 302
Giant goldenrod • 114
Giant hyssop • 225
Giant ragweed • 62, 131
Giant Solomon's-seal • 233
Giant St. John's-wort • 211
Giant sunflower • 83
Gill-over-the-ground • 225
Glade mallow • 238
Glaucous campion • 147
Glaucous white-lettuce • 103
Glechoma hederacea • 225
Globular coneflower • 105
Gloriosa daisy • 106
Gnaphalium obtusifolium • 104
Goat's-beard • 135
Goat's-rue • 193
Goblet aster • 124
Golden alexanders • 50-51
Golden cassia • 144
Golden ragwort • 100-101
Goldenrod • 6, 9, 14, 22, 80, 90, 113-120, 122, 134
GRAMINEAE • 257-275
Grass • 2-3, 5-7, 9-11, 14, 16-18, 25, 257-275, 345
GRASS FAMILY • 159, 257-275
Grass pink • 24, 246
Grass-leaved goldenrod • 7, 80, 118
Gray dogwood • 7, 157
Gray goldenrod • 116
Great blue lobelia • 234-235
Great bulrush • 173
Great Indian-plantain • 65
Great Plains fringed gentian • 206
Great Plains lady's-tresses • 251
Great ragweed • 131
Great St. John's-wort • 211
Greater fringed gentian • 206
Greater sand goat's-beard • 135
Greater straw sedge • 163, 167
Greek-valerian • 278
Ground-ivy • 225
Groundsel • 100-101
Habenaria
    flava • 249
    leucophaea • 250
Hackelia virginiana • 139
Hair grass • 273
Hairy aster • 128
Hairy grama grass • 260

Hairy hawkweed • 89
Hairy puccoon • 137-138
Hairy sunflower • 84-85
Hairy vetch • 11, 195
Hairy white vervain • 337
Hairy wood violet • 343
Halberd-leaved tear-thumb • 285
Handsome-Harry • 240
Hardhack • 314
Harebell • 145
Hawk's-beard • 78
Hay sedge • 165
Hayden's sedge • 162
HAY-SCENTED FERN FAMILY • 200
Heal-all • 220
Heart-leaved aster • 127
Heart-leaved golden alexanders • 50-51
Heart-leaved groundsel • 101
Heart's-ease • 283, 287
Heath aster • 7, 120, 124, 128
Hedge bindweed • 156
Hedge-nettle • 223-224
Hedyotis caerulea • 319
Helenium autumnale • 82
Helianthemum • 154
    bicknellii • 152
    canadense • 153
Helianthus • 82, 88, 106, 110, 345
    decapetalus • 85
    divaricatus • 85
    giganteus • 83
    grosseserratus • 83
    hirsutus • 84
    occidentalis • 86
    pauciflorus • 86-87
    strumosus • 85
    tuberosus • 85, 87
    ×*laetiflorus* • 87
Heliopsis • 110
    helianthoides • 88
Hemerocallis fulva • 230
Hemlock-parsley • 43
Hemp dogbane • 53
Herb-Robert • 208
Hesperis matronalis • 142
Hesperostipa spartea • 265
Heuchera richardsonii • 321
Hieracium
    aurantiacum • 89
    caespitosum • 89, 133
    longipilum • 89
Hill's oak • 6
Hillside sedge • 165
Hoary false madwort • 142
Hoary frostweed • 152
Hoary puccoon • 137-138
Hoary vervain • 337-338
Hoary-alyssum • 142
Hog-peanut • 41, 179

Hogweed • 72
Honeysuckle • 7
Horse-cane • 131
Horsemint • 219
HORSETAIL FAMILY • 174-175
Horseweed • 72, 78
Houstonia
    caerulea • 319
    longifolia • 319
Howell's pussy-toes • 63
Hybrid cat-tail • 335
Hybrid prairie sunflower • 87
Hybrid sumac • 39
HYPERICACEAE • 209-211
Hypericum
    kalmianum • 209
    perforatum • 210
    pyramidatum • 211
Hypoxis hirsuta • 229
Hystrix patula • 273
Illinois tick-trefoil • 184-185
Impatiens
    biflora • 136
    capensis • 136
    pallida • 136
Indian breadroot • 192
Indian grass • 3, 5, 7, 9, 11, 25, 269-270
Indian hemp • 53
Indian paintbrush • 10, 324
Indian-tobacco • 235
Indigo • 6, 9, 180-181
Indigofera tinctoria • 180
Inland New Jersey tea • 300
Inland rush • 214
Inland sedge • 173
Inland star sedge • 173
Innocence • 319
Interrupted fern • 199
Intermediate pinweed • 153-154
Ionactis linariifolia • 90
IRIDACEAE • 212-213
IRIS FAMILY • 212-213
Iris
    versicolor • 212
    virginica • 212
Ironweed • 130
Jack pine • 6
Jacob's-ladder • 278
Japanese hedge-parsley • 44
Japanese spiraea • 314
Jerusalem-artichoke • 85, 87
Jewelweed • 24, 136
JUNCACEAE • 159, 214-215
Juncus • 159, 214-215
Juncus
    dudleyi • 214-215
    interior • 214
    tenuis • 214-215
June grass • 6-7, 266
Kalm's brome • 275

Kalm's St. John's-wort • 209
Kentucky bluegrass • 7, 257, 274
King-devil • 89, 133
Klamath-weed • 210
Koeleria
    macrantha • 266
    pyramidata • 266
Krigia
    biflora • 91
    virginica • 91
Kuhnia eupatorioides • 92
Labrador marsh bedstraw • 318
Lactuca
    biennis • 93
    canadensis • 93
    sativa • 93
    scariola • 134
    serriola • 93, 134
LAMIACEAE • 216-225
Lance-leaf tickseed • 73
Lance-leaved ground-cherry • 333
Lance-leaved loosestrife • 289
Large cotton-weed • 36
Large-flowered beard-tongue • 329
Large-flowered yellow false foxglove • 323
Large-leaved wild indigo • 180
Large-pod pinweed • 154
Large yellow lady's-slipper • 247-248
Late goldenrod • 7, 113-115
Lathyrus • 179
    maritimus • 186
    odoratus • 186
    palustris • 186
    venosus • 187
Lead-plant • 6-7, 178, 182
Leafy spurge • 7, 23, 177
Lechea intermedia • 153-154
Leiberg's panic grass • 6, 262
Leonard's skullcap • 222
Leonurus cardiaca • 225
Lepidium
    campestre • 142, 213
    densiflorum • 142
Lespedeza capitata • 188
Leucanthemum vulgare • 94
Liatris
    aspera • 95, 97
    cylindracea • 96
    ligulistylis • 97
    pycnostachya • 98-99
    spicata • 98-99
LILIACEAE • 226-233
Lilium
    canadense • 230
    michiganense • 230-231
    philadelphicum • 231
    superbum • 230
LILY FAMILY • 19, 155, 226-233
Linaria vulgaris • 331

Lion's-foot • 103
Lion's-tail • 225
Lithospermum
    canescens • 37
    caroliniense • 138
Little bluestem • 5-7, 9, 11, 258, 269
Little-leaved mint • 217
Lobed blue violet • 341
Lobelia
    cardinalis • 236
    inflata • 235
    silphilitica • 234
    spicata • 235
LOBELIA FAMILY • 234-236
LOBELIACEAE • 234-236
Locoweed • 297
Locust-weed • 144
Long-awned bracted sedge • 7
Long-beard hawkweed • 89
Long-bracted green orchid • 249
Long-bracted spiderwort • 155
Long-bracted wild indigo • 181
Long-branch frostweed • 153
Long-haired hawkweed • 89
Long-headed anemone • 292
Long-headed coneflower • 105
Long-leaved bluets • 319
Long-scaled tussock sedge • 162, 166
Long-spike evening-primrose • 243
Long-stalked panic grass • 7
LOOSESTRIFE FAMILY • 237
Lotus corniculatus • 196
Louisiana sage-wort • 68
Lousewort • 326
Lupinus perennis • 189
Luzula • 214
Lychnis alba • 147
Lycopus
    americanus • 216
    uniflorus • 216
    virginicus • 216
Lyrate rock-cress • 140
Lysimachia
    ciliata • 290
    lanceolata • 289
    quadriflora • 289
    terrestris • 289
    thyrsiflora • 289
LYTHRACEAE • 237
Lythrum
    alatum • 237
    salicaria • 237
MADDER FAMILY • 317-319
Maianthemum racemosum • 233
MALLOW FAMILY • 238-239
Malva neglecta • 239
MALVACEAE • 238-239
Marguerite • 94
Marsh bellflower • 145
Marsh blazing-star • 99
Marsh fern • 201

Marsh gay-feather • 98-99
Marsh-marigold • 299
Marsh pea • 186
Marsh-pepper knotweed • 284
Marsh skullcap • 222
Marsh straw sedge • 167
Marsh vetchling • 186
Matricaria discoidea • 94
Mayweed • 131
Mead's sedge • 7, 168
Meadow anemone • 291
MEADOW-BEAUTY FAMILY • 240
Meadow garlic • 226
Meadow hawkweed • 133
Meadow horsetail • 174
Meadow-parsnip • 51
Meadow sundrops • 244
Meadow willow • 7
Medicago lupulina • 196
MELASTOMATACEAE • 240
Melilotus
    alba • 190
    officinalis • 190
Mentha
    arvensis • 216-217
    spicata • 217
    X*gracilis* • 217
Michigan lily • 23, 230-231
Midwestern white-lettuce • 103
Milfoil • 60
Milk-vetch • 195
MILKWEED FAMILY • 15, 54-59
MILKWORT FAMILY • 279-282
Milky wild indigo • 180
Mimulus ringens • 331
MINT FAMILY • 68, 216-225
Missouri goldenrod • 113, 115
Monarda
    fistulosa • 218
    punctata • 219
Monkey-flower • 331
MORNING-GLORY FAMILY • 156
Morning-glory • 156
Motherwort • 225
Mountain mint • 6-7, 221
Mountain wood-sorrel • 252
Mouse-ear chickweed • 149
Mullein • 210, 331
Multiflora rose • 311
MUSTARD FAMILY • 140-142
Mustard • 141
Naked-stemmed sunflower • 86
Napaea dioica • 238
Narrow-leaf purple coneflower • 75
Narrow-leaved cat-tail • 335
Narrow-leaved hedge-nettle • 223
Narrow-leaved loosestrife • 289
Narrow-leaved oval sedge • 167
Narrow-leaved plantain • 256
Narrow-leaved vervain • 338
Needle grass • 6-7, 9, 25, 265
Nepeta cataria • 225
Nettle-leaved vervain • 339

New England aster • 6-7, 120, 125
New Jersey tea • 7, 300
NIGHTSHADE FAMILY • 332-334
Nightshade • 334
Nodding beggar-ticks • 69
Nodding bur-marigold • 69
Nodding lady's-tresses • 251
Nodding wild onion • 227
Northern bedstraw • 317
Northern blue flag • 212
Northern bog aster • 123
Northern bog violet • 343
Northern bugleweed • 216
Northern meadow groundsel • 100
Northern Plains blazing-star • 97
Northern ragwort • 100-101
Northern slender lady's-tresses • 251
Northern small yellow lady's-slipper • 248
Northern swamp dogwood • 157
Northern sweet grass • 7
Northern three-lobed bedstraw • 318
Northern water-horehound • 216
Norwegian cinquefoil • 306
Nut sedge • 7
Oblong sunflower • 84
Oenothera • 345
    biennis • 242
    clelandii • 243
    parviflora • 242
    perennis • 244
    pilosella • 244
    rhombipetala • 243
Ohio goldenrod • 7, 117-118
Oklahoma grass pink • 246
Old-field balsam • 104
Old-field cinquefoil • 308
Old-field cudweed • 104
Old-field five-fingers • 308
Old-field goldenrod • 7, 116
Old-man's-whiskers • 302
ONAGRACEAE • 241-245, 345
Onoclea sensibilis • 198
Ontario aster • 124
Open-field sedge • 172
Opuntia
    compressa • 143
    fragilis • 143
    humifusa • 143
    rafinesquei • 143
Orange-cup lily • 231
Orange dwarf-dandelion • 91
Orange hawkweed • 89
Orange jewelweed • 136
Orange milkweed • 58
Orange touch-me-not • 136
ORCHID FAMILY • 246-251
ORCHIDACEAE • 246-251
Oriental bittersweet • 9, 150-151
OROBANCHACEAE • 323-324, 326

Osmunda
   cinnamomea • 199
   claytoniana • 199
   regalis • 199
OXALIDACEAE • 252-253
Oxalis
   dillenii • 253
   montana • 252
   stricta • 253
   violacea • 252
Ox-eye • 88, 110
Ox-eye daisy • 94, 210
Oxypolis rigidior • 46
Packera
   aurea • 101
   paupercula • 100
   plattensis • 101
   pseudaurea • 101
Pale beard-tongue • 328
Pale green orchid • 249
Pale Indian-plantain • 9, 11, 65
Pale-leaved woodland sunflower • 85
Pale purple coneflower • 74-75
Pale-spike lobelia • 235
Pale sunflower • 85
Pale touch-me-not • 136
Panic grass • 6, 262-263, 353, 355-357
Panicled aster • 121, 123-124
Panicled dogwood • 157
Panicum • 25, 257
   leibergii • 262
   oligosanthes • 263
   praecocius • 263
   virgatum • 267
Parasol aster • 133
Parlin's pussy-toes • 64
PARSLEY FAMILY • 42-51
Parthenium integrifolium • 102
Parthenocissus • 41
Partridge pea • 144, 288
Pasqueflower • 6-7, 9, 293, 302
Pastinaca sativa • 47, 51
Pasture rose • 310, 312
Pasture thistle • 70
Path rush • 214
PEA FAMILY • 23, 144, 178-197, 279
Pearly everlasting • 104
Pedicularis • 323
   canadensis • 326
   lanceolata • 326
   palustris • 326
Pediomelum esculentum • 192
Pennsylvania buttercup • 299
Pennsylvania sedge • 7, 164
Penstemon
   digitalis • 327-329
   gracilis • 328
   grandiflorus • 329
   pallidus • 328

Pentaphylloides
   floribunda • 303
   fruticosa • 303
Pepper-grass • 7
Persicaria
   amphibia • 283
   arifolia • 285
   hydropiper • 284
   hydropiperoides • 284
   maculosa • 287
   punctata • 284
   sagittata • 285
Pest sedge • 163
Petalostemon
   candidum • 182
   purpureum • 183
Phalaris arundinacea • 268
Phleum pratense • 274
PHLOX FAMILY • 276-278
Phlox • 10
   glaberrima • 276
   pilosa • 277
Phragmites communis • 261
Physalis
   alkekengi • 332
   heterophylla • 332
   longifolia • 333
   virginiana • 333
Pineapple-weed • 94
PINK FAMILY • 146-149, 276
Plains oval sedge • 7, 172
Plains puccoon • 138
Plains snake-cotton • 36
Plains wild indigo • 181
PLANTAGINACEAE • 254-256, 325, 327
Plantago • 66
   lanceolata • 256
   major • 255-256
   patagonica • 254
   purshii • 254
   rugelii • 255-256
PLANTAIN FAMILY • 254-256, 325, 327, 330
Plantain-leaved pussy-toes • 64
Plantain pussy-toes • 63-64
Platanthera
   flava • 249
   huronensis • 249
   leucophaea • 250
   praeclara • 250
Plumeless thistle • 71, 132
Poa
   compressa • 274
   pratensis • 274
POACEAE • 159, 257-275
Pogonia ophioglossoides • 246
Poison-hemlock • 43
Poison-ivy • 40-41, 125
POLEMONIACEAE • 276-278
Polemonium reptans • 278

Polygala
   ambigua • 282
   polygama • 279
   sanguinea • 280
   senega • 281
   verticillata • 282
   vulgaris • 280
POLYGALACEAE • 279-282
POLYGONACEAE • 283-287
Polygonatum
   biflorum • 233
Polygonum
   amphibium • 283
   coccineum • 283
   hydropiper • 284
   natans • 283
   persicaria • 287
   sagittatum • 285
Polytaenia nuttallii • 48
Pomme-de-prairie • 192
Porcupine grass • 265
Potentilla
   anserina • 304-305
   argentea • 304
   arguta • 305
   fruticosa • 303
   norvegica • 306-307
   recta • 306-307
   simplex • 308
Poverty rush • 214
Prairie alumroot • 321
Prairie blazing-star • 98
Prairie blue-eyed-grass • 6, 213
Prairie brome • 275
Prairie cinquefoil • 305
Prairie-clover • 7, 10, 182-3
Prairie coneflower • 74
Prairie cord grass • 7, 271
Prairie coreopsis • 73
Prairie dock • 9, 110, 112
Prairie dropseed • 5-7, 272
Prairie fleabane • 78
Prairie gentian • 203-204
Prairie gray sedge • 168, 172
Prairie hawkweed • 89
Prairie heart-leaved aster • 127
Prairie Indian-plantain • 65-66
Prairie June grass • 266
Prairie larkspur • 9, 297
Prairie lily • 231
Prairie milkweed • 56
Prairie onion • 227
Prairie panic grass • 7
Prairie-parsley • 48
Prairie pepper-weed • 142
Prairie phlox • 277
Prairie ragwort • 101
Prairie red-root • 300
Prairie rose • 310, 312-313
Prairie rosinweed • 109, 112
Prairie sagewort • 68
Prairie-smoke • 6, 293, 302

Prairie thistle • 6, 70
Prairie tickseed • 6-7, 73
Prairie tick-trefoil • 185
Prairie-turnip • 192
Prairie violet • 6, 341
Prairie white fringed orchid • 250
Prairie willow • 7
Prenanthes
    alba • 103
    aspera • 103
    crepidinea • 103
    racemosa • 103
Prickly lettuce • 93, 134
Prickly sow-thistle • 134
PRIMROSE FAMILY • 288-290
Primula • 288
PRIMULACEAE • 242, 288-290
Prunella vulgaris • 220
Pseudognaphalium
    obtusifolium • 104
    saxicola • 104
Psoralea esculenta • 192
Pteridium aquilinum • 200
Pulsatilla patens • 293
Purple clematis • 296
Purple coneflower • 74-75, 106
Purple giant hyssop • 225
Purple Joe-Pye-weed • 81
Purple loosestrife • 237
Purple love grass • 274
Purple meadow-rue • 298
Purple milkweed • 56-57
Purple milkwort • 279
Purple prairie-clover • 6, 183
Purple rattlesnake-root • 103
Purple-stem aster • 135
Pussy-toes • 7, 9, 63-64, 104
Pussy willow • 7
Pycnanthemum virginianum • 221
Quackgrass • 7, 257, 273
Quaker-ladies • 319
Quaking aspen • 7
Queen Anne's-lace • 7, 44, 210
Rabbit-pea • 193
Rabbit-tobacco • 104
Racemed milkwort • 279
Ragwort • 100-101
RANUNCULACEAE • 291-299
Ranunculus • 14, 299, 306, 308
Ranunculus pensylvanicus • 299
Ratibida • 82, 106
    columnifera • 105
    pinnata • 105
Rattlesnake-master • 6-7, 9-10, 45, 66
Red-cedar • 6, 11
Red clover • 197
Red columbine • 295
Red osier dogwood • 157-158
Red-root • 300
Red sorrel • 286

Red-stalked plantain • 255
Redtop • 7, 257, 273
Reed canary grass • 7, 10, 257, 268
Reed grass • 261
RHAMNACEAE • 300
Rhexia virginica • 240
Rhus
    glabra • 38
    hirta • 39
    radicans • 40
    typhina • 39
    ✕*borealis* • 39
    ✕*pulvinata* • 39
Ribgrass • 256
Richardson's sedge • 7
Riddell's goldenrod • 117-118
Rigid goldenrod • 118
Rigid sedge • 168
Roadside rush • 214
Robin's-plantain • 77
Rock-rose • 152-154
ROCK-ROSE FAMILY • 152-154
Rosa
    arkansana • 310-312
    blanda • 311
    carolina • 310, 312
    multiflora • 311
    palustris • 311
    setigera • 313
    woodsii • 311
    ✕*rudiuscula* • 312
ROSACEAE • 301-316
ROSE FAMILY • 301-316
Rose pogonia • 246
Rosinweed • 9, 88, 109, 112
Rough avens • 302
Rough barnyard grass • 7
Rough bedstraw • 318
Rough blazing-star • 6-7, 95, 97
Rough cinquefoil • 306-307
Rough fleabane • 78
Rough-fruited cinquefoil • 307
Rough-leaved sunflower • 84-85
Rough rose • 312
Rough sunflower • 84
Rough white-lettuce • 103
Round-headed bush-clover • 105, 188
Royal fern • 199
RUBIACEAE • 317-319
Rubus • 41
    allegheniensis • 316
    occidentalis • 316
Rudbeckia • 82
    hirta • 106, 108
    laciniata • 107
    subtomentosa • 107
    triloba • 107-108
Rugel's plantain • 255

Rumex
    acetosella • 286
    crispus • 287
Running savanna sedge • 7, 165
Rush • 25, 214-215
RUSH FAMILY • 159, 214-215
Rydberg's poison-ivy • 40
Sand cherry • 7
Sand coreopsis • 73
Sand cress • 140
Sand evening-primrose • 243
Sand milkweed • 54
Sand reed grass • 261
SANDALWOOD FAMILY • 320
SANTALACEAE • 320
Saponaria officinalis • 149
Saw-leaf mugwort • 68
Saw-tooth sunflower • 6-7, 83
Saxifraga pensylvanica • 322
SAXIFRAGACEAE • 321-322
SAXIFRAGE FAMILY • 321-322
Scarlet gaura • 241
Scarlet painted-cup • 324
Schizachyrium scoparium • 269
Schoenoplectus tabernaemontani • 173
Scirpus validus • 173
Scleria triglomerata • 171
Scribner's panic grass • 263
SCROPHULARIACEAE • 323-331
Scutellaria
    galericulata • 222
    lateriflora • 222
    leonardii • 222
    parvula • 222
Sedge • 6-7, 17-18, 25, 159-173
SEDGE FAMILY • 159-173, 257
Self-heal • 220
Seneca snakeroot • 281
Senecio
    pauperculus • 100
    pseudaureus • 101
SENNA FAMILY • 144
Sensitive fern • 198
Sessile blazing-star • 99
Sessile-headed blazing-star • 99
Shaggy prairie-turnip • 192
Shasta daisy • 94
Sheep sorrel • 286
Shining aster • 121
Shiny-leaved aster • 121
Shooting-star • 6, 288
Short green milkweed • 14, 59
Short ragweed • 62
Short's aster • 122
Showy blazing-star • 97
Showy goldenrod • 6, 119
Showy sunflower • 87
Showy tick-trefoil • 184
Shreve's iris • 212
Shrubby cinquefoil • 303
Shrubby five-fingers • 303

Side-flowering aster • 124
Side-oats grama • 6-7, 259-260, 269
Silene
    csereii • 147
    latifolia • 147-148
    noctiflora • 147
    pratensis • 147
    stellata • 49, 148
    vulgaris • 147
Silky aster • 6-7, 129
Silky dogwood • 157-158
Silphium • 88
    integrifolium • 109
    laciniatum • 110
    perfoliatum • 109, 111
    terebinthinaceum • 110, 112
Silver-weed • 304-305
Silvery cinquefoil • 304
Silvery five-fingers • 304
Simpler's-joy • 337
Sisymbrium
    altissimum • 142
Sisyrinchium
    albidum • 213
    campestre • 213
Sium suave • 43, 46
Sky-blue aster • 7, 122, 127
Sleeping-plant • 144
Slender beard-tongue • 328
Slender cotton-weed • 36
Slender goldentop • 80
Slender-stem pea-vine • 186
Slender wedge grass • 266
Slough grass • 6, 271
Small evening-primrose • 244
Small-flowered evening-primrose • 242
Small pepper-grass • 142
Small skullcap • 222
Small sundrops • 244
Small white lady's-slipper • 24, 247-248
Small yellow lady's-slipper • 247-248
SMARTWEED FAMILY • 283-287
SMILACACEAE • 233
Smilacina racemosa • 233
Smilax herbacea • 233
Smooth aster • 6-7, 122-123, 127
Smooth blue aster • 122
Smooth brome • 7, 275
Smooth goldenrod • 114
Smooth horsetail • 175
Smooth ironweed • 130
Smooth long-leaved ground-cherry • 333
Smooth loosestrife • 289
Smooth milkweed • 54, 56
Smooth phlox • 276-277
Smooth rose • 7, 311
Smooth scouring rush • 175
Smooth Solomon's-seal • 233
Smooth spiderwort • 155

Smooth sumac • 6-7, 38-39
Soapwort • 149
Soft-stem bulrush • 173
SOLANACEAE • 332-334
Solanum
    americanum • 334
    caroliniense • 333
    dulcamara • 334
    nigrum • 334
    ptychanthum • 334
Solidago • 122
    altissima • 113
    canadensis • 113-115
    gigantea • 113-115
    graminifolia • 80
    juncea • 115
    missouriensis • 113, 115
    nemoralis • 116
    ohioensis • 117-118
    riddellii • 117-118
    rigida • 118
    speciosa • 119
    ulmifolia • 134
Solomon's-plume • 233
Solomon's-seal • 233
Sonchus asper • 134
Sorghastrum nutans • 270
Sour dock • 287
Southern blue flag • 212
Southern three-lobed bedstraw • 318
Southern yellow wood-sorrel • 253
Spanish broom • 271
Spartina pectinata • 271
Spartium junceum • 271
Spearmint • 217
Sphenopholis • 266
SPIDERWORT FAMILY • 155
Spiked lobelia • 235
Spiny plumeless thistle • 132
Spiny sow-thistle • 134
Spiraea
    alba • 314
    japonica • 314
    tomentosa • 314
    X*vanhouttei* • 314
Spiranthes
    cernua • 251
    lacera • 251
    magnicamporum • 251
Sporobolus heterolepis • 272
Spotted bee balm • 219
Spotted geranium • 208
Spotted Joe-Pye-weed • 79, 81
Spotted knapweed • 71, 132
Spotted lady's-thumb • 287
Spotted touch-me-not • 136
Spotted water-hemlock • 43
Spreading dogbane • 52-53
Spreading Jacob's-ladder • 278
SPURGE FAMILY • 176-177
ST. JOHN'S-WORT FAMILY • 209-211

Stachys
    palustris • 223
    tenuifolia • 223
Staghorn sumac • 38-39
Starry campion • 49, 148
Stickseed • 139
Sticky cockle • 147
Stiff aster • 90
Stiff cowbane • 7, 46
Stiff gentian • 205
Stiff goldenrod • 6, 118
Stiff-leaved panic grass • 263
Stiff sunflower • 6-7, 87
Stiff tickseed • 73
Stinking chamomile • 94, 131
Stipa spartea • 265
Strawberry • 41, 301, 306
Strawberry-weed • 306
Straw-colored cyperus • 173
Sullivant's milkweed • 56
Sulphur cinquefoil • 306-307
Sulphur five-fingers • 307
Sumac • 6, 38-39
Sundial lupine • 189
Sunflower • 6, 17-18, 22, 61, 77, 83-88, 98, 106, 110
SUNFLOWER FAMILY • 17-18, 60-135
Sunflower-everlasting • 88
Sunshine rose • 310
Swamp aster • 121, 123, 135
Swamp-betony • 326
Swamp-candles • 289
Swamp loosestrife • 289
Swamp milkweed • 54-55
Swamp rose • 311
Swamp saxifrage • 322
Swamp thistle • 71
Sweet Betty • 149
Sweet black-eyed Susan • 107
Sweet-clover • 7, 190-1
Sweet coneflower • 107
Sweet everlasting • 104
Sweet pea • 186
Switch grass • 3, 5, 7, 267
Symphyotrichum • 76
    boreale • 123
    ericoides • 120, 124, 128
    firmum • 121
    laeve • 122-123, 127
    lanceolatum • 121, 123-124
    lateriflorum • 123-124
    novae-angliae • 120, 125
    oblongifolium • 90, 126
    ontarionis • 124
    oolentangiense • 122, 127
    pilosum • 120, 124, 128
    puniceum • 123, 135
    sericeum • 129
    shortii • 122
    X*amethystinum* • 120
Taenidia integerrima • 49
Tall anemone • 292, 294

Tall beard-tongue • 327-329
Tall blue lettuce • 93
Tall boneset • 92
Tall cinquefoil • 305
Tall flat-topped white aster • 133
Tall gay-feather • 95
Tall hairy agrimony • 315
Tall lettuce • 93
Tall meadow-rue • 7, 298
Tall northern bog orchid • 249
Tall nut-rush • 7, 171
Tall potentilla • 305
Tall thistle • 70
Tall tumble mustard • 142
Tall wild lettuce • 93
Tall wood-sorrel • 253
Tap-rooted valerian • 336
Taraxacum officinale • 135
Tephrosia virginiana • 193
Teucrium canadense • 223-224
Thalictrum
    dasycarpum • 298
    revolutum • 298
Thaspium trifoliatum • 51
Thelypteris palustris • 201
Thick-spike blazing-star • 98-99
Thick-spike gay-feather • 6-7, 98-99
Thimbleweed • 292, 294
Thoroughwort • 79
Three-lobed coneflower • 108
Tickle grass • 273
Tickseed • 73, 88, 110
Tick-trefoil • 7, 184, 188
Timothy • 274
Torilis japonica • 44
TOUCH-ME-NOT FAMILY • 136
Toxicodendron
    radicans • 40
    rydbergii • 40
Tradescantia
    bracteata • 155
    occidentalis • 155
    ohiensis • 155
Tragopogon dubius • 135
Trembling aspen • 10
Trifolium
    hybridum • 197
    pratense • 197
    repens • 197
Troublesome sedge • 163
Tubercled orchid • 249
Tuberous Indian-plantain • 65-66
Tumble grass • 274
Turk's-cap lily • 230
Turkey-foot • 258
Turtlehead • 325
Tussock sedge • 7, 161-162, 166

Twiggy spurge • 177
Two-flowered Cynthia • 91
Typha
    angustifolia • 335
    latifolia • 335
    X*glauca* • 335
TYPHACEAE • 335
Umbrella sedge • 173
Upland wild-timothy • 6-7
VALERIAN FAMILY • 336
Valeriana
    edulis • 336
    officinalis • 336
VALERIANACEAE • 336
Veiny pea • 187
Veiny pea-vine • 187
Verbascum thapsus • 331
Verbena
    bracteata • 338
    hastata • 337-338
    simplex • 338
    stricta • 337-338
    urticifolia • 337, 339
VERBENACEAE • 337-339
Vernonia fasciculata • 130
Veronicastrum virginicum • 330
VERVAIN FAMILY • 337-339
Vicia • 187
    americana • 194
    cracca • 195
    sativa • 195
    villosa • 195
Viola
    cucullata • 343
    nephrophylla • 343
    papilionacea • 343
    pedata • 340-341
    pedatifida • 341
    sagittata • 341-342
    sororia • 341, 343
    X*palmata* • 341
    X*subsinuata* • 341
VIOLACEAE • 340-343
VIOLET FAMILY • 340-343
Violet wood-sorrel • 252
Virgin's-bower • 41, 293, 296
Virginia creeper • 41
Virginia dwarf-dandelion • 91
Virginia ground-cherry • 333
Virginia meadow-beauty • 240, 288
Virginia water-horehound • 216
Virginia wild-rye • 264
Viscid grass-leaved goldenrod • 80
Waldstein's spurge • 177
Water drop-wort • 46
Water heart's-ease • 283
Water-hemlock • 43

Water-horehound • 216
Water-parsnip • 43, 46
Water-pepper • 7, 284
Water smartweed • 283
Wax-leaf meadow-rue • 298
Wedge grass • 266
Western golden ragwort • 100-101
Western heart-leaved groundsel • 101
Western mugwort • 68
Western poison-ivy • 40
Western prairie fringed orchid • 250
Western ragweed • 62
Western rose • 311
Western silvery aster • 129
Western spiderwort • 155
Western sunflower • 86
Whip-grass • 171
White avens • 315
White camas • 228, 232
White campion • 147-148
White clover • 197
White cockle • 147
White lady's-slipper • 247-248
White meadowsweet • 7, 314
White old-field aster • 128
White panicle aster • 123
White prairie aster • 120
White prairie-clover • 182-183
White sage • 68
White snakeroot • 61
White sweet-clover • 190
White turtlehead • 325
White vervain • 337, 339
White wild indigo • 9, 180
Whole-leaf rosinweed • 109
Whorled milkweed • 7, 59
Whorled milkwort • 282
Widow's-frill • 148
Wild beet • 322
Wild bergamot • 9, 218
Wild carrot • 44
Wild columbine • 295
Wild comfrey • 139
Wild garlic • 226
Wild geranium • 208
Wild golden-glow • 107
Wild lettuce • 93
Wild lupine • 189
Wild madder • 318
Wild mint • 23, 216-217
Wild onion • 226-227
Wild parsnip • 7, 47, 51
Wild quinine • 11, 102
Wild rose • 311
Wild strawberry • 41, 301

358

Wild-hyacinth • 228
Winged loosestrife • 237
Wing-stem meadow-pitcher • 240
Winter vetch • 195
Winter-cress • 141
Wiregrass • 274
Wolf's-milk • 177
Woman's-tobacco • 64
Wood-betony • 326
Wood lily • 231
Wood sage • 224
Wood sedge • 172
Woodbine • 41
Woodland strawberry • 301
WOOD-SORREL FAMILY • 252-253
Woolly plantain • 254
Woundwort • 60, 223
Yellow avens • 302, 315
Yellow coneflower • 6-7, 105-106
Yellow false foxglove • 323
Yellow-headed fox sedge • 169, 172
Yellow king-devil • 133
Yellow lady's-slipper • 248
Yellow lake sedge • 25
Yellow marsh-marigold • 299
Yellow salsify • 135
Yellow star-grass • 7, 229
Yellow sweet-clover • 190
Yellow-pimpernel • 49
Yellow-rocket • 141
Zigadenus
   elegans • 228, 232
   glaucus • 232
   venenosus • 228
Zizia
   aptera • 50
   aurea • 51

NOTES

NOTES

NOTES

Notes

NOTES

Notes

Arboretum bouquet